PCR Protocols

Methods in Molecular Biology

John M. Walker, SERIES EDITOR

Methods in Molecular Biology • 15

PCR Protocols

Current Methods and Applications

Edited by

Bruce A. White

*University of Connecticut Health Center,
Farmington, CT*

Humana Press ☀ Totowa, New Jersey

© 1993 Humana Press Inc.
999 Riverview Drive, Suite 208
Totowa, New Jersey 07512

Printed in the United States of America. 9 8 7 6 5 4 3

Library of Congress Cataloging in Publication Data

Main entry under title:
Methods in molecular biology.

PCR protocols ; current methods and applications / edited by Bruce A. White.
      p. cm. – (Methods in molecular biology : 15)
   Includes index.
   ISBN 0-89603-244-2
   1. Polymerase chain reaction–Methodology. I. White, Bruce Alan.
II. Series: Methods in molecular biology (Totowa, NJ) ; 15.
QP606.D46P363 1993
574.87'322–dc20   92-34874
                                                                          CIP

Preface

PCR has been successfully utilized in every facet of basic, clinical, and applied studies of the life sciences, and the impact that PCR has had on life science research is already staggering. Concomitant with the essentially universal use of PCR has been the creative and explosive development of a wide range of PCR-based techniques and applications. These increasingly numerous protocols have each had the general effect of facilitating and accelerating research. Because PCR technology is relatively easy and inexpensive, PCR applications are well within the reach of every research lab. In this sense, PCR has become the "equalizer" between "small" and "big" labs, since its use makes certain projects, especially those related to molecular cloning, now far more feasible for the small lab with a modest budget.

This new volume on *PCR Protocols* does not attempt the impossible task of representing all PCR-based protocols. Rather, it presents a range of protocols, both analytical and preparative, that provide a solid base of knowledge on the use of PCR in many common research problems. The first six chapters provide some basic information on how to get started. Chapters 7–19 represent primarily analytical uses of PCR, both for simple DNA and RNA detection, as well as for more complex analyses of nucleic acid (e.g., DNA footprinting, RNA splice site localization). The remaining chapters represent "synthetic," or preparative, uses of PCR. The use of PCR for aspects of cloning, including obtaining full-length cDNA sequence, site-directed mutagenesis, and production of synthetic genes has been emphasized in these chapters. Some duplication of important topics (e.g., sequencing, cDNA cloning, the use of degenerate oligonucleotides, and site-directed mutagenesis) has been introduced purposely to offer the reader several approaches to the same problem. As has been done in previous volumes of the *Methods in Molecular Biology* series, an emphasis has been placed on generally

v

applicable protocols. The description of specific experimental systems has been deemphasized, and used only when the provision of a specific example is helpful. As part of the *Methods in Molecular Biology* series, the chapters each include a "Notes" section, whose purpose is to provide a discussion of problems, tips, and alternatives. This type of discussion, not usually available in the original publications, should enhance the ability of the reader to get a procedure up and running, as well as increase experimenters' understanding of its strengths, limitations, and pitfalls. It is hoped that this collection of PCR protocols will be especially useful to young investigators, or to those new to PCR, by providing a knowledge base and encouraging the design of novel approaches and applications of PCR.

I am indebted to Jennifer Swanson for her superb secretarial skills. I also wish to thank John Walker and Humana Press for their assistance and support in putting this volume together. Finally, I want to express my appreciation to the contributing authors, who displayed a remarkable degree of enthusiasm for this volume and provided such excellent material for it.

Bruce A. White

Contents

Contents

Contributors

PUJA AGARWAL • *Department of Anatomy, University of Connecticut Health Center, Farmington, CT*

ANTHONY ALBINO • *Memorial Sloan-Kettering Cancer Center, New York*

ROBERT C. ALLEN • *Department of Pathology and Laboratory Medicine, Medical University of South Carolina, Charleston, SC*

GIOVANNA FERRO-LUZZI AMES • *Department of Molecular and Cell Biology, University of California, Berkeley, CA*

STEPHEN C. BAIN • *University of Oxford, John Radcliffe Hospital, Headington, Oxford, UK*

DEBABRATA BANERJEE • *Memorial Sloan-Kettering Cancer Center, New York*

SAILEN BARIK • *Department of Molecular Biology, The Cleveland Clinic Foundation, Cleveland, OH*

JAMES BATTEY • *NCI-Navy Medical Oncology, Bethesda, MD*

JOSEPH R. BERTINO • *Cornell University Graduate School of Medical Sciences and the Laboratory of Molecular Pharmacology, Memorial Sloan-Kettering Cancer Center, New York*

LAURA J. BLOEM • *Department of Medical and Molecular Genetics, Indiana University School of Medicine, Indianapolis, IN*

DAVID A. BRIAN • *Department of Microbiology, University of Tennessee, Knoxville, TN*

THOMAS R. BROKER • *Department of Biochemistry, University of Rochester School of Medicine, Rochester, NY*

BRUCE BUDOWLE • *Forensic Science Research and Training Center, Federal Bureau of Investigation Academy, Quantico, VA*

BING CAI • *Division of Medical Genetics, University of Southern California, Childrens Hospital of Los Angeles, CA*

SHIZHONG CHEN • *Molecular Genetics Laboratory, Center for Human Genome Research, The Salk Institute for Biological Sciences, San Diego, CA*

CHENG-MING CHIANG • *Department of Biochemistry, University of Rochester School of Medicine, Rochester, NY (Present address: Laboratory of Biochemistry and Molecular Biology, The Rockefeller University, New York)*

LOUISE T. CHOW • *Department of Biochemistry, University of Rochester School of Medicine, Rochester, NY*

DAVID L. COOPER • *Department of Pathology, University of Pittsburgh School of Medicine, Pittsburgh, PA*

KATHLEEN DANENBERG • *Kenneth Norris Jr. Cancer Cancer Hospital and Research Institute, University of Southern California Comprehensive Cancer Center, Los Angeles, CA*

PETER V. DANENBERG • *Kenneth Norris Jr. Cancer Cancer Hospital and Research Institute, University of Southern California Comprehensive Cancer Center, Los Angeles, CA*

BEVERLY C. DELIDOW • *Department of Anatomy, University of Connecticut Health Center, Farmington, CT*

JACQUES DELORT • *Laboratoire de Neurobiologie Cellulaire et Moleculaire, Gif sur Yvette, Cedex, France (Present address: Howard Hughes Medical Institute, University of Utah, Salt Lake City, UT)*

ADAM P. DICKER • *Cornell University Graduate School of Medical Sciences and the Laboratory of Molecular Pharmacology, Memorial Sloan-Kettering Cancer Center, New York*

PATRICK J. DILLON • *Department of Gene Regulation, Roche Institute of Molecular Biology, Nutley, NJ*

CRAIG A. DIONNE • *Rhone-Poulenc Rorer Central Research, Collegeville, PA*

JEAN BAPTISTE DUMAS MILNE EDWARDS • *Laboratoire de Neurobiologie Cellulaire et Moleculaire, Gif sur Yvette, Cedex, France*

GLEN A. EVANS • *Molecular Genetics Laboratory, Center for Human Genome Research, The Salk Institute for Biological Sciences, San Diego, CA*

RENATO FANIN • *Memorial Sloan-Kettering Cancer Center, New York*

MARTIN A. HOFMANN • *Department of Microbiology, University of Tennessee, Knoxville, TN*

JOHN HOLCENBERG • *Division of Hematology-Oncology, Department of Pediatrics, University of Southern California, Childrens Hospital of Los Angeles, CA*

TETSURO HORIKOSHI • *Kenneth Norris Jr. Cancer Cancer Hospital and Research Institute, University of Southern California Comprehensive Cancer Center, Los Angeles, CA*

ROBERT M. HORTON • *Department of Biochemistry, College of Biological Sciences, University of Minnesota, St. Paul, MN*

SHENG-HE HUANG • *Division of Hematology-Oncology, Department of Pediatrics, University of Southern California, Childrens Hospital of Los Angeles, CA*

NARAYANA R. ISOLA • *Department of Pathology, University of Pittsburgh School of Medicine, Pittsburgh, PA*

MICHAEL JAYE • *Rhone-Poulenc Rorer Central Research, Collegeville, PA*

DOUGLAS H. JONES • *Department of Pediatrics, College of Medicine, University of Iowa, Iowa City, IA*

AMBROSE Y. JONG • *Division of Hematology–Oncology, Departments of Pediatrics and Microbiology, University of Southern California, Childrens Hospital of Los Angeles, CA*

ELENA D. KATZ • *Biotechnology Department, Perkin-Elmer, Norwalk, CT*

CHRISTOPH KESSLER • *Department of Molecular Genetics, Boehringer Mannheim GmbH, Penzberg/Obb., Germany*

W. MICHAEL KUEHL • *NCI-Navy Medical Oncology, Bethesda, MD*

JOHN P. LYNCH • *Department of Anatomy, University of Connecticut Health Center, Farmington, CT*

JACQUES MALLET • *Laboratoire de Neurobiologie Cellulaire et Moleculaire, Gif sur Yvette, Cedex, France*

TIM MCDANIEL • *University of Maryland Hospital, Baltimore, MD*

STEPHEN J. MELTZER • *University of Maryland Hospital, Baltimore, MD*

JOHN J. PELUSO • *Department of Ob/Gyn, University of Connecticut Health Center, Farmington, CT*

RICCARDO PERFETTI • *Johns Hopkins University School of Medicine and National Institute on Aging, National Institutes of Health, Department of Medicine, Baltimore, MD*

GERD P. PFEIFER • *Department of Biology, Beckman Research Institute of the City of Hope, Duarte, CA*

GREGORY M. PRESTON • *Departments of Medicine and Cell Biology/Anatomy, Johns Hopkins University School of Medicine, Baltimore, MD*

UDO REISCHL • *Department of Molecular Genetics, Boehringer Mannheim GmbH, Penzberg/Obb., Germany*

ARTHUR D. RIGGS • *Department of Biology, Beckman Research Institute of the City of Hope, Duarte, CA*

CRAIG A. ROSEN • *Department of Gene Regulation, Roche Institute of Molecular Biology, Nutley, NJ*

JESSE ROTH • *Johns Hopkins University School of Medicine and National Institute on Aging, National Institutes of Health, Department of Medicine, Baltimore, MD*

RÜDIGER RÜGER • *Department of Molecular Genetics, Boehringer Mannheim GmbH, Penzberg/Obb., Germany*

WOJCIECH RYCHLIK • *National Biosciences, Plymouth, MN*

LAWRENCE B. SCHOOK • *Laboratory of Molecular Immunology, Department of Animal Sciences, University of Illinois, Urbana, IL*

ALAN R. SHULDINER • *Johns Hopkins University School of Medicine and National Institute on Aging, National Institutes of Health, Department of Medicine, Baltimore, MD*

VENKATAKRISHNA SHYAMALA • *Chiron Corporation, Emeryville, CA*

THOMAS STADLBAUER • *Kenneth Norris Jr. Cancer Cancer Hospital and Research Institute, University of Southern California Comprehensive Cancer Center, Los Angeles, CA*

KEITH TANNER • *Johns Hopkins University School of Medicine, Department of Medicine, Baltimore, MD*

JOHN A. TODD • *University of Oxford, John Radcliffe Hospital, Headington, Oxford, UK*

MATTHIAS VOLKENANDT • *Department of Dermatology, University of Munich, Germany*

BRUCE A. WHITE • *Department of Anatomy, University of Connecticut Health Center, Farmington, CT*

STANLEY C. WINISTORFER • *Department of Pediatrics, College of Medicine, University of Iowa, Iowa City, IA*

ALICE L. WITSELL • *Laboratory of Molecular Immunology, Department of Animal Sciences, University of Illinois, Urbana, IL*

CHUN-HUA WU • *Division of Hematology-Oncology, Department of Pediatrics, University of Southern California, Childrens Hospital of Los Angeles, CA*

WU YANG • *Department of Molecular Pharmacology, University of Southern California, Childrens Hospital of Los Angeles, CA*

LEI YU • *Department of Medical and Molecular Genetics, Indiana University School of Medicine, Indianapolis, IN*

CHAPTER 1

Polymerase Chain Reaction

Basic Protocols

Beverly C. Delidow, John P. Lynch, John J. Peluso, and Bruce A. White

1. Introduction

The melding of a technique for repeated rounds of DNA synthesis with the discovery of a thermostable DNA polymerase has given scientists the very powerful technique known as polymerase chain reaction (PCR). PCR is based on three simple steps required for any DNA synthesis reaction: (1) *denaturation* of the template into single strands; (2) *annealing* of primers to each original strand for new strand synthesis; and (3) *extension* of the new DNA strands from the primers. These reactions may be carried out with any DNA polymerase and result in the synthesis of defined portions of the original DNA sequence. However, in order to achieve more than one round of synthesis, the templates must again be denatured, which requires temperatures well above those that inactivate most enzymes. Therefore, initial attempts at cyclic DNA synthesis were carried out by adding fresh polymerase after each denaturation step *(1,2)*. The cost of such a protocol becomes rapidly prohibitive.

The discovery and isolation of a heat-stable DNA polymerase from a thermophilic bacterium, *Thermus aquaticus (Taq)*, enabled Saiki et al. *(3)* to synthesize new DNA strands repeatedly, exponentially amplifying a defined region of the starting material, and allowing the birth of a new technology that has virtually exploded into prominence. Not

From: *Methods in Molecular Biology, Vol. 15: PCR Protocols: Current Methods and Applications*
Edited by: B. A. White Copyright © 1993 Humana Press Inc., Totowa, NJ

since the discovery of restriction enzymes has a new technique so revolutionized molecular biology. There are scores of journal articles published *per month* in which PCR is used, as well as an entire journal (at least one) devoted to it. To those who use and/or read about PCR every day, it is remarkable that this method is not yet 10 years old.

One of the great advantages of PCR is that, although some laboratory precaution is called for, the equipment required is relatively inexpensive and very little space is needed. The only specialized piece of equipment needed for PCR is a thermal cycler. Although it is possible to perform PCR without a thermal cycler—using three water baths at controlled temperatures—the manual labor involved is tedious and very time-consuming. A number of quality instruments are now commercially available. A dedicated set of pipets is useful, but not absolutely necessary. If one purchases oligonucleotide primers, all of the other equipment required for PCR is readily found in any laboratory involved in molecular biology. Thus, a very powerful method is economically feasible for most research scientists.

The versatility of PCR will become clear in later chapters, which demonstrate its use in a wide variety of applications. Additionally, the reader is referred to several recent reviews *(4,5)*. In this chapter, we outline the preparations required to carry out PCR, the isolation of DNA and RNA as templates, the basic PCR protocol, and several common methods for analyzing PCR products.

2. Materials

2.1. Preparation for PCR

2.1.1. Obtaining Primers

1. Prepared oligonucleotide on a cartridge. Cap ends with parafilm and store horizontally (the columns contain fluid, which can leak) at –20°C until the oligo is to be purified.
2. Ammonium hydroxide, reagent grade. Ammonium hydroxide should be handled in a fume hood, using gloves and protective clothing.
3. 1-mL tuberculin syringes (needles are not required).
4. 1.25-mL screw-cap vials, with O-rings (e.g., Sarstedt #D-5223, Sarstedt, Inc., Pennsauken, NJ).
5. Parafilm.
6. Sterile water, filter deionized distilled water through a 0.2-μm filter, store at room temperature.

7. $1M$ MgSO$_4$. Filter through a 0.2-μm filter and store at room temperature.
8. 100% Ethanol.
9. 95% Ethanol; for precipitations store at –20°C.

2.1.2. Isolation of DNA

1. Source of tissue or cells from which DNA will be extracted.
2. Dounce homogenizer.
3. Digestion buffer: 100 mM NaCl, 10 mM Tris-HCl, pH 8.0, 25 mM EDTA, 0.5% SDS.
4. Proteinase K, 20 mg/mL.
5. a. Buffered phenol *(6,7)*: Phenol is highly corrosive, wear gloves and protective clothing when handling it. Use only glass pipets and glass or polypropylene tubes. Phenol will dissolve polystyrene plastics.
   b. Buffering solutions: $1M$ Tris base; 10X TE, pH 8.0 = 100 mM Tris-HCl, pH 8, 10 mM EDTA; 1X TE, pH 8 = 10 mM Tris, pH 8, 1 mM EDTA. To a bottle of molecular biology grade recrystallized phenol add an equal volume of $1M$ Tris base. Place the bottle in a 65°C water bath and allow the phenol to liquify (approx 1 h). Transfer the bottle to a fume hood and allow it to cool. Cap the bottle tightly and shake to mix the phases, **point the bottle away** and vent. Transfer the mix to 50-mL screw-top tubes by carefully pouring or using a glass pipet. Centrifuge at 2000 rpm for 5–10 min at room temperature to separate the phases. Remove the upper aqueous phase by aspiration. To the lower phase (phenol) add an equal volume of 10X TE, pH 8. Cap tubes tightly, shake well to mix, and centrifuge again. Aspirate the aqueous phase. Reextract the phenol two or three more times with equal volumes of 1X TE, pH 8.0, until the pH of the upper phase is between 7 and 8 (measured using pH paper). Aliquot the buffered phenol, cover with a layer of 1X TE, pH 8, and store at –20°C.
6. CHCl$_3$.
7. 100% Ethanol.
8. 70% Ethanol.
9. TE buffer, pH 8.0: 10 mM Tris-HCl, pH 8.0, 1mM EDTA.
10. Phosphate-buffered saline (PBS): 20X stock = 2.74M NaCl, 53.6 mM KCl, 166 mM Na$_2$HPO$_4$, 29.4 mM KH$_2$PO$_4$, pH 7.4. Make up in deionized distilled water, filter through a 0.2-μm filter, and store at room temperature. For use, dilute 25 mL of 20X stock up to 500 mL with deionized distilled water and add 250 μL of $1M$ MgCl$_2$. Sterile-filter and store at 4°C.
11. 7.5M Ammonium acetate.
12. RNase A. Prepare at 10 mg/mL in 10 mM Tris-HCl, pH 7.5, 15 mM

NaCl. Incubate at 100°C for 15 min and allow to cool to room temperature. Store at –20°C.
13. 20% SDS.

2.1.3. Isolation of RNA

2.1.3.1. ISOLATION OF RNA
BY CsCl CENTRIFUGATION (*SEE* NOTE 1)

1. Source of tissue or cells from which RNA will be extracted.
2. PBS (*see* Section 2.1.2., item 10).
3. 2-mL Wheaton glass homogenizer.
4. Guanidine isothiocyanate/β-mercaptoethanol solution (GITC/ßME): 4.2M guanidine isothiocyanate, 0.025M sodium citrate, pH 7.0, 0.5% N-laurylsarcosine (Sarkosyl), 0.1M β-mercaptoethanol. Prepare a stock solution containing everything except β-mercaptoethanol in deionized distilled water. Filter-sterilize using a Nalgene 0.2-μm filter (Nalge Co., Rochester, NY) (*see* Note 2). Store in 50-mL aliquots at –20°C. To use, thaw a stock tube, transfer the required volume to a fresh tube, and add 7 μL of β-mercaptoethanol/mL of buffer. Guanidine isothiocyanate and β-mercaptoethanol are strong irritants, handle them with care.
5. 1-mL tuberculin syringes, with 21-g needles.
6. Ultraclear ultracentrifuge tubes, 11 × 34 mm (Beckman #347356).
7. Diethylpyrocarbonate, 97% solution, store at 4°C.
8. Diethylpyrocarbonate (DEPC)-treated water *(6,7)*. Fill a baked glass autoclavable bottle to two-thirds capacity with deionized distilled water. Add diethyl pyrocarbonate to 0.1%, cap and shake. Vent the bottle, cap loosely, and incubate at 37°C for at least 12 h (overnight is convenient). Autoclave on liquid cycle for 15 min to inactivate the DEPC. Store at room temperature.
9. 200 mM EDTA, pH 8.0. Use molecular biology grade disodium EDTA. Make up in deionized distilled water and filter through a 0.2-μm filter. Place in an autoclavable screw-top bottle. Treat with DEPC as described in the preceding step for DEPC water. Store at room temperature.
10. CsCl: molecular biology grade. For 20 mL, place 20 g of solid CsCl in a sterile 50-mL tube. Add 10 mL of 200 mM EDTA, pH 8.0 (DEPC-treated). Bring volume to 20 mL with DEPC water. Mix to dissolve. Filter through a 0.2-μm filter and store at 4°C.
11. TE buffer, pH 7.4: 10 mM Tris-HCl, pH 7.4, 1 mM EDTA. Make a solution of 10 mM Tris-HCl and 1 mM EDTA, pH 7.4, in DEPC water (*see* Note 3). Filter through a 0.2-μm filter, autoclave 15 min on liquid cycle, and store at room temperature.

12. TE-SDS: Make fresh for each use. From a stock solution of 10% SDS in DEPC water, add SDS to a concentration of 0.2% to an aliquot of TE, pH 7.4.
13. Buffered phenol (*see* Section 2.1.2., item 5).
14. CHCl$_3$.
15. 4M NaCl. Make up in deionized distilled water and DEPC treat. Autoclave 15 min on liquid cycle and store at room temperature.
16. 95% Ethanol, stored at –20°C.
17. Polyallomer 1.5-mL microcentrifuge tubes, for use in an ultracentrifuge (Beckman #357448, Beckman Instrument Inc., Fullerton, CA).
18. RNasin RNase inhibitor, 40 U/µL (Promega, Madison, WI). Store at –20°C.
19. Beckman TL-100 table-top ultracentrifuge, TLS 55 rotor, and TLA-45 rotor.

2.1.3.2. ISOLATION OF RNA BY GUANIDINE/PHENOL (RNAZOL™) EXTRACTION

1. RNAzol reagent (TEL-TEST, Inc., Friendswood, TX). This reagent contains guanidine isothiocyanate, β-mercaptoethanol, and phenol; handle with care.
2. Glass-Teflon homogenizer.
3. Disposable polypropylene pellet pestle and matching microfuge tubes (1.5 mL) (Kontes Life Science Products, Vineland, NJ).
4. CHCl$_3$ (ACS grade).
5. Isopropanol (ACS grade). Store at –20°C.
6. 80% Ethanol. Dilute 100% ethanol with DEPC-treated H$_2$O and store at –20°C.
7. TE buffer, pH 7.4, in DEPC-treated water (*see* Section 2.1.3.1.).

2.1.4. Synthesis of Complementary DNAs (cDNAs) from RNA

1. RNA in aqueous solution.
2. Oligo dT$_{18-20}$ primer (Pharmacia, Piscataway, NJ). Dissolve 5 OD U in 180 µL of sterile water to give a concentration of 1.6 µg/µL.
3. Specific primer, optional. Choose sequence and obtain as for PCR primers (*see* Section 3.1.1.).
4. MMLV reverse transcriptase (200 U/µL) with manufacturer-recommended buffer and 0.1M DTT.
5. Deoxynucleotides dATP, dCTP, dGTP, and dTTP. Supplied as 10 mg solids. To make 10 mM stocks: Resuspend 10 mg of dNTP in 10% less

sterile water than is required to give a 10 mM solution. Adjust the pH to approximate neutrality using sterile NaOH and pH paper. Determine the exact concentration by OD, using the wavelength and molar extinction coefficient provided by the manufacturer for each deoxynucleotide. For example, the A$_m$ (259 nm) for dATP is 15.7×10^3; therefore a 1:100 dilution of a 10 mM solution of dATP will have an A$_{259}$ of $(0.01M \times 15.7 \times 10^3$ OD U/$M) \times 1/100 = 1.57$. If the actual OD of a 1/100 dilution of the dATP is 1.3, the dATP concentration is $1.3/1.57 \times 10$ m$M = 8.3$ mM. Store deoxynucleotides at –20°C in 50- to 100-µL aliquots. Make a working stock containing 125 µM of each dNTP in sterile water for cDNA synthesis or for PCR. Unused working stock may be stored at –20°C for up to 2 wk.
6. RNasin, 40 U/µL (Promega) or other RNase inhibitor. Store at –20°C.

2.2. Performing PCR

2.2.1. Basic PCR Protocol (see Note 4)

1. Genomic DNA or cDNA to be amplified in aqueous solution.
2. Oligonucleotide primers complementary to the 5' and 3' ends of the sequence to be amplified.
3. Sterile UV-irradiated water (*see* Note 5). Sterile-filter deionized distilled water. UV irradiate for 2 min in a Stratagene (La Jolla, CA) Stratalinker UV crosslinker (200 mJ/cm^2) (8) or at 254 and 300 nm for 5 min (9). Store at room temperature.
4. PCR stock solutions: Dedicate these solutions for PCR use only. Prepare the following three solutions, filter-sterilize, and autoclave 15 min on liquid cycle: 1M Tris-HCl, pH 8.3; 1M KCl; and 1M MgCl2.
5. 10X PCR buffer: 100 mM Tris-HCl, pH 8.3; 500 mM KCl; 15 mM MgCl$_2$; 0.01% (w/v) gelatin. This buffer is available from Perkin-Elmer/ Cetus. Per milliliter of 10X buffer combine 100 µL of 1M Tris-HCl, pH 8.3, 500 µL of 1M KCl, 15 µL of 1M MgCl$_2$ and 375 µL of UV-irradiated sterile water. Make up a 1% solution of gelatin in UV-irradiated sterile water. Heat at 60–70°C, mixing occasionally, to dissolve the gelatin. Filter the gelatin solution while it is still warm through a 0.2-µm filter, and add 10 µL of gelatin to each milliliter of 10X PCR buffer. Store PCR buffer in small aliquots (300–500 µL) at –20°C. As an extra precaution, the 10X buffer may be UV-irradiated before each use.
6. 10 mM Deoxynucleotide stocks (dATP, dCTP, dGTP, and dTTP), made up in UV-irradiated sterile water; *see* Section 2.1.4.5.
7. 1.25 mM Deoxynucleotide working stock. Make a solution 1.25 mM in each nucleotide, in UV-irradiated sterile water.
8. Light mineral oil.

9. CHCl₃.
10. 7.5*M* Ammonium acetate, filter through a 0.2-μm filter and store at room temperature.
11. 95% Ethanol. Store at –20°C.
12. *Taq* DNA polymerase.

2.3. Analysis of PCR Products

2.3.1. Agarose Gel Electrophoresis

2.3.1.1. DETECTION OF PCR PRODUCTS BY ETHIDIUM BROMIDE STAINING

1. DNA grade agarose.
2. E buffer, for running agarose gels (40X stock): 1.6*M* Tris-HCl, 0.8*M* anhydrous sodium acetate, 40 m*M* EDTA. Adjust pH to 7.9 with glacial acetic acid and filter through a 0.2-μm filter. To make 1X buffer, dilute 25 mL of stock up to 1 L in distilled water. Store at room temperature.
3. 6X Agarose gel-loading dye: 0.25% bromophenol blue, 0.25% xylene cyanol, 30% glycerol. Prepare in sterile water and store at room temperature.
4. DNA markers. Several are available. We routinely use a *BstE* II digest of lambda DNA (New England Biolabs, Beverly, MA). This preparation contains 14 DNA fragments, ranging from 8454–117 bp. Store at –20°C.
5. Ethidium bromide (10 mg/mL) in sterile water. Store at 4°C in a dark container. **Ethidium bromide is a potent mutagen**. Use a mask and gloves when weighing powder. Clean up spills immediately. Wear gloves when handling solutions. Dispose of wastes properly.

2.3.1.2. DETECTION OF PCR PRODUCTS BY SOUTHERN BLOT HYBRIDIZATION ANALYSIS

1. Materials for agarose gel electrophoresis (Section 2.3.1.1., items 1–5).
2. Gel denaturation buffer: Make fresh. 1.5*M* NaCl, 0.5*M* NaOH.
3. Gel neutralizing buffer: 1*M* Tris-HCl, pH 8, 1.5*M* NaCl.
4. Nitrocellulose, 0.45 μm pore size.
5. 20X SSC: 3*M* NaCl, 0.3*M* sodium citrate, pH 7.0. Make up a bulk stock, unfiltered for use in transfers and blot washes. Make up a sterile 0.2-μm filtered stock for presoaking nitrocellulose (*see* Note 6). Store at room temperature.
6. 10X SSC: 1.5*M* NaCl, 0.15*M* sodium citrate, pH 7.0. Make by diluting 20X SSC 1:2.
7. 50X Denhardt's solution: 1% Ficoll, 1% polyvinylpyrollidine, 1% BSA. Make up in deionized distilled water and filter through a 0.2-μm filter. Aliquot and store at –20°C.

8. Deionized formamide, molecular biology grade *(6,7)*: Place the formamide to be deionized in a clean baked glass beaker. Add 10 g of mixed-
   bed ion exchange resin (e.g., Biorad AG 501 X8, BioRad Laboratories,
   Richmond, CA) per 100 mL of formamide. Stir at room temperature for
   30 min. Filter twice through Whatman #1 filter paper and store aliquots
   at −70°C.
9. 20X SSPE: 3.6M NaCl, 200 mM NaH$_2$PO$_4$, pH 7.4, 20 mM EDTA.
   Filter through a 0.2-μm filter and store at room temperature.
10. Denatured salmon sperm DNA: 10 mg/mL in water. Dissolve the DNA
    in water by stirring at room temperature for several hours. Shear the
    DNA by passing it through an 18-g needle, then denature it by incubat-
    ing it in a boiling water bath for 10 min. Aliquot and store at −20°C.
    Sonicate each aliquot for 30 s before using it for the first time.
11. 10% SDS.
12. Prehybridization solution: 50% formamide, 5X Denhardt's, 5X SSPE,
    100 μg/mL of denatured salmon sperm DNA, and 0.1% SDS.
13. Plasmid containing desired probe sequences.
14. Nick translation kit or random primer kit for labeling nucleic acids.
15. α^{32}P-dCTP, 3000 Ci/mmol.
16. Blot washing buffers:
    a. High salt: 2X SSC, 0.1% SDS
    b. Low salt: 0.1X SSC, 0.1% SDS
17. X-ray film.

2.3.1.3. ANALYSIS OF PCR PRODUCTS BY NESTED PCR *(10)*

1. Products of an initial round of PCR.
2. Low-melting-point agarose.
3. Agarose gel electrophoresis reagents (Section 2.3.1.1., items 2–5).
4. Oligonucleotide primers complementary to internal portions of the DNA
   amplified (nested primers).
5. PCR reagents (Section 2.2.1., items 3–11).
6. DNA grade agarose.

2.3.2. Analysis of PCR Products
by Acrylamide Gel Electrophoresis

2.3.2.1. ACRYLAMIDE GEL ELECTROPHORESIS
WITH ETHIDIUM BROMIDE STAINING

1. 30% Acrylamide: 0.8% *bis*. **Acrylamide in its powdered and liquid
   forms is a neurotoxin**. Always wear gloves when handling acrylamide.
   Weigh powder in a fume hood wearing gloves and a mask. For 400 mL,
   dissolve 116.8 g acrylamide and 3.2 g *bis*-acrylamide in water. Stir to
   dissolve and filter through a 0.2-μm filter. Store at 4°C.

2. 10X TBE buffer: 0.89M Tris, pH 8.0, 0.89M boric acid, 2 mM EDTA. Filter through a 0.2-µm filter and store at room temperature.
3. 10% Ammonium persulfate. Make up fresh weekly in deionized distilled water.
4. TEMED (*N,N,N',N'*-Tetramethylethylenediamine).
5. 6X Acrylamide gel-loading dye: 0.125% bromophenol blue, 0.125% xylene cyanol, 25% glycerol (v/v), 2.5% SDS, 12.5 mM EDTA. This dye may be made in two parts.
   a. 250 µL of 1% bromophenol blue, 250 µL of 1% xylene cyanol, and 500 µL glycerol. Mix well by pipetting up and down.
   b. 5% SDS, 25 mM EDTA.
   To make the 6X gel loading dye, mix equal parts of a and b. Store at room temperature.
5. DNA markers.
6. 10 mg/mL ethidium bromide (*see* Section 2.3.1.1.).

2.3.2.2. Acrylamide Gel Electrophoresis
     of Directly Labeled PCR Products

1. α^{32}P-dCTP, 3000 Ci/mmol.
2. PCR reagents (Section 2.2.1.).
3. Acrylamide gel reagents (Section 2.3.2.1., items 1–6).
4. 3MM Filter paper.
5. X-ray film.

3. Methods
3.1. Preparation for PCR
3.1.1. Obtaining Primers

Determine the primer sequences required (*see* Chapter 2 for selection of primers). Double-check sequence and orientation of primers. Once the sequence is determined, synthesize primers locally or order them from commercial suppliers. Our primers are synthesized locally by the β-cyanoethyl phosphoramidite method on a Cyclone machine (MilliGen/Biosearch, Burlington, MA) and delivered to us in the form of protected oligomers covalently linked to a CPG support cartridge. The following procedure is used to deprotect, release, and purify the primers.

1. Wear gloves when handling PCR primers to avoid inadvertent contamination.
2. **In a fume hood**, draw 0.5 mL of ammonium hydroxide into each of two 1-mL tuberculin syringes (without needles), making sure there are no air bubbles.

3. Attach the syringes to either side of the oligo cartridge. Make sure there is a good seal at each end.
4. Holding a syringe in each hand so that the cartridge is horizontal, slowly wash the ammonium hydroxide back and forth across the cartridge by pushing alternately on the syringe plungers. Go back and forth 20 times.
5. After the final wash, adjust the plungers so that each is halfway down, lay the whole apparatus on a clean surface, and allow it to sit for 45 min at room temperature.
6. At the end of the incubation, wash the ammonium hydroxide back and forth, as in step 4, another 20 times.
7. To remove the solution now containing the released oligo, push all of the solution into one syringe. Gently detach the full syringe from the cartridge while pulling back on the plunger of the other syringe to preserve the fluid still in the cartridge.
8. Empty the full syringe into a screw-cap, O-ring vial. Pull back on the plunger of the syringe still attached to the cartridge to retrieve all of the remaining fluid. Empty the second syringe into the O-ring vial (*see* Note 7).
9. Tightly cap the vial and transfer it to a heated water bath. Incubate at 70°C for 3 h, or at 55°C overnight.
10. Poke a well into a container of ice. Carefully transfer the heated vial into this well and allow it to cool before handling it further.
11. Spin the vial briefly in a table-top microfuge to collect all of the condensate.
12. Place the vial back on ice. Remove the cap carefully and cover the vial with two layers of Parafilm. Poke 10–12 holes in the Parafilm with a 21-g needle.
13. To remove the solvent, place the vial and a balance tube in the rotor of a SpeedVac evaporator (Savant, Hicksville, NY). Close the lid, turn on the rotor, and wait for it to reach top speed before slowly applying the vacuum. **Do not use heat**. Evaporate to dryness. This takes 3–4 h.
14. Resuspend the pellet in 200 µL of sterile water.
15. Precipitate the oligo by addition of 2 µL of $1M$ $MgSO_4$ and 1 mL of 100% ethanol. Mix well and spin at $12,000g$ for 15 min in a table-top microfuge.
16. After precipitation, a large white pellet should be visible. Decant the supernatant and add 200 µL of 80% ethanol to the side of the tube. Spin briefly and decant again. Allow the pellet to air-dry.
17. Resuspend the pellet in 500 µL of sterile water.
18. To quantitate the oligo, take the OD_{260} of 5 µL of oligo in 1 mL of sterile water. Multiply the reading by 20 and divide by 5 to obtain the

concentration in µg/µL. The expected yield is 1–2 µg/µL, or a total of between 0.5 and 1 mg.

19. To determine the mol wt of the primer, use the following approximate nucleotide monophosphate mol wt: dAMP, 313.2; dCMP, 289.2; dGMP, 329.2; dTMP, 304.2. Multiply each mol wt by the number of residues of that nucleotide in the primer and add all four together. A 20-mer will have a mol wt in the range of 6000 dalton; therefore, approx 0.6 µg will equal 100 pmol.

20. The oligo can be stored at –20°C. However, it may be helpful to aliquot it and to store aliquots not meant for immediate use at –70°C.

3.1.2. Isolation of DNA (7)

Several chapters in this volume contain methods for treating small samples of cells or tissue so that the DNA may be PCR amplified (*see* Chapter 7) or for isolating DNA from small samples (*see* Chapter 11). The following method works well for isolation of DNA from larger tissue samples or for bulk preparations of DNA from cultured cells.

1. Remove tissue into ice-cold PBS. Weigh tissue and mince with a razor blade. For cultured cells, collect by centrifugation, wash once in ice-cold PBS, and resuspend in 1 pellet vol of PBS.
2. Transfer tissue or cells to a Dounce homogenizer containing 12 mL of digestion buffer/g of tissue (per mL of packed cells).
3. Homogenize by 20 gentle strokes using a B pestle. Keep on ice.
4. Transfer the sample into a test tube, add proteinase K to a final concentration of 100 µg/mL, and incubate at 50°C overnight.
5. Extract sample twice with an equal volume of phenol/CHCl$_3$ (1:1 by volume).
6. Extract twice with an equal volume of CHCl$_3$.
7. Add 0.5 vol of 7.5M ammonium acetate and 2 vol of 100% ethanol. Mix gently. DNA should immediately form a stringy precipitate.
8. Recover the DNA by centrifugation at 12,000g for 15 min at 4°C.
9. Rinse pellet with 70% ethanol, decant, and air-dry.
10. Resuspend DNA in TE buffer, pH 8.0 (7–10 mL/g of tissue). Resuspension can be facilitated by incubation of sample at 65°C with gentle agitation.
11. Add SDS to final concentration of 0.1% and RNase A to 1 µg/mL. Incubate at 37°C for 1 h.
12. Reextract with phenol/CHCl$_3$, precipitate, and resuspend DNA as described above in steps 5–10. Keep the DNA in ethanol at 4°C for long-term storage.

3.1.3. Isolation of RNA

There are a number of protocols now available for the isolation of RNA from cells or tissues (*see* Chapters 16–19). The following are two procedures we routinely use to isolate RNA from small tissue samples or from cultured cells. One procedure more rigorously removes DNA by centrifugation of the RNA through a CsCl cushion. The other relies on the extraction of RNA out of a guanidine solution and is less time-consuming.

3.1.3.1. ISOLATION OF RNA BY CsCL CENTRIFUGATION *(11)*

We have used this procedure for isolating RNA from whole rat ovaries (up to six ovaries, or about 150 mg of tissue, per sample), from ovarian granulosa cells and from nuclei of GH_3 pituitary tumor cells (nuclei from up to 5×10^7 cells). This procedure requires more time than the following guanidine/phenol extraction, but we found it gives cleaner RNA preparations from ovarian tissue, which contains not only DNA, but also substantial lipid deposits. The procedure is also recommended for preparing nuclear RNA because of the much higher DNA content of nuclei as opposed to whole cells or tissues.

1. Remove the tissue from the animal within several minutes of death. Place in ice-cold PBS and trim off fat and/or fascia if necessary. Cut large pieces of tissue into smaller pieces (2- to 3-mm cubes) (*see* Note 8).
2. Place the tissue in a 2-mL Wheaton glass homogenizer containing 1 mL of GITC/βME buffer. Homogenize by hand until no visible clumps remain (*see* Notes 9 and 10).
3. Transfer the sample to a 5-mL or 15-mL Falcon tube. To shear the DNA, draw the homogenate up into 1-mL tuberculin syringe with an 18-g needle. Pass the homogenate up and down through the needle, avoiding foaming, until it becomes less viscous and can be released in individual drops (*see* Note 11).
4. Rinse Beckman Ultraclear centrifuge tubes with 0.3 mL of GITC/βME buffer and allow to dry inverted. Turn dried tubes up and place 875 µL of CsCl solution into the bottom of each tube.
5. Add 300 µL of GITC/βME to each tissue sample. Mix. Layer each entire sample (1.3 mL) on top of a CsCl cushion, taking care not to disrupt the boundary.
6. Fill each tube with sample and/or GITC/βME to within 2 mm of the top. Balance tubes to within 0.01 g with GITC/βME.

7. Load the tubes into the buckets of a Beckman TLS-55 rotor and centrifuge at 40,000 rpm for 3 h at 16°C. This pellets the RNA, but not DNA (*see* Note 12).

8. Remove the tubes from the rotor buckets. Empty by rapid inversion and immediately place the inverted tubes in a rack or on a clean paper towel to drain-dry for about 15 min. Do not right the tubes until they dry.

9. Using a clean Kimwipe, remove the last traces of liquid from the sides of the tube, without touching the bottom. The RNA pellet will not be visible.

10. Add 400 µL of TE-SDS to the bottom of each tube, without allowing the solution to run down the sides. Cover the tubes and place in a rack on a rotary platform. Solubilize the RNA pellets by gently rocking for 20 min at room temperature.

11. Using a pipettor set at 200 µL, transfer each sample to a 1.5-mL microfuge tube (*see* Note 13). This requires two transfers. During each transfer, pipet the sample up and down in the Ultraclear tube and scrape the pipet tip across the bottom to ensure that the RNA is solubilized. Avoid foaming of the SDS during this procedure.

12. To the RNA sample in a 1.5-mL microfuge tube add 200 µL of buffered phenol and 200 µL of chloroform. Mix well.

13. Separate the phases by centrifuging at top speed in a table-top microfuge for 2 min.

14. Transfer the upper aqueous phase to a clean microfuge tube, add 400 µL of chloroform, mix, and spin as in step 13.

15. Again, transfer the aqueous phase to a clean tube and repeat the chloroform extraction.

16. Transfer the final clean aqueous phase to a Beckman ultramicrocentrifuge tube. Add 25 µL of 4M NaCl and mix. Add 1 mL of cold 95% ethanol and mix again. Precipitate at –20°C overnight (*see* Note 14).

17. Collect the RNA by centrifuging in a Beckman TLA-45 rotor at 15,000 rpm for 30 min at 4°C.

18. Decant the supernatant and invert the tubes over a clean tissue (e.g., Kimwipe) to air-dry. The RNA should be visible as a translucent white pellet at the bottom of the tube.

19. Resuspend the pellet in 25–100 µL of TE, pH 7.4. The volume used will be determined by the size of the pellet. To prevent degradation, add 1 U/µL of RNasin ribonuclease inhibitor and mix gently.

20. Measure the OD_{260} and OD_{280} of 3–5 µL of RNA in a total of 0.4–1 mL of sterile water. The ratio of OD_{260}/OD_{280} should be close to 2.0. If this ratio is <1.7, the sample may contain residual phenol or proteins and should be reextracted and precipitated. To obtain the concentration of RNA, use the following formula:

$$[RNA] (\mu g/\mu L) = (OD_{260} \times 40) \times total\ vol\ OD'd\ (mL)/\mu L\ RNA\ OD'd$$

21. For short-term storage (several weeks), store RNA in aqueous solution at –20°C. For more stable long-term storage, store RNA in ethanol. Add NaCl to 0.25M to RNA in aqueous solution, add 2.5 vol of 95% ethanol, mix well, and store at –20°C. To recover the RNA, centrifuge it as in step 17.

3.1.3.2. ISOLATION OF RNA BY RNAZOL METHOD

RNA can be isolated quickly and with great purity using the RNAzol technique (TEL-TEST, Inc.), based on the method of Chomczynski and Sacchi *(12)*. This procedure is most useful for isolating RNA from many samples, especially small tissue specimens (<500 mg). The following protocol is from TEL-TEST *(13)*, with minor modifications we commonly employ.

1. Homogenize tissue samples in RNAzol (2 mL for each 100 mg of tissue) with several strokes of a glass-Teflon homogenizer. Samples of <50 mg should be homogenized directly in 1.5-mL Eppendorf tubes using Kontes polypropylene pestles. *Brief* sonication is helpful to break up any residual tissue clumps, but do not allow the homogenate to become heated (*see* Note 11).
2. Cells grown in suspension should be pelleted in culture media (5 min, 200g_{max}). After pouring off the supernatant, add 0.2 mL RNAzol/10^6 cells and completely lyse the pellet by repeated pipetting and vortexing.
3. Cells grown on culture dishes can be lysed in the dish. After removing the medium, add RNAzol until the dish is well covered (e.g., 1.5 mL/ 3.5-cm culture dish). Scraping and/or repipetting will ensure complete lysis. Alternatively, attached cells can be collected by scraping them from the dish, then pelleted and lysed as in step 2.
4. Add 0.1 mL of CHCl$_3$ for each 1 mL of homogenate. Vortex rapidly for at least 15 s, until the homogenate is completely frothy white, and incubate on ice for 15 min. After the incubation, vortex again as before, then centrifuge for 15 min at 10,000g at 4°C.
5. There should now be two liquid phases visible in the tube. Carefully remove and save the upper aqueous phase that contains the RNA. The volume of this aqueous phase is approx half of the volume of the homogenate. Do not transfer any of the interface. Pour the lower organic phase into a waste bottle and dispose of properly.
6. Precipitate the RNA by adding an equal volume of ice-cold isopropanol to the aqueous phase and incubate at –20°C for 45 min (*see* Note 15). Pellet the RNA by centrifuging at 12,000g at 4°C for 15 min (or 10,000g

at 4°C for 30 min in a table-top microfuge). A white pellet of RNA is often (but not always) visible after this step.

7. Carefully decant the supernatant and wash the pellet with 80% ethanol (0.8 mL/100 µg of RNA). Vortex briefly to loosen pellet, then centrifuge for 10 min at 12,000g at 4°C. Remove supernatant and repeat the ethanol wash. The RNA pellet is often not well attached to the wall of the tube, so the decanting should be performed gently.

8. Allow the pellet to air-dry until just damp (completely dried pellets are difficult to resuspend). Resuspend the pellet in approx 50 µL of TE buffer, pH 7.4, for each 100 µg of RNA by vortexing and by repipetting. A room temperature incubation (15–30 min) can help resuspend difficult pellets. Incubation at 60°C (10–15 min) may also be used for resuspension, but only if all else fails. We often obtain an OD 260/280 ratio of 2.0:2.1 by this method. Samples with a ratio of <1.7, should be reextracted and precipitated, as described in Section 3.1.3.1. RNA isolated by this method should also be reprecipitated prior to enzymatic manipulation.

3.1.4. Synthesis of Complementary DNAs from RNA

In order to perform PCR on RNA sequences using *Taq* DNA polymerase, it is necessary to first convert the sequence to a complementary DNA (cDNA) because *Taq* has limited reverse transcriptase activity *(14)* (*see* Note 16). Several different kinds of primers can be used to make cDNAs. Oligo-dT will prime cDNA synthesis on all polyadenylated RNAs and is most often used for convenience, as these cDNAs can be used for amplification of more than one species of RNA. Random-primed cDNA synthesis similarly gives a broad range of cDNAs and is not limited to polyadenylated RNAs. Lastly, oligonucleotide primers complementary to the RNA(s) of interest may be used to synthesize highly specific cDNAs. We developed the following procedure for use with oligo-dT or RNA-specific primers. A procedure for using random primers to synthesize cDNAs may be found in Chapter 19.

1. Place up to 20 µg of RNA in a microfuge tube containing 4 µg of oligo dT or 200 pmol of specific primer and 5 µL of 10X RT buffer in a total volume of 36.5 µL (*see* Note 17). Mix gently.
2. Incubate at 65°C for 3 min. Cool on ice.
3. Add 5 µL of 100 m*M* DTT, 1 µL (40 U) of RNasin, and 5 µL of a deoxynucleotide mix containing 1.25 m*M* of each dNTP (final concen-

tration 125 μ*M* each). Add 2.5 μL (500 U) of MMLV reverse transcriptase and mix gently. The final volume is 50 μL.
4. Incubate at 37°C for 1 h.
5. This cDNA may be used directly in PCR reactions or may be modified further (*see* Note 18).

3.2. Performing PCR
3.2.1. Basic PCR Protocol

The ideal way to perform PCR is in a dedicated room, using reagents and equipment also dedicated only to PCR. Such luxuries are often not available. Dedicated PCR reagents are essential. A set of dedicated pipets is very helpful, as are filter-containing pipet tips now available from several manufacturers. Gloves should always be worn when handling PCR reagents. An attempt should be made to keep concentrated stocks of target sequences (e.g., recombinant plasmids) away from PCR areas and equipment. Chance contamination can be very difficult to trace and to get rid of.

PCR cycles consist of three basic steps:

1. Denaturation, to melt the template into single strands and to eliminate secondary structure; this step is carried out at 94°C for 1–2 min during regular cycles. However, amplification of genomic DNA requires a longer initial denaturation of 5 min to melt the strands.
2. Annealing, to allow the primers to hybridize to the template. This step is carried out at a temperature determined by the strand-melting temperature of the primers (*see* Chapter 2) and by the specificity desired. Typical reactions use an annealing temperature of 55°C for 1–2 min. Reactions requiring greater stringency may be annealed at 60–65°C. Reactions in which the primers have reduced specificity may be annealed at 37–45°C.
3. Extension, to synthesize the new DNA strands. This step is usually carried out at 72°, which is optimal for *Taq* polymerase. The amplification time is determined by the length of the sequence to be amplified. At optimal conditions, *Taq* polymerase has an extension rate of 2–4 kb/min (manufacturer's information). As a rule of thumb, we allow 1 min/kb to be amplified, with extra time allowed for each kb >3 kb (*see* Note 19).

Between 20 and 30 cycles of PCR are sufficient for many applications. DNA synthesis will become less efficient as primers and deoxynucleotides are used up and as the number of template molecules surpasses the supply of polymerase. Therefore, following the last cycle, the enzyme is allowed to finish any incomplete synthesis by including a final extension of 5–15 min at 72°C. Following completion of the program, many cycling blocks have a convenient feature allowing an indefinite hold at 15°C, to allow preservation of the samples, particularly during overnight runs.

Ideally, PCR conditions should be optimized for each template and primer combination used. Practically, most researchers will use the manufacturer's recommended conditions unless the results obtained fall far short of expectations. Other than primer sequence, which is discussed in Chapter 2, there are six variables that may be optimized for a given amplification reaction: annealing temperature, primer concentration, template concentration, $MgCl_2$ concentration, extension time, and cycle number (e.g., *see* ref. 15). Standard conditions are described in the following.

1. Prepare a master mix of PCR reagents containing (per 100 µL of PCR reaction): 10 µL of 10X PCR buffer, 100 pmol of upstream primer, 100 pmol of downstream primer, and 16 µL of 1.25 m*M* dNTP working stock (*see* Note 20). Bring to volume with sterile UV-irradiated water, such that, after addition of the desired amount of sample and *Taq* polymerase, the total reaction volume will be 100 µL. Make up a small excess (an extra 0.2–0.5 reaction's worth) of master mix to ensure that there is enough for all samples.

2. Aliquot the desired amount of sample to be amplified into labeled 0.5-mL microfuge tubes. We routinely amplify 5 µL (1/10) of a 50-µL cDNA made using up to 10 µg of RNA. Genomic DNA is usually amplified in amounts of 100 ng to 1 µg. Adjust the volumes with sterile UV-irradiated water so that all are equal.

3. To the master mix, add 0.5 µL of *Taq* polymerase (2.5 U) for each reaction. Mix well and spin briefly in a microfuge to collect all of the fluid (*see* Note 21).

4. Add the correct volume of master mix to each sample tube so that the total volume is now 100 µL. Cap and vortex the tubes to mix. Spin briefly in a microfuge.

5. Reopen the tubes and cover each reaction with a few drops of light mineral oil to prevent evaporation.

6. Put a drop of mineral oil into each well of the thermal cycler block that will hold a sample. Load the sample tubes (*see* Note 22).

7. Amplify the samples, according to the principles previously oulined.
   a. A typical cycling program for a cDNA with a 1-kb amplified region is 30 cycles of:

94°C, 2 min (denaturation)
55°C, 2 min (hybridization of primers)
72°C, 1 min (primer extension)
Followed by:
72°C, 5 min (final extension)
15°C, indefinite (holding temperature until the samples are removed)
b. A typical cycling program for genomic DNA with a 2-kb amplified region is:
94°C, 5 min (initial denaturation)
Followed by 30 cycles of:
94°C, 2 min
60°C, 2 min
72°C, 2 min
Final extension: 72°C, 10 min
Hold: 15°C.
8. Following PCR, remove the samples from the block and add 200 μL of chloroform to each tube. The mineral oil will sink to the bottom.
9. Without mixing, centrifuge the tubes for 30 s at top speed in a table-top microfuge.
10. Transfer the upper phase to a clean microfuge tube. Add 50 μL of 7.5*M* ammonium acetate and mix well (*see* Note 23).
11. Add 375 μL of 95% ethanol and mix. Precipitate for 10 min at room temperature for concentrated samples, or for 30 min on ice for less concentrated products.
12. Centrifuge at top speed in a table-top microfuge for 15 min.
13. Decant the supernatant (*see* Note 24), air-dry the pellet, and resuspend in 20 μL of sterile water.

3.3. Analysis of PCR Products

Both agarose and acrylamide gel electrophoresis may be used to analyze PCR products, depending on the resolution required and whether the sample is to be recovered from the gel. Agarose gel electrophoresis on minigels is fast and easy and allows quick estimates of the purity and concentration of a PCR product. DNA may be recovered much more quickly and efficiently out of agarose gels than out of acrylamide. On the other hand, acrylamide gel electrophoresis provides better resolution and a much more precise estimate of product size (*see* Chapter 11). This is the method of choice for detecting directly labeled PCR products. Denaturing acrylamide gels containing urea may be used to analyze single-stranded products, as from asymetric PCR (*see* Chapter 4).

3.3.1. Agarose Gel Electrophoresis

3.3.1.1. AGAROSE GEL ELECTROPHORESIS
WITH ETHIDIUM BROMIDE STAINING *(6,7)*

This is the method of choice for checking the size and purity of a PCR product before using it in other applications, such as cloning or labeling. Agarose gel electrophoresis may also be used to separate a specific PCR fragment from contaminating sequences. A number of products are now commercially available for extracting DNA out of agarose gels with recoveries of up to 95%.

Never use PCR-dedicated pipets to aliquot concentrated PCR products! Always use filter-containing pipet tips if reamplification of PCR products is desired.

1. To prepare a minigel (5 × 7.5 cm), place 0.25 g of DNA grade agarose in a flask or bottle of at least a 50 mL vol. Add 25 mL of 1X E buffer and swirl. Heat the mixture in a beaker of just boiling water, or in a microwave at about 85% of full power, swirling about every 30 s. It will take 3–5 min for the agarose to dissolve completely, at which point it will no longer be visible as small transparent globules. Cool the solution on the bench-top for a minute, then pour onto a glass plate in a gel-casting stand with a well comb in place (*see* Note 25). Allow about 20 min for the gel to set. Remove the comb. Transfer the solid gel to a gel tank and add enough 1X E buffer to cover the gel by at least several millimeters.

2. To prepare samples for electrophoresis, aliquot the equivalent of at least 1/10 of the PCR product from each reaction to be analyzed into a microfuge tube (2 µL of a precipitated sample resuspended in 20 µL). For a 2-µL sample, add 8 µL of sterile water and 2 µL of 6X agarose gel-loading dye. Mix well and spin briefly in a table-top microfuge to collect all of the fluid.

3. Load the samples into the wells and run the gel on constant voltage at 40 V for about 2 h. The lower dye front will be one-half to two-thirds of the way down the gel.

4. Place the gel in 100 mL of distilled water and add 10 µL of 10 mg/mL ethidium bromide. Shake gently on a rocker or rotating platform for 10 min to stain the DNA. View the DNA by placing the stained gel on a UV lightbox. If the gel is overstained (overall pink background), destain it in 100 mL of distilled water, with gentle shaking for 10–30 min. Destaining can be extended to several hours and can dramatically improve visualization of bands. If the DNA bands are not well resolved, place

the gel back in the gel tank and electrophorese it further. Usually the DNA will not require restaining when electrophoresis is completed.
5. Photograph the gel on the UV illuminator using a Polaroid camera with a yellow filter and Polaroid Polaplan 52 film. A 1-s exposure with the aperture all the way open (f4.5) is usually sufficient.

3.3.1.2. Detection of PCR Products by Southern Blot Hybridization Analysis (6,7)

Agarose gel electrophoresis, followed by Southern blotting and hybridization of a specific probe, allows the detection of a given PCR product in a background of high nonspecific amplification (3). It is also a means of proving that the amplified fragment is related to a known sequence (3,16). Finally, this method can be used to detect PCR products that are still not abundant enough to be detected by ethidium bromide staining (16).

1. Prepare a minigel or an 11×16 cm 1% agarose gel. Follow the instructions for a minigel (Section 3.3.1.1., step 1), but prepare 100 mL of 1% agarose using 1 g of agarose and 100 mL of 1X E buffer.
2. For this application, load half to all of the PCR product of each reaction onto the gel. To prepare the samples, add 1/5 vol of 6X agarose gel-loading dye to each. Mix well and spin briefly in a microfuge to collect all the fluid. Include one sample containing DNA markers of sizes near those expected for the sample bands.
3. Place the gel in a tank, cover with 1X E buffer, and load the samples. Run the gel at 4 V/cm, constant voltage, until the lower dye is about two-thirds of the way down the gel (40 V for about 4 h for large gels). Alternatively, the gel may be run at 40 V for 15 min to allow the samples to run into the agarose, then turned down to 12–15 V and allowed to run overnight.
4. Carefully remove the gel from the tank and place it in enough distilled water to cover it well. Stain the gel with ethidium bromide and photograph it, as in Section 3.3.1.1., steps 5 and 6. It is convenient to align a ruler along one side of the gel in the photograph so that the sizes of bands appearing on autoradiograms of the blotted gel may be estimated by comparing their positions to those of the markers.
5. To denature the DNA, soak the gel in enough denaturing buffer to cover it for 30 min, with gentle shaking (*see* Note 26).
6. Pour off the denaturing buffer and cover the gel in neutralizing solution. Again, soak for 30 min with gentle shaking.
7. While the gel is soaking, prepare a blotting apparatus. Across the middle

of a pan or an unused gel box, lay a glass plate or gel box cover from edge to edge, sideways. Into the pan, or into each side of the gel box, pour 10X SSC to a depth of about 4 cm. Cut three large strips of 3MM filter paper big enough to cover the platform created by the plate and long enough to dip well into the SSC on either side. One at a time, wet the strips in 10X SSC and lay them across the platform, one on top of the other, making sure there are no air bubbles.

8. With a clean, sharp razor blade, cut the gel across the wells and remove the upper piece. Also remove 1–2 mm from the sides and bottom of the gel, where the agarose slopes upwards. The lane containing DNA markers may also be removed.

9. Place the neutralized gel **face down** on the blotting apparatus in the center of the platform. Make sure there are no air bubbles under the gel.

10. Cut a piece of nitrocellulose to the exact dimensions of the gel. Wet the nitrocellulose in deionized distilled water, then soak it for a few minutes in 2X SSC. Carefully lay the nitrocellulose on the gel so that it fits exactly. Make sure there are no bubbles under the nitrocellulose.

11. Cut two pieces of 3MM paper to fit the gel. Moisten them in 2X SSC and lay them on top of the nitrocellulose. Get rid of bubbles.

12. Cut a stack of paper towels to the size of the gel. Pile them on top of the filter papers to a height of about 6–8 cm. Place a glass or plastic plate on top of the towels and center a weight on top of it.

13. Add more 10X SSC to the pan if necessary. Allow the transfer to continue overnight.

14. Remove the weight, the paper towels, and the filter paper. Very carefully lift the nitrocellulose off of the gel by peeling it from one corner. Lay it on a clean piece of filter paper so that the side that was facing the gel is now face up. This is the surface now holding the DNA.

15. To fix the DNA to the nitrocellulose, place the blot on the filter paper in a Stratagene Stratalinker UV crosslinker and crosslink it on the automatic program. (Alternatively, nitrocellulose blots may be placed between two layers of 3MM filter paper and baked at 80°C for 2 h in a vacuum oven.)

16. Put the blot with the immobilized DNA into a hybridization bag and seal three sides. Add 10 mL of prepared prehybridization buffer and squeeze out all the bubbles before sealing the last side of the bag.

17. Place the bag in a 42°C water bath and prehybridize for at least 6 h, with gentle shaking.

18. Prepare a ^{32}P-labeled probe, using 200 ng of plasmid DNA containing the sequence of interest, and following the manufacturer's instructions in the labeling reagent kit. Separate the probe from unincorporated nucle-

otides following the manufacturer's recommendations. Count 3 µL of probe in scintillation fluid to determine the cpm/µL.

19. After the prehybridization, take an aliquot of probe containing 10 million cpm and transfer it to a microfuge tube with a firm-sealing cap. To denature the probe, incubate it in a boiling water bath for 5 min. Chill the tube on ice.
20. Cut open a corner of the hybridization bag containing the blot, add the probe, and reseal the bag, as before.
21. Return the blot to the 42°C bath and hybridize for 2 d.
22. To wash the blot, remove it from the bag (the probe in hybridization buffer may be stored at –20°C and reused once, without reboiling). Wash twice for 15 min at room temperature in 100–200 mL high-salt wash buffer, with gentle shaking. Wash once in low-salt wash buffer at 55°C for 1 h, with shaking.
23. Wrap the washed blot in plastic wrap and expose to X-ray film for 16 h to 1 wk.

3.3.1.3. ANALYSIS OF PCR PRODUCTS BY NESTED PCR

This technique allows the definition of PCR products by reamplification of an internal portion of the DNA. The method takes advantage of the fact that DNA bands in a low-melting agarose gel may be excised and used directly in PCR reactions without further purification *(17,18)*.

1. Resolve amplified product(s) on a gel of 0.7–1% low-melting agarose such as NuSieve (FMC BioProducts). Resolution may be improved by running the gel slowly (12–25 V) at 4°C.
2. Locate the band of interest using UV illumination of the ethidium bromide-stained gel. Excise the band and transfer it to a microfuge tube.
3. Melt the gel slice by incubation at 68°C for 5–10 min.
4. Transfer 10 µL directly into a tube containing the second PCR mix with 100 pmol each of the nested primers and reamplify.
5. Examine the product(s) of the second PCR by agarose gel electrophoresis as described in Section 3.3.1.1.

3.3.2. Detection of PCR Products by Acrylamide Gel Electrophoresis

3.3.2.1. ACRYLAMIDE GEL ELECTROPHORESIS WITH ETHIDIUM BROMIDE STAINING

1. Set up glass plates with 1.5-mm spacers in an acrylamide gel-casting stand. Make sure the seals along the sides and bottom are tight.

2. Make up a 5% acrylamide gel solution in 1X TBE. This solution must be made up fresh for each gel. An 11 × 16 cm gel that is 1.5-mm thick (*see* Note 27) requires 40 mL of this solution, made up as follows: 6.7 mL of 30% acrylamide: 0.8% *bis*, 4 mL of 10X TBE, 29.3 mL of deionized distilled water, 250 µL of 1% ammonium persulfate, and 25 µL of TEMED. Mix the solution well and carefully pipet it between the glass plates. When the plates are almost full, insert a 1.5-mm comb with 10–12 wells. Finish filling the plates. Allow 20–30 min for the gel to polymerize.
3. Prepare the samples by adding 4 µL of 6X acrylamide gel-loading dye per each 20 µL sample (*see* Note 28).
4. Remove the comb from the gel and rinse the wells with distilled water. Assemble the gel in the tank and add enough 1X TBE buffer to the buffer reservoirs to cover the electrodes. The sample wells should be filled with buffer. Make sure there are no bubbles in the wells or along the bottom of the gel.
5. Using a long gel-loading pipet tip, load the samples into the wells (*see* Note 29).
6. Cover the gel tank and attach cables from a power source. Run the gel at 100–125 V (constant voltage) for 3–4 h (*see* Note 30). The lower gel dye should be 2–3 cm from the bottom of the gel when it is done.
7. Prepare a staining tank by lining a shallow dish larger than the gel with a single sheet of plastic wrap larger than the dish. Add about 300 mL of deionized distilled water into the plastic wrap.
8. Remove the gel from the tank. Remove the side clamps and carefully push out the spacers from between the glass plates. Very gently pry open the plates; note which one the gel adheres to and keep that plate facing upwards. Place the plate with the gel face down in the staining tank. Rock gently back and forth to allow the gel to come off the plate. Lift out the plate, leaving the gel in the water.
9. Add 30 µL of 10 mg/mL ethidium bromide and shake gently for 10 min. Discard the staining solution properly.
10. To view the stained gel, lift it carefully onto a UV illuminator on the plastic wrap. Gently smooth out any folds. The gel may be photographed as for stained agarose gels (Section 3.3.1.1., step 6).

3.3.2.2. ACRYLAMIDE GEL ELECTROPHORESIS FOR DETECTION OF DIRECTLY LABELED PCR PRODUCTS

For very sensitive detection and relative quantitation of PCR products, the DNA fragments may be labeled by inclusion of radiolabeled nucleotide in the PCR mix, followed by acrylamide gel electrophoresis and autoradiography. To quantitate the bands, the autoradiograms may be scanned by densitometry, or the labeled bands themselves may be cut out of the gel and counted. We have used autoradiography of directly labeled PCR products to measure the relative levels of several mRNAs in rat ovarian granulosa cells *(19)* and in a pituitary tumor cell line *(11)*.

1. Prepare the gel and samples, run the gel, stain and photograph it, as in Section 3.3.1.1-10). Remember that the gel is radioactive and handle it accordingly.
2. Cut a piece of 3MM filter paper larger than the gel. Lift the gel on the plastic wrap and place it on a flat surface. Smooth out any wrinkles in the gel.
3. Lay the filter paper on top of the gel, it should begin to wet immediately as the gel adheres to it. Turn over the gel, plastic wrap, and filter paper all at once. The gel now has a filter backing.
4. Dry the gel on a gel dryer for 30–45 min. To avoid contamination of the gel dryer, place a second layer of filter paper below the gel.
5. Wrap the dried gel in fresh plastic wrap. Place in a film cassette with X-ray film and expose for 4 h to 2 d.

4. Notes

1. In order to protect RNA from ubiquitous RNases, the following precautions should be followed during the preparation of reagents for RNA isolation and during the isolation procedure: Wear gloves at all times. Use the highest quality molecular biology grade reagents possible. Bake all glassware. Use sterile, disposable plasticware.
2. Guanidine solutions must be sterilized using Nalgene filters because they dissolve Corning filters *(6)*.
3. DEPC breaks down in the presence of Tris buffers and cannot be used to treat them *(6)*.
4. Wear gloves when preparing PCR reagents or performing PCR to prevent contamination.
5. UV irradiation of all solutions used for PCR that do not contain nucleotides or primers is recommended to reduce the chance that accidental contamination of stocks with PCR target sequences will interfere with sample amplifications *(8,9)*.
6. If the prehybridization and blotting solutions (SSC, Denhardt's, SSPE) are to be used for RNA blots as well, the solutions should be made up with precautions as for RNA-grade solutions, and the SSC and SSPE should be DEPC-treated. The 50X Denhardt's may be made up in DEPC-treated water.

7. It is critical that the screw caps of the vials fit flat and tight and that they have O-rings. Ammonium hydroxide is both volatile and corrosive, so the vials must be well sealed to contain the solvent during the heating step. Ill-fitting caps may pop off during the incubation, or worse, when the heated vials are handled.

8. To collect cultured cells for RNA isolation, pour the desired volume of cells in suspension into 50-mL tubes, or remove attached cells from plates by scraping. Avoid enzymatic detachment of plated cells because the enzyme preparations may contain contaminating nucleases. Spin the cells down at 1000 rpm for 4 min at 4°C. Resuspend in 1/10 the original volume of cold PBS and spin down again. For isolation of whole cell RNA, proceed as in Note 10. For nuclear RNA, lyse the cells in 1 mL of PBS plus 0.5% NP-40, incubate for 3 min on ice, then collect the nuclei by centrifugation at 2000 rpm for 5 min at 4°C. Proceed to Note 10.

9. Avoid foaming of the sample during homogenization.

10. For cell or nuclear pellets, gently flick the side of the tube to loosen the pellet, then add 1 mL of GITC/βME, incubate on ice for several minutes, and allow the pellet to dissolve. Proceed as for tissue with shearing of the DNA (Section 3.1.3.1., step 3).

11. An alternative to shearing the DNA is to sonicate the sample in a 1.5-mL microfuge tube. We have used a Virsonic 50 cell disruptor (Virtis, Gardiner, NY) at a setting that delivers a pulse of 40–50% of maximal, for 10–30 s, depending on the viscosity of the sample. To use a sonicator, make sure the tip of the probe is placed all the way at the bottom of the sample tube to prevent foaming. Activate the sonicator only when the probe is immersed, and cool the sample between 10-s pulses if more than one is necessary. Rinse the probe in sterile water prior to use to protect the RNA sample, as well as afterwards to remove the guanidine solution. Ear protection is recommended for the user.

12. The TLS-55 rotor holds four buckets. Although it can be used containing only two samples, all four buckets must be in place during centrifugation.

13. The samples cannot be extracted in the Ultraclear tubes because these tubes are not resistant to organic solvents.

14. Precipitation may also be carried out at −70°C for 1 h.

15. Prolonged isopropanol precipitation at −20°C can precipitate contaminants with the RNA. If the procedure must be halted here, store the samples at 4°C. Resume the isolation at step 4 by incubating the samples at −20°C for 45 min.

16. Several dual-function thermostable enzymes that have both reverse transcriptase and DNA polymerase activities are now commercially available (TetZ, Amersham, Arlington Hts., IL; TTh; *20*). The different

activities rely on differing divalent cations and therefore can be regulated by buffer changes.

17. The lower limit for the amount of RNA required to synthesize a PCR-amplifiable cDNA is beyond the limit of normal detection. We have made cDNAs starting with as little as 0.4 µg of measurable RNA, and with even less of RNAs too dilute to obtain an OD measurement *(11,19)*. Others have used PCR to detected specific mRNAs in RNA/cDNA samples prepared from single cells *(21)*.

18. For some applications, such as tailing, synthesis of a double-stranded cDNA may be required. Second-strand synthesis is achieved by addition of RNase H and DNA polymerase. Procedures for this may be found in manufacturers' information in kits sold for this purpose and in the following references *(6,7)*. Homopolymeric tailing is used to anchor a DNA sequence so that amplification may be performed from a known region out to the tailed end which is unknown *(18)*. Procedures for tailing may also be found in the following references *(6,7)*. Finally, a procedure has been described for addition of a linker to the 3' end of a single-stranded cDNA (to the unknown 5' end of an RNA) using RNA ligase (*see* Chapter 35). This allows subsequent PCR to be carried out between a known region of the cDNA and the linker, and is used to help define the 5' ends of mRNAs. Other PCR applications require removal of the primers used to prepare the cDNA. This may be achieved by several rounds of ammonium acetate precipitation or by purification out of gels.

19. Although the efficiency may be reduced for very large target sequences, we have successfully amplified sequences >6 kb in length.

20. To label PCR products directly, alter the nucleotide mix to contain 625 µM dCTP, 1.25 mM dATP, 1.25 mM dGTP, and 1.25 mM dTTP. To this mix add 5–10 µCi (0.5–1 µL) of α^{32}P-dCTP (3000 Ci/mmol)/100 µL.

21. For PCR programs that are allowed an initial denaturation before addition of the polymerase ("hot-started," *22)*, the reactions are made up lacking a small volume and the *Taq*. The samples are loaded into the block and put through an initial denaturation step, then held at annealing temperature for 2 min. The temperature is then raised to 65°C and the *Taq* is added in a small volume of sterile UV water; the timing of this step is set to allow addition of enzyme to all of the samples. The samples are then allowed to extend at 72°C for the appropriate time and then are cycled normally.

22. More uniform heating and cooling may be achieved if the samples are clustered in the block and surrounded by blank tubes.

23. Ammonium acetate precipitation removes primers and unincorporated nucleotides *(23)*.
24. If the samples are radioactive, be sure to dispose of wastes properly.
25. If a casting stand is not available, place the glass plate on a smooth level surface and position an adjustable well comb over it so that the teeth are *just* above the plate. If a comb with adjustable height is not available, a comb for a larger gel may be used by attaching a black binder clip to each end. Set this apparatus over the gel plate, resting the clips on their sides on the counter-top. Release the clips to allow the teeth of the comb to rest squarely on the plate, then reattach the clips. To bring the teeth to the right height, place a layer or two of paper or toweling under each clip. Using a 10-mL pipet, slowly add the molten agarose to the top until the plate is covered and the solution is gently bowed outward by surface tension. A minigel should hold about 12–15 mL. The total sample volume that can be loaded onto a gel poured in this fashion is about half of what can be loaded onto a cast gel with a similar comb size. (A comb with 6 × 1 mm teeth will leave wells that hold about 20–25 µL in a cast gel.)
26. Smaller gels (minigels) require less time for denaturation, neutralization, and transfer.
27. 0.75-mm Gels are a little more brittle, but save on reagents.
28. 1.5-mm Sample wells will hold up to 100 µL.
29. If a sample starts to float upwards, immediately stop loading and return it to the tube. Floating is usually caused by residual ethanol in the samples. The ethanol cannot be removed, but you can add another 3–5 µL of gel-loading dye, the glycerol content of which effectively "weights" the sample.
30. After running the samples into the gel at 100 V, the gel may be run overnight at 15–20 V.

Acknowledgments

We thank Puja Agarwal and Anna Pappalardo for expert technical assistance. Supported by Grant DK43064 from the NIH.

References

1. Saiki, R., Scharf, S., Faloona, F., Mullis, K. B., Horn, G. T., Erlich, H. A., and Arnheim, N. (1985) Enzymatic amplification of ß-globin genomic sequences and restriction site analysis for diagnosis of sickle cell anemia. *Science* **230,** 1350–1354.
2. Mullis, K. B. and Faloona, F. A. (1987) Specific synthesis of DNA in vitro via a polymerase-catalyzed chain reaction. *Methods Enzymol.* **155,** 335–350.

3. Saiki, R. K., Gelfand, D. H., Stoffel, S., Scharf, S. J., Higuchi, R., Horn, G. T., Mullis, K. B., and Erlich, H. A. (1988) Primer-directed enzymatic amplification of DNA with a thermostable DNA polymerase. *Science* **239**, 487–491.
4. Erlich, H. A., Gelfand, D., and Sninsky, J. J. (1991) Recent advances in the polymerase chain reaction. *Science* **252**, 1643–1651.
5. Robert, S. S. (1991) Amplification of nucleic acid sequences: The choices multiply. *J. NIH Res.* **3(2)**, 81–94.
6. Maniatis T., Fritsch, E. F., and Sambrook, J. (1982) *Molecular Cloning: A Laboratory Manual.* Cold Spring Harbor Laboratory, Cold Spring Harbor, NY.
7. Ausubel, F. M., Brent, R., Kingston, R. E., Moore, D. D., Smith, J. A., Seidman, J. G., and Struhl, K. (eds.) (1987) *Current Protocols in Molecular Biology.* Wiley Interscience, New York.
8. Dycaico, M. and Mather, S. (1991) Reduce PCR false positives using the Stratalinker UV crosslinker. *Stratagene Strategies* **4(3)**, 39,40.
9. Sarkar, G. and Sommer, S. S. (1990) Shedding light on PCR contamination. *Nature* **343**, 27.
10. Zintz, C. B. and Beebe, D. C. (1991) Rapid re-amplification of PCR products purified from low melting point agarose gels. *Biotechniques* **11**, 158–162.
11. Delidow, B. C., Peluso, J. J. and White, B. A. (1989) Quantitative measurement of mRNAs by polymerase chain reaction. *Gene Anal. Tech.* **6**, 120–124.
12. Chomczynski, P. and Sacchi, N. (1987) Single-step method of RNA isolation by acid guanidinium thiocyanate-phenol-chloroform extraction. *Anal. Biochem.* **162**, 156–159.
13. Chomczynski, P. (1991) Isolation of RNA by the RNAzol method. *TEL-Test, Inc., Bulletin* **2**, 1–5.
14. Myers, T. W. and Gelfand, D. H. (1991) Reverse transcription and DNA amplification by a Thermus thermophilus DNA polymerase. *Biochemistry* **30**, 7661–7666.
15. Zimmer, A. and Gruss, P. (1991) Use of polymerase chain reaction (PCR) to detect homologous recombination in transfected cell lines, in *Methods in Molecular Biology, vol. 7: Gene Transfer and Expression Protocols* (Murray, E. J., ed.) Humana, Clifton, NJ, pp. 411–418.
16. Ohara, O., Dorit, R. L., and Gilbert, W. (1989) One-sided polymerase chain reaction: The amplification of cDNA. *Proc. Natl. Acad. Sci. USA* **86**, 5673–5677.
17. Belyavsky, A. (1989) Polymerase chain reaction in the presence of NuSieve™ GTG Agarose. *FMC Resolutions* **5**, 1,2.
18. Belyavsky, A., Vinogradova, T., and Rajewsky, K. (1989) PCR-based cDNA library construction: general cDNA libraries at the level of a few cells. *Nucleic Acids Res.* **17**, 2919–2932.
19. Delidow, B. C., White, B. A., and Peluso, J. J. (1990) Gonadotropin induction of c-fos and c-myc expression and deoxyribonucleic acid synthesis in rat granulosa cells. *Endocrinology* **126**, 2302–2306.
20. Tse, W. T. and Forget, B. G. (1990) Reverse transcription and direct amplification of cellular RNA transcripts by Taq polymerase. *Gene* **88**, 293–296.

21. Rappolee, D. F. A., Mark, D., Banda, M. J., and Werb, Z. (1988) Wound macrophages express TGF-α and other growth factors in vivo: Analysis by mRNA phenotyping. *Science* **241,** 708–712.
22. Rauno, G., Brash, D. E., and Kidd, K. K. (1991) PCR: The first few cycles. *Perkin-Elmer Cetus Amplifications* **7,** 1–4.
23. Crouse, J. and Amorese, D. (1987) Ethanol precipitation: Ammonium acetate as an alternative to sodium acetate. *BRL Focus* **9,** 3–5.

CHAPTER 2

Selection of Primers
for Polymerase Chain Reaction

Wojciech Rychlik

1. Introduction

One of the most important factors affecting the quality of polymerase chain reaction (PCR) is the choice of primers. Several rules
should be observed when designing primers and, in general, the more
DNA sequence information available, the better the chance of finding
an "ideal" primer pair. Fortunately, not all primer selection criteria
need be met in order to synthesize a clean, specific product, since the
adjustment of PCR conditions (such as composition of the reaction
mixture, temperature, and duration of PCR steps) may considerably
improve the reaction specificity. Amplification of 200–400-bp DNA
is the most efficient and, in these cases, one may design efficient
primers simply by following a few simple rules described in this chapter. It is more difficult to choose primers for efficient amplification of
longer DNA fragments, and use of an appropriate primer analysis
software is worthwhile.

The important parameters to be considered when selecting PCR
primers are the ablility of the primer to form a stable duplex with the
specific site on the target DNA, and no duplex formation with another
primer molecule or no hybridization at any other target site. The primer
stability can be measured in the length (base pairs) of a DNA duplex,
the GC/AT ratio, kcal/mol (duplex formation free energy), or in °C
(melting temperature). The most accurate methods for computing helix

From: *Methods in Molecular Biology, Vol. 15: PCR Protocols: Current Methods and Applications*
Edited by: B. A. White Copyright © 1993 Humana Press Inc., Totowa, NJ

Table 1
Free Energy Values of a Nearest Neighbor Nucleotide[a]

First (5') nucleotide	Second nucleotide			
	dA	dC	dG	dT
	ΔG (kcal/mol)			
dA	−1.9	−1.3	−1.6	−1.5
dC	−1.9	−3.1	−3.6	−1.6
dG	−1.6	−3.1	−3.1	−1.3
dT	−1.0	−1.6	−1.9	−1.9

[a]Calculated according to Eq. (1) in 25°C.

stability are based on nearest neighbor thermodynamic parameters
(1). Calculation of T_m according to the nearest neighbor method is
complicated, and therefore not practical for use without computer
software. Similar duplex stability accuracy, however, may be achieved
by calculating the free energy of duplex formation (ΔG). This calculation is simple and can be performed manually.

The Methods section describes the following: an example of ΔG
calculation, needed for accurate determination of duplex stability;
general rules for PCR primer selection; primer design based on a peptide
sequence; and primer design for subcloning PCR products.

2. Methods

2.1. Calculations of DNA Duplex Stability

The method of predicting free energy of duplex formation (ΔG) for
DNA oligomers, described in the following, is a simplified method of
Breslauer et al. *(1)*. It is based on the equation

$$\Delta G = \Delta H - T\Delta S \tag{1}$$

where ΔH and ΔS are the enthalpy and entropy of duplex formation,
respectively, and *T* is the temperature in K. Table 1 lists the ΔG values
of nucleotide pairs.

For simplicity, all calculations are made with *T* set to 298.15 K
(25°C). The relative stability of base-pairing in a duplex is dependent
on the neighboring bases *(1)*. Thus, for example, to calculate the ΔG
of the d(ACGG/CCGT) duplex formation, add the ΔG values of the
three nucleotide pairs as follows:

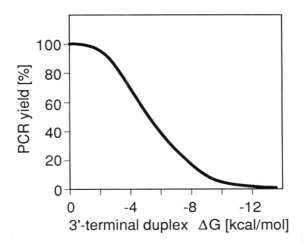

Fig. 1. Dependence of PCR yield on the ΔG of 3'-terminal primer duplexes. The ΔG values were calculated as described in Section 2.1.

$$\Delta G \, (ACGG) = \Delta G \, (AC) + \Delta G \, (CG) + \Delta G \, (GG) \qquad (2)$$

$$\Delta G \, (ACGG) = -(1.3 + 3.6 + 3.1) = -8.0 \, (kcal/mol) \qquad (3)$$

This method is especially useful for determination of primer compatibility owing to formation of 3'-terminal duplexes, discussed in the following section. Use the same approach when calculating the ΔG of a hairpin loop structure, except that the ΔG increment for loop must be added. For loops, sizes 3–8 nucleotide, I use the following values (averaged from refs. 2 and 3): 3 nucleotide, 5.2 kcal/mol; 4 nucleotide, 4.5; 5 nucleotide, 4.4; 6 nucleotide, 4.3; 7 and 8 nucleotide, 4.1 kcal/ mol. More data can be found in ref. 2.

2.2. Selection of PCR Primers

2.2.1. General Rules

2.2.1.1. DIMER FORMATION

PCR primers should be free of significant complementarity at their 3' termini as this promotes the formation of primer–dimer artifacts that reduce product yield. Formation of primer–dimer artifacts may also cause more serious problems, such as nonspecific DNA synthesis owing to an unbalanced primer ratio (asymmetric PCRs fail more frequently than "standard" reactions). Figure 1 illustrates the PCR yield dependence on the ΔG of 3'-terminal duplexes.

These values are approximate, since the yield also depends on the annealing temperature, the specificity of primers, and other parameters not considered here. The high dependence of yield on dimer formation tendency is the result of the very high processivity of *Taq* polymerase. Duplexes need not be stable to prime DNA synthesis. Very little time is required for the enzyme to recognize a 3'-terminal duplex and start polymerization.

2.2.1.2. SELF-COMPLEMENTARITY

In general, oligonucleotides forming intramolecular duplexes with negative ΔG should be avoided. Although self-complementary PCR primers with hairpin loop ΔG approaching –3 kcal/mol (at 25°) are suitable in certain cases, a hairpin loop forming primer is troublesome when its 3' end is "tied up," since this can cause internal primer extension, thus eliminating a given primer from the reaction. Hairpins near the 5' end, however, do not significantly affect the PCR.

2.2.1.3. MELTING TEMPERATURE: STABILITY

There is a widely held assumption that PCR primers should have about a 50% GC/AT ratio. This is not correct. An 81% AT-rich primer (with a second primer of a similar composition and human genomic DNA as substrates) produced a single, specific, 250-bp PCR product (70% AT-rich). Without getting into the complex calculations of product and primer T_m values, PCR primers should have a GC/AT ratio similar to or higher than that of the amplified template.

A more important factor is the T_m difference between the template and the less stable primer. PCR is efficient if this difference is minimized. Note that the T_m of DNA also depends on its length. This is the reason why researchers typically design primers that are too long and unnecessarily too stable. Longer oligos, however, are less likely to be suitable in terms of dimer formation and self-complementarity and, therefore, generally scarce in a given sequence. If the expected PCR product is ≤500 bp, select short (16–18 nucleotide) primers. For the synthesis of a 5-kb fragment, choose about 24-mers. In the latter case, however, it is difficult to choose a compatible primer pair without the aid of primer selection software to check dimer formation, self-complementarity, and the specificity of primers. When amplifying a long DNA fragment, there is a good chance that an oligonucleotide selected "by eye" will prime from other than the intended target site,

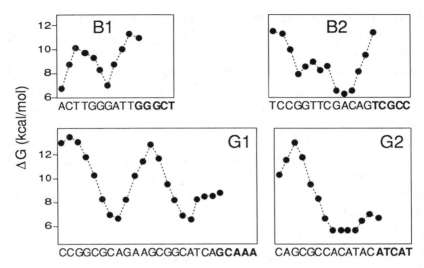

Fig. 2. Internal stability of two poorly functioning (B1, B2) and two efficient (G1, G2) sequencing primers. Primer G1 and G2 performed above average (with almost any other compatible primer) in PCR. The ΔG values were calculated for all pentamers in each primer. The last symbol in each inset represents the ΔG value of the subsequence written in bold (the 3'-terminal pentamer).

yielding nonspecific product(s). The likelihood of false priming can be significantly reduced by observing the internal stability rule, as described in the following.

2.2.1.4. INTERNAL STABILITY

Primers that are stable at their 5' termini but somewhat unstable on their 3' ends perform best in sequencing and PCR as well. This primer structure effectively eliminates false priming. These recent findings, based on primer internal stability, are supported by the experimental data presented in Fig. 2. A primer with low stability on its 3' end will function well in PCR because the base pairings near and at the 3' end with nontarget sites are not sufficiently stable to initiate synthesis (false priming). Therefore, the 5' and central parts of the primer must also form a duplex with the target DNA site in order to prime efficiently. Conversely, oligonucleotides with stable, GC-rich, 3' termini need not anneal with the target along their entire length in order to efficiently prime, resulting often in nonspecific product synthesis. Examples of efficient PCR (and sequencing) primers are presented in Fig. 2 (primers G1 and G2).

Notice the high 3'-end stability of nonspecific primers (B1 and B2) and low stability of specific primers. The optimal annealing temperature range is unusually broad when primers exhibiting low 3'-terminal stability are used. This improves the chances of running the PCR at optimal conditions without preliminary optimization experiments. It is worth noting that the quality of the PCR product depends on the template (substrate complexity, product length, and T_m), as well as on the annealing time and temperature *(4)*. In certain conditions, primers with high 3' terminal stability perform satisfactory in PCR. Nevertheless, oligonucleotides with 3' terminal pentamers less stable than –9 kcal/mol (check Section 2.1. for calculations) are more likely to be specific primers.

2.2.1.5. UNIQUE PRIMERS

In order to amplify a single, specific DNA fragment, the primer's sequence should not repeat in the template *(5)*. Although it is highly unlikely that the entire primer matches perfectly at more than one site on the template, primers with 6–7 nucleotide-long nonunique 3' termini are not uncommon. This may create problems when a "false" priming site is located inside the amplified region. In these cases, a nonspecific product formation is observed (especially in later cycles), because the PCR of shorter DNA fragments is usually more efficient. Note that the more unstable the primer's 3' end, the lower the likelihood of false priming *(see* Section 2.2.1.4.). When working with mammalian genomic sequences, it is helpful to check the primer of interest for complementarity with *Alu* sequences or with other short repetitive elements. For a similar reason, homooligomers (like –AAAAAA–) and dinucleotide repeats (like –ATATAT–) should rather be avoided.

2.2.2. *Specific Applications*

2.2.2.1. PRIMER DESIGN BASED ON PEPTIDE SEQUENCES

When designing primers from peptide sequences, the use of degenerate primers rather than "guessmers" is preferred. Although it has been reported that up to 1024-degenerate primers have been used successfully *(6)*, regions of high degeneracy should be avoided. There are many (unreported) cases in which less degenerate primers have not worked. It is generally assumed that PCR is acceptably efficient when using primers with 15–20% bp mismatches with the template. Mismatches at a primer's 3' end, however, cause more serious problems than the same mismatch ratio at the 5' end. The PCR yield using a primer with two mismatches within the last four bases is drastically reduced. Studies of Kwok et al. *(7)* indicate, however, that primers with 3'-terminal "T"-mismatches can be efficiently utilized by *Taq* polymerase when the nucleotide concentration is high. At 0.8 m*M*, most 3'-end mismatches are acceptable *(7)*, although nonspecific product formation is high, and the fidelity DNA synthesis is reduced *(8)*. There is a low level of priming from mismatched bases even at low nucleotide concentrations *(9)*, and therefore, increasing the annealing time to 3–5 min in the initial PCR cycles may yield a desired product of a better quality than when using standard annealing times and high dNTP concentrations. A total nucleotide concentration of 0.2 m*M*, or below, is recommended when unique primers are used, since high concentrations increase the misincorporation rate *(8,10)*. When degenerate oligonucleotides are used, PCRs should be run at higher primer concentrations (1–3 µ*M* instead of 0.2 µ*M*) because most oligos in the mixture will not prime specifically and only contribute to high background. More information on optimizing the reaction mixture and the use of degenerate primers can be found in Chapters 30 and 31.

2.2.2.2. PRIMER DESIGN FOR SUBCLONING

The addition of a (mismatched) restriction site at the 5' terminus is the most useful method. Add a few "dummy" 5'-terminal bases beyond the recognition site, so that the restriction endonucleases can cut the DNA. Try not to extend a potential dimer structure (inherent to restriction sites) beyond the recognition site. There are no general rules as to how many nucleotides to add. A list of cleavage efficiencies of short oligonucleotides has been published *(11)*; the summary is listed in Table 2.

An alternative to incorporating a full restriction enzyme recognition site is to use oligonucleotide primers with only half a palindromic recognition site at the 5' termini of each phosphorylated primer. After amplification, the PCR product should be concatamerized with ligase and then digested with the appropriate enzyme *(12)*. This is an efficient method, actually forcing a researcher to use high fidelity synthesis conditions *(8,13)*, i.e., low nucleotide concentration, low number of cycles, short extension times, and no "final extension." In these conditions, the formation of 3' overhangs, preventing efficient ligation, is minimal.

If the amplified product is to be subcloned, and the restriction site not needed, use unphosphorylated primers for the reaction and then

Table 2
Cleavable Efficiencies of Short DNA Fragments

Enzyme	Excess bp[a]	%Cleavage after 2 h	20 h	Enzyme	Excess bp[a]	%Cleavage after 2 h	20 h
Acc I	3	0	0	*Not* I	8	25	90
Afl III	1	0	0		10	25	>90
	2	>90	>90	*Nsi* I	3	10	>90
Asc I	1	>90	>90	*Pst* I	1	0	0
BamH I	1	10	25		4	10	10
	2	>90	>90	*Pvu* I	1	0	0
Bgl II	1	0	0		2	10	25
	2	75	>90		3	0	10
	3	25	>90	*Sac* I	1	10	10
BssH II	2	0	0	*Sac* II	1	0	0
	3	50	>90		3	50	90
BstE I	1	0	10	*Sca* I	1	10	25
Cla I	1	0	0		3	75	75
	2	>90	>90	*Sma* I	0	0	10
	3	50	50		1	0	10
EcoR I	1	>90	>90		2	10	50
Hae III	1	>90	>90		3	>90	>90
Hind III	2	0	0	*Spe* I	1	10	>90
	3	10	75		2	10	>90
Kpn I	1	0	0		3	0	50
	2	>90	>90		4	0	50
Mlu I	1	0	0	*Sph* I	1	0	0
	2	25	50		3	0	25
Nco I	1	0	0		4	10	50
	4	50	75	*Stu* I	1	>90	>90
Nde I	1	0	0	*Xba* I	1	0	0
Nhe I	1	0	0		2	>90	>90
	2	10	25		3	75	>90
	3	10	50		4	75	>90
Not I	2	0	0	*Xho* I	1	0	0
	4	10	10		2	10	25
	6	10	10		3	10	75

[a]Number of base pairs added on each side of the recognition sequences.

ligate the product with a *Sma* I-digested vector in the presence of low concentrations of *Sma* I (a blunt-end cutter compatible with the ligation conditions). Again, high fidelity PCR conditions should be used, as mentioned earlier, to minimize formation of 3' overhangs.

When high fidelity synthesis is less essential, one may utilize the template-independent activity of *Taq* polymerase to create 3'-"A" overhangs in the PCR product and use a vector with 3'-"T"-overhangs *(14,15)*. This method is very efficient when high concentration of nucleotides and long extension times are used, followed by prolonged incubation at the extension temperature after the last cycle.

References

1. Breslauer, K. J., Frank, R., Blocker, H., and Markey, L. A. (1986) Predicting DNA duplex stability from the base sequence. *Proc. Natl. Acad. Sci. USA* **83**, 3746–3750.
2. Freier, S. M., Kierzek, R., Jaeger, J. A., Sugimoto, N., Caruthers, M. H., Neilson, T., and Turner, D. H. (1986) Improved free-energy parameters for predictions of RNA duplex stability. *Proc. Natl. Acad. Sci. USA* **83**, 9373–9377.
3. Groebe, D. R. and Uhlenbeck, O. C. (1988) Characterization of RNA hairpin loop stability. *Nucleic Acids Res.* **16**, 11,725–11,735.
4. Rychlik, W., Spencer, W. J., and Rhoads, R. E. (1990) Optimization of the annealing temperature for DNA amplification in vitro. *Nucleic Acids Res.* **18**, 6409–6412.
5. Rychlik, W. and Rhoads, R. E. (1989) A computer program for choosing optimal oligonucleotides for filter hybridization, sequencing and in vitro amplification of DNA. *Nucleic Acids Res.* **17**, 8543–8551.
6. Lee, C. C. and Caskey, C. T. (1990) cDNA cloning using degenerate primers, in *PCR Protocols* (Innis, M. A., Gelfand, D. H., Sninsky, J. J., and White, T. J. eds.), Academic, New York, pp. 46–53.
7. Kwok, S., Kellogg, D. E., McKinney, N., Spasic, D., Goda, L., Levenson, C., and Sninsky, J. J. (1990) Effects of primer-template mismatches on the polymerase chain reaction: human immunodeficiency virus type 1 model studies. *Nucleic Acids Res.* **18**, 999–1005.
8. Eckert, K. A. and Kunkel, T. A. (1990) High fidelity DNA synthesis by the *Thermus aquaticus* DNA polymerase. *Nucleic Acids Res.* **18**, 3739–3744.
9. Petruska, J., Goodman, M. F., Boosalis, M. S., Sowers, L. C., Cheong, C., and Tinoco, I., Jr. (1988) Comparison between DNA melting thermodynamics and DNA polymerase fidelity. *Proc. Natl. Acad. Sci. USA* **85**, 6252–6256.
10. Kawasaki, E. (1990) Amplification of RNA, in *PCR Protocols* (Innis, M. A., Gelfand, D. H., Sninsky, J. J., and White, T. J. eds.), Academic, New York, pp. 21–27.

11. New England BioLabs, 1990–1991 Catalog, "Cleavage close to the end of DNA fragments," p. 132.
12. Jung, V., Pestka, S. B., and Pestka, S. (1990) Efficient cloning of PCR generated DNA containing terminal restriction endonuclease recognition sites. *Nucleic Acids Res.* **18**, 6156.
13. Eckert, K. A. and Kunkel, T. A. (1991) The fidelity of DNA polymerase used in PCR, in *Polymerase Chain Reaction: A Practical Approach* (McPherson, M. J., Quirke, P., and Taylor, G.R. eds.), IRL, Oxford, UK, pp. 227–246.
14. Marchuk, D., Drumm, M., Saulino, A., and Collins, F. S. (1991) Construction of T-vectors, a rapid and general system for direct cloning of unmodified PCR products *Nucleic Acids Res.* **19**, 1154.
15. Holton, T. A. and Graham, M. W. (1991) A simple and efficient method for direct cloning of PCR products using ddT-tailed vectors *Nucleic Acids Res.* **19**, 1156.

CHAPTER 3

Direct Radioactive Labeling of Polymerase Chain Reaction Products

Tim McDaniel and Stephen J. Meltzer

1. Introduction

Radioactively labeled polymerase chain reaction (PCR) products are being used in an increasing number of molecular biology research techniques. Among these are PCR-based polymorphism assays such as linkage analysis *(1)* and detecting allelic loss in cancer cells *(2)*. Other uses of radioactive PCR include generating probes for Southern and Northern blotting *(3)* and screening for polymorphisms and point mutations by the single-strand conformation polymorphism (SSCP) technique *(4)*.

The two most common methods for radioactively labeling PCR products, adding radioactive deoxynucleotide triphosphates (dNTPs) to the PCR mixture *(1)* and end-labeling primers prior to the reaction *(5)*, are problematic in that they risk radioactive contamination of the thermal cycler and extend the radioactive work area. The technique described here *(6)*, based on the methods of O'Farrell et al. *(7)* and Nelkin *(8)*, overcomes these problems because labeling is carried out after the PCR is completed, totally separate from the PCR machine. The technique is rapid, requires no purification of PCR products (e.g., phenol-chloroform extraction), and allows visualization of a tiny fraction of the product in under 2 h after labeling.

From: *Methods in Molecular Biology, Vol. 15: PCR Protocols: Current Methods and Applications*
Edited by: B. A. White Copyright © 1993 Humana Press Inc., Totowa, NJ

The method employs the Klenow fragment of *E. coli* DNA polymerase I, exploiting both the 5'–3' synthetic and 3'–5' excision (proofreading) functions of the enzyme. Briefly, unlabeled PCR product is mixed with Klenow fragment in the absence of dNTPs. Under these conditions, the enzyme lacks necessary substrates for its 5'–3' synthetic function, and thus engages solely in its 3'–5' excision activity, nibbling away bases at the 3' end of the PCR fragments. After allowing this reaction to proceed a short time, dNTPs, including [^{32}P]-dCTP, are added, allowing Klenow fragment to fill in and thereby label the recessed 3' ends.

2. Materials

1. PCR product.
2. Klenow fragment (1 U/μL).
3. 10X Klenow buffer: 500 mM Tris-HCl, pH 7.8, 100 mM MgCl$_2$, 10 mM ß-mercaptoethanol.
4. [α-^{32}P]dCTP (deoxycytidine 5'-[α-^{32}P]triphosphate, triethyl ammonium salt, aqueous solution, 10 mCi/mL, 3000 Ci/mmol).
5. dNTP solution (400 μM each: dGTP, dATP, and dTTP).
6. Sequenase stop solution: 95% formamide, 20 mM EDTA, .05% bromophenol blue, 0.05% xylene cyanol FF.
7. DNA sequencing apparatus (including glass plates, spacers, and combs).

3. Methods

1. Digest the 5' ends of the PCR product by mixing 2 μL of PCR product (from a 100 μL reaction), 2 μL of 10X Klenow buffer, 1 μL of Klenow fragment, and 15 μL of H$_2$O in a microfuge tube.
2. Let stand for 10 min at room temperature.
3. Initiate fill-in of the digested ends by adding 0.5 μL of [α-^{32}P]dCTP and 1 μL of dNTP solution.
4. Let stand 10 min at room temperature (*see* Note 1).
5. Destroy the enzyme by incubating at 70°C for 5 min.
6. At this point, labeled DNA can be incorporated into whatever assay requires it, or can be visualized by running on a denaturing polyacrylamide gel as follows.
7. Mix 8 μL of the final labeling reaction mixture with 2 μL of sequenase stop solution.
8. Denature double-stranded DNA by heating for 5 min at 90°C.
9. Load samples into prerun 8% acrylamide, 50% urea sequencing gel and run 1-1/2 h at a voltage that maintains a surface temperature of 50°C (this voltage varies among sequencing apparatuses).

10. Autoradiograph the gel 1-1/2 h at −70°C with an intensifying screen. *See* Note 2.

4. Notes

1. To ensure complete fill-in of the 3' ends, it may be desirable to chase the reaction with 1 µL of 400 µ*M* dCTP for 5 min just before destroying with the 70°C incubation. However, we have found this step to be unnecessary.
2. To verify the size of the labeled PCR products, run them along with DNA size markers that have been labeled by the same method. To label marker DNA, run the aforementioned protocol substituting 2 µL (2 µg) of marker DNA for the PCR product.

Acknowledgments

Supported by ACS grant #PDT-419, the Crohn's and Colitis Foundation of America, and the Department of Veterans Affairs.

References

1. Peterson, M. B., Economou, E. P., Slaugenhaupt, S. A., Chakravarti, A., and Antonarakis, S. E. (1990) Linkage analysis of the human HMG 14 gene on chromosome 21 using a GT dinucleotide repeat as polymorphic marker. *Genomics* **7**, 136–138.
2. Bookstein, R., Rio, P., Madreperla, S.A., Hong, F., Allred, C., Grizzle, W. E., and Lee, W.-H. (1990) Promoter deletion and loss of retinoblastoma gene expression in human prostate carcinoma. *Proc. Nat. Acad. Sci. USA* **87**, 7762–7766.
3. Schowalter, D. B. and Sommer, S. S. (1989) The generation of radiolabeled DNA and RNA probes with polymerase chain reaction. *Anal. Biochem.* **177**, 90–94.
4. Orita, M., Suzuki, Y., Sekiya, T., and Hayashi, K. (1989) Rapid and sensitive detection of point mutations and DNA polymorphisms using the polymerase chain reaction. *Genomics* **5**, 874–879.
5. Hayashi, S., Orita, M., Suzuki, Y., and Sekiya, T. (1989) Use of labeled primers in polymerase chain reaction (LP-PCR) for a rapid detection of the product. *Nucleic Acids Res.* **18**, 3605.
6. McDaniel, T. K., Huang, Y., Yin, J., Needleman, S. W., and Meltzer, S. J. (1991) Direct radiolabeling of unpurified PCR product using Klenow fragment. *BioTechniques* **11**, 7,8.
7. O'Farrell, P. H., Kutter, E., and Nakanishi (1980) A restriction map of the bacteriophage T4 genome. *Mol. Gen. Genet.* **179**, 421–435.
8. Nelkin, B. D. (1990) Labeling of double stranded oligonucleotides to high specific activity. *BioTechniques* **8**, 616–618.

CHAPTER 4

Use of Arithmetic Polymerase Chain Reaction for Synthesis of Single-Stranded Probes for S1 Nuclease Assays

Puja Agarwal and Bruce A. White

1. Introduction

The S1 nuclease protection assay involves the hybridization in solution of a single-stranded DNA probe to RNA, followed by digestion with S1 nuclease, which is specific for single-stranded nucleic acid *(1)*. The protected probe is measured by first resolving the sample by denaturing gel electrophoresis, followed by autoradiography and densitometry. The amount of protected hybridized probe is proportional to the complementary RNA in the sample. The S1 nuclease assay is sensitive (usually 10- to 100-fold more sensitive than Northern blot hybridization), reproducible, and quantitative. The assay is also more rapid than Northern blot hybridization, since hybridization is performed in a small volume and is thus completed during an overnight incubation, and there is no prehybridization or transfer. The S1 nuclease assay is also more reliable than dot blot assays because the size of the product is determined. Despite these advantages, the S1 nuclease assay is often avoided by investigators who do not wish to invest the time in generating the single-stranded probes. In this chapter, we describe a simple method for generating single-stranded DNA probes by arithmetic polymerase chain reaction (PCR), which we have used in an S1 nuclease assay *(2)* to measure PRL mRNA. This procedure is similar to one reported by Blakeley and Carman *(3)*.

From: *Methods in Molecular Biology, Vol. 15: PCR Protocols: Current Methods and Applications*
Edited by: B. A. White Copyright © 1993 Humana Press Inc., Totowa, NJ

In addition to generating single-stranded probes for S1 nuclease assays, we have used arithmetic PCR for synthesis of single-stranded unlabeled probes, which were used to measure strand-specific transcription in a nuclear run-on transcription assay *(4)*.

2. Materials

2.1. Generation of Single-Stranded Probe

1. Plasmid DNA template: For our studies, we used a PRL cDNA cloned into Bluescript SK+ vector (Stratagene, La Jolla, CA). This construct is digested with *Pst* 1 before use, which removes PRL cDNA insert.
2. Buffered phenol (*see* Chapter 1 for preparation of phenol).
3. $CHCl_3$.
4. *E. coli* tRNA (20 mg/mL in sterile water).
5. 95% Ethanol (cold).
6. 7.5*M* Ammonium acetate.
7. 10X PCR amplification buffer: 100 m*M* Tris-HCl, pH 8.3; 500 m*M* KCl; 15 m*M* $MgCl_2$; 0.01% gelatin (Sigma [St. Louis, MO] G2500), stored at –20°C.
8. Nucleotide mix: 2 m*M* each of dATP, dGTP, dTTP, and 50 µ*M* dCTP prepared in sterile water. Prepare this mix fresh from concentrated stocks of each dNTP.
9. $[\alpha\text{-}^{32}P]dCTP$ (3000 Ci/mmol).
10. 200 pmol of selected primer. We used a 20-bp oligonucleotide that primed the synthesis of an antisense probe from exon 3. The fragment synthesized was approx 250 bp in length.
10. PCR thermal cycler.
11. Sodium acetate, pH 5.2.
12. Mineral oil.
13. TE buffer: 10 m*M* Tris-HCl, pH 7.4, 1 m*M* EDTA.
14. TL-100 Beckman (Columbia, MD) table-top ultracentrifuge and TLA 45 rotor.
15. 1.5-mL Beckman polyallomer microfuge tubes (#357448).
16. Amplitaq *Taq* DNA polymerase (Perkin-Elmer Cetus, Norwalk, CT).

2.2. S1 Nuclease Assay (2)

All reagents should be RNA grade, made (wearing gloves) with sterile, DEP-treated H_2O (*see* Chapter 1), DEP treated, and stored in RNase-free containers.

1. Single-stranded DNA probe freshly prepared by arithmetic PCR (*see* Note 1).
2. Cytoplasmic RNA (*see* Chapter 1).

3. TL-100 Beckman table-top ultracentrifuge and TLA 45 rotor.
4. 1.5-mL Beckman polyallomer microfuge tubes (#357448).
5. 3*M* Sodium acetate, pH 5.2.
6. Cold 95% ethanol.
7. *E. coli* tRNA (10 mg/mL).
8. 400 m*M* PIPES (piperazine-*N,N'-bis*[2-ethanesulfonic acid]). Prepare by adding 12.1 g of PIPES to 75 mL of sterile H_2O. Adjust pH to 6.4 with 10*N* NaOH, and bring volume to 100 mL.
9. 100 m*M* EDTA, pH 8.
10. Hybridization buffer: 80% deionized formamide, 40 m*M* PIPES, pH 6.4, 400 m*M* NaCl, 1 m*M* EDTA, pH 8.0. Store at −70°C in 1-mL aliquots.
11. 1*M* Sodium acetate, pH 4.5.
12. 90 m*M* $ZnSO_4$.
13. S1 nuclease (400 U/µL).
14. 2X S1 nuclease buffer: 0.56*M* NaCl, 0.1*M* sodium acetate, pH 4.5, 9 m*M* $ZnSO_4$. Filter through a 0.2-µm filter and store at 4°C.
15. S1 nuclease stop buffer: 4*M* ammonium acetate, 20 m*M* EDTA, pH 8, and 40 µg/mL of tRNA. Store at 4°C.
16. Salmon sperm DNA. Prepare as described in Chapter 1.
17. Acrylamide solution. Add 116.8 g of acrylamide and 3.2 g of *N,N'*-methylene *bis*-acrylamide to 400 mL of sterile H_2O. Filter through Whatman #1 paper and store at 4°C.
18. Urea.
19. 10% Ammonium persulfate.
20. TEMED.
21. 10X TBE buffer. Prepare by adding 108 g of Tris base, 55 g of boric acid, and 10 mL of 200 m*M* EDTA, pH 8.0, to 1 L of dd H_2O.
22. Formamide loading buffer. Prepare from 98 µL of formamide, 1 µL of 2% xylene cyanol, and 1 µL of 2% bromophenol blue.

3. Methods

3.1. Synthesis of Single-Stranded Probe by Arithmetic PCR

1. Assemble a 50-µL PCR in a 500-µL microfuge tube containing the following:

10X PCR buffer	5.0 µL
dNTP mix	8.0 µL
^{32}P dCTP (5 µCi)	5.0 µL
primer	200 pmol
DNA template	100 ng of PRL cDNA insert (*see* Note 2)
sterile H_2O	to 45 µL final vol

2. Vortex and spin the sample for 5 s. Overlay two drops of mineral oil.
3. Perform PCR using the following cycle profile:
   Step 1   94°C, 2 min (denaturing)
   Step 2   60°C, 2 min (annealing)
   Step 3   65°C, 1 min (HOT START)
4. Add DNA Ampli*Taq* polymerase mix containing 4.5 µL of sterile H_2O and 0.5 µL of *Taq* polymerase during step 3.
5. Continue PCR using the following cycle profile:
   48 main cycles   72°C, 1 min (extension)
                    94°C, 1 min (denaturation)
                    60°C, 1 min (annealing)
   Final extension 72°C, 5 min
6. Extract sample from under the oil layer by pipet and transfer to a clean 1.5-mL Beckman centrifuge tube.
7. Precipitate with 0.5 vol of 7.5*M* ammonium acetate, 2.5 vol of 95% cold ethanol, and 1 µL of 10 mg/mL *E. coli* tRNA. Incubate on ice for 1 h, and spin for 30 min at 12,000*g*, at 4°C in a TLA 45 rotor.
8. Invert tube, dry pellet at room temperature for about 10 min, and resuspend pellet in 50 µL of DEP-treated H_2O.
9. Determine the total amount of incorporated radioactivity by counting a small aliquot (2 µL) in a scintillation counter.
10. A second precipitation may be desirable in order to remove the unincorporated radioactivity from the probe. Use probe on same day of synthesis (*see* Note 3).

3.2. S1 Nuclease Assay

1. Dilute probe in DEP-treated H_2O to 50,000 cpm/10 µL and add 10 µL to 1.5-mL Beckman microfuge tube.
2. Add RNA (0.1–5 µg) and adjust volume to a total volume of 50 µL (*see* Note 4).
3. Add 5 µL of 3*M* sodium acetate, pH 5.2, 1 µL of 10 mg/mL of *E. coli* tRNA, and 137.5 µL of cold 95% ethanol. Incubate at –20°C for 1 h.
4. Centrifuge for 30 min in a TLA-45 rotor in 4°C at 12,000*g*.
5. Decant supernatant and dry pellet at room temperature with tubes inverted.
6. Resuspend pellet in 20 µL of hybridization buffer. Pipet the solution up and down about 50 times until the pellet is completely dissolved and vortex vigorously.
7. Heat the hybridization reaction for 10 min in a 85°C water bath to denature the nucleic acids. Quickly transfer the reaction to a 42°C shaking water bath to incubate overnight for 12–16 h. *See* Note 5.
8. To each hybridization reaction add:

2X S1-nuclease buffer	150 µL
2 mg/mL salmon sperm DNA	3 µL
DEP-treated H_2O	145 µL
600 U of S1 Nuclease	1.6 µL

See Note 6.

9. Incubate in a shaking water bath at 42°C for 2 h (*see* Note 7).
10. During this incubation, prepare a 6% polyacrylamide-urea denaturing gel. Add 16.8 g of urea, 4 mL of 10X TBE, 8 mL of acrylamide solution, and 10 mL of DEP-treated H_2O. Dissolve urea by stirring on a warm hot plate/magnetic stirrer. Adjust volume to 40 mL with H_2O. Degas for 5 min. Add 200 µL of fresh 10% ammonium persulfate and 20 µL of TEMED. Mix gently and pour into 1.5 mm × 18 cm × 16 cm vertical gel mold (assembled according to manufacturer's directions). Let gel polymerize (about 30 min). Wash wells with the running buffer and prerun gel for 1–2 h at 200 V.
11. Chill the S1 nuclease reactions to 0°C by transferring tubes on ice.
12. Add 80 µL of S1-nuclease stop buffer and mix.
13. Add 1 mL of 100% ethanol and precipitate at –20°C for 60 min.
14. Centrifuge at 12,000*g* at 4°C for 30 min.
15. Wash pellet with 100 µL of 70% ethanol and recentrifuge as in step 14. Decant the supernatant, invert tube, and dry pellet at room temperature until ethanol has evaporated.
16. Resuspend the pellet completely in 4 µL of TE buffer and 6 µl of formamide loading buffer.
17. Heat denature the reactions at 95°C for 5 min and transfer the tubes to ice.
18. Resolve samples on polyacrylamide/urea gel at 150 V for about 3–4 h. Dry gel and expose to film.

4. Notes

1. We have purified probes by resolving them on a urea-acrylamide gel, followed by elution of the band. This is done by incubating the gel slice containing probe in 400 µL of 300 m*M* sodium acetate, pH 5.2, at 37°C for 1 h. Transfer liquid to a Ultrafree-MC filter (Millipore, Bedford, MA) and centrifuge in cold room according to manufacturer's instructions. The elution step can be repeated, and the samples pooled before centrifugation through filter unit. Add 1 µL of 10 mg/mL of tRNA and precipitate probe. We have found that gel purification is not usually required, and often results in multiple bands anyway because of degradation of probe during purification.
2. We do not purify the cDNA insert from the plasmid DNA after digestion with *Pst* I. Thus, our calculations of the quantity of DNA used as a template in the PCR is based on the amount of cDNA insert only.

3. The probe should be used as soon as possible after its synthesis to avoid degradation.
4. In the S1 nuclease assay, include an undigested probe with no RNA as a marker for the correct size and digested probe with no RNA as a negative control.
5. The hybridization temperature needs to be optimized for different probes.
6. Prepare solutions as master mixes (with a few extra reactions' worth) and aliquot the appropriate amount into individual tubes.
7. The temperature for S1 nuclease digestion should be optimized.

Acknowledgment

Supported by grant DK43064 from the NIH.

References

1. Berk, A. J. and Sharp, P. A. (1977) Sizing and mapping of early adenovirus mRNAs by gel electrophoresis of S1 endonuclease-digested hybrids. *Cell* **12,** 721–732.
2. Ausebel, F. M., Brent, R., Kingston, R. E., Moore, D. D., Smith, J. A., Seidman, J. G., and Struhl, K. (eds.) (1987) *Current Protocols in Molecular Biology.* Wiley Interscience, New York, Section 4.6.1.
3. Blakeley, M. S. and Carman, M. D. (1991) Generation of an S1 probe using arithmetic polymerase chain reaction. *BioTechniques* **10,** 52,53.
4. Billis, W. M., Delidow, B. C., and White, B. A. (1992) Posttranscriptional regulation of prolactin gene expression in prolactin-deficient pituitary tumor cells. *Mol. Endocrinol.* **6,** in press.

CHAPTER 5

Nonradioactive Labeling
of Polymerase Chain Reaction
Products

Udo Reischl, Rüdiger Rüger,
and Christoph Kessler

1. Introduction

Polymerase chain reaction (PCR) was originally introduced to amplify in vitro particular DNA sequences by the application of temperature cycles *(1)*. In a modification, RNA molecules also may serve as templates by an additional reverse transcription step converting RNA in complementary DNA sequences *(2)*.

In addition to PCR, a variety of alternative amplification systems have been recently developed, which either are also based on temperature cycles (LCR *[3]*, RCR *[4]*) or are designed as isothermal amplification reactions; e.g., transcription amplification (TAS *[5]*, 3SR *[6]*, NASBA *[7]*, LAT *[8]*) or replication approaches (Qß *[9]*).

However, PCR remains the most widely applied amplification method at the moment because of its great potential regarding different modes of applications in molecular biology as well as in other fields like the medical area *(10)*.

In this chapter we describe direct nonradioactive labeling during PCR resulting in labeled amplification products that directly can be quantified (Fig. 1). However, we also focus on the use of PCR for generation of nonradioactively labeled hybridization probes. The advantage of this method is that hybridization probes of strongly reduced vector

From: *Methods in Molecular Biology, Vol. 15: PCR Protocols: Current Methods and Applications*
Edited by: B. A. White Copyright © 1993 Humana Press Inc., Totowa, NJ

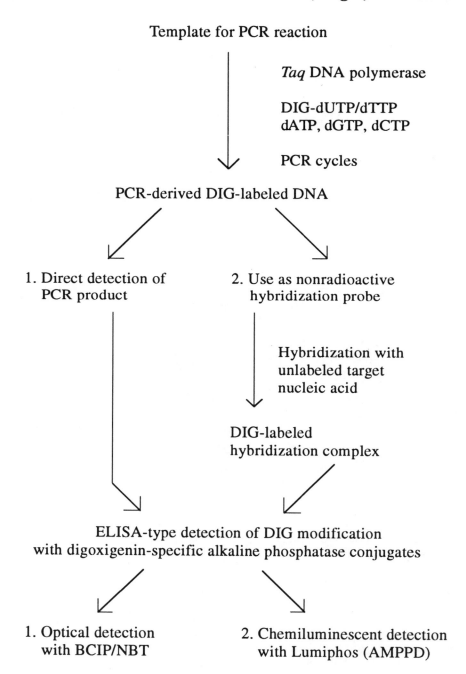

Fig. 1. Flow diagram for detection of PCR-derived DIG-labeled DNA.

content can be synthesized in a one-step procedure. The PCR-generated probes can be applied in hybridization experiments reducing nonspecific background reactions reflecting cross-hybridizations between vector sequences.

PCR amplification is combined with the integration of the nonradioactive modification group digoxigenin (DIG) *(11)*. Digoxigenin is used as a nonradioactive marker because of its high sensitivity and specificity as compared, e.g., with the biotin:streptavidin system. Applying digoxigenin-labeled hybridization probes, sensitivities in the 0.1-pg range are obtained *(12–15)*. Using the digoxigenin label for direct quantitation of the amplified DNA PCR products, detection of single molecules is possible in a nonradioactive way by applying digoxigenin-specific antibodies in an ELISA-type detection reaction *(16)*.

Digoxigenin-labeled hybridization probes can be used under standard hybridization conditions. Hybridized filters are either detected immediately or are stored dry for later detection. After blocking of the membrane with blocking reagent the antibody conjugate is bound to hapten-labeled DNA hybrids.

The indicator reaction is mediated either by the optical substrates BCIP and NBT resulting in blue color products or by use of the chemiluminescent substrate Lumiphos (AMPPD) *(17,18)*. Rehybridization especially with the latter chemiluminescent substrate, is possible after removing the hybridized probe by standard procedures; i.e., 2×15 min in $0.2M$ NaOH and 0.1% SDS at 37°C followed by a short wash in 2X SSC *(19)*.

Thus, the combination of high efficient PCR amplification and high sensitive nonradioactive digoxigenin technology may serve as powerful tool for the isotope-free detection of single DNA molecules as well as for the convenient generation of modified hybridization probes.

2. Materials

2.1. PCR Amplification

1. Automated PCR thermal cycler (Perkin Elmer, Norwalk, CT).
2. Reaction tubes: The tubes must be adapted to the respective kind of PCR thermal cycler to ensure optimal temperature transfer.
3. Oligonucleotides: Single-strand oligonucleotides are synthesized on an automated DNA synthesizer. After deblocking, extraction *(see* Note 1) and ethanol precipitation/washing, no further purification is required prior to use as PCR primers.
4. *Taq* DNA polymerase.

5. Mineral oil, light (Sigma Diagnostics, St. Louis, MO).
6. A set of sterile, autoclavable pipets (Eppendorf, Hamburg, Germany).
7. 10X PCR buffer: 100 mM Tris-HCl, pH 8.5 at 25°C, 500 mM KCl, 15 mM MgCl$_2$, 0.1 mg/mL gelatine (adjust the volume with autoclaved water).
8. dNTP labeling mixture: 1 mM dATP, 1 mM dGTP, 1 mM dCTP, 0.65 mM dTTP, 0.35 mM DIG-[11]-dUTP, or 0.35 mM bio-[16]-dUTP respectively. Use commercially available pH 7.5 preadjusted dNTP solutions.

2.2. Southern Blotting and Hybridization

1. Agarose: SeaPlaque or NuSieve GTG (FMC BioProducts, Rockland, ME).
2. Electrophoresis unit for submarine gels and a constant voltage power supply (Hölzel, Dorfen, Germany).
3. Gel trays (11 × 13 cm) and a 14-teeth comb bridge (Hölzel, Dorfen, Germany).
4. Nylon membrane (11 × 13 cm) (Boehringer Mannheim GmbH, Mannheim, Germany).
5. Whatman 3MM paper (Whatman International, Ltd., Maidstone, England).
6. Glass tray and glass plates or alternatively a vacuum-blotting device (VakuGene, Pharmacia P-L Biochemicals, Milwaukee, WI).
7. UV transilluminator (300 nm) (UVP, San Gabriel, CA).
8. Heatable water bath (GFL, Burgwedel, Germany).
9. Plastic hybridization bags (Life Technologies, Gaithersburg, MD).
10. Plastic bag sealing apparatus.
11. 50X TAE buffer: 2M Tris-HCl, pH 7.8 at 25°C, 250 mM sodium acetate, 50 mM EDTA.
12. Ethidium bromide stock solution: 10 mg/mL (w/v) in water (store in a light-proof container at room temperature). **Caution:** Ethidium bromide is a powerful mutagen and is moderately toxic!
13. Gel loading buffer: 0.4 % (w/v) bromophenol blue, 60% (w/v) sucrose in water.
14. 20X SSC: 3M NaCl, 0.3M Na-citrate, pH 7.0 at 25°C.
15. Denaturation solution: 0.5M NaOH, 0.5M NaCl.
16. Renaturation solution: 1M Tris-HCl, pH 7.4, 1.5M NaCl.
17. Hybridization solution: 5X SSC, 1% (w/v) blocking reagent (Boehringer Mannheim), 0.5% (w/v) N-lauroylsarcosine Na-salt, 0.02% (w/v) SDS, pH 7.0 at 25°C. The blocking reagent does not dissolve very rapidly. Therefore, prepare the solution 1 h in advance by dissolving at 50–70°C. The solution remains turbid (*see* Note 2).
18. Alternative hybridization solution for RNA probing: 5X SSC; 50% (v/v) formamide, 5% (w/v) blocking reagent, 0.1% (w/v) N-lauroyl-sarcosine Na-salt, 0.02% (w/v) SDS, pH 7.0 at 25°C. Dissolve the blocking reagent as previously described.

19. Washing solution 1: 2X SSC; 0.1% (w/v) SDS.
20. Washing solution 2: 0.1X SSC; 0.1% (w/v) SDS.

2.3. Immunological Detection

The chemicals and enzyme conjugates used in this section are available from Boehringer Mannheim, GmbH, Mannheim, Germany.

1. <DIG>:AP-conjugate: polyclonal sheep antidigoxigenin Fab fragments, conjugated to alkaline phosphatase from calf intestine.
2. Streptavidin:AP-conjugate: streptavidin, conjugated with alkaline phosphatase from calf intestine.
3. Lumi-Phos (AMPPD): 3-(2'-spiroadamantane)-4-methoxy-4(3"-phosphoryloxy)-phenyl-1,2-dioxetane (Lumi-Phos is a Trademark of Lumigen Inc., Detroit, MI).
4. Film cassette.
5. X-ray film (Du Pont de Nemours, Bad Homburg, Germany).
6. Buffer 1 (maleate buffer): 100 mM maleic acid, pH 7.5 at 20°C, 50 mM NaCl. Adjust pH with concentrated or solid NaOH.
7. Blocking stock solution: Dissolve the blocking reagent in buffer 1 (maleate buffer) to a final concentration of 10% (w/v) by shaking and heating. Autoclave this stock solution and store at 4°C subsequently.
8. Buffer 2 (blocking solution). Prepare by dissolving the blocking stock solution 1:10 to a final concentration of 1% blocking reagent in buffer 1 (maleate buffer).
9. Buffer 3: 100 mM Tris-HCl, pH 9.5 at 20°C, 100 mM NaCl, 50 mM MgCl$_2$.
10. Buffer 4: 10 mM Tris-HCl, pH 8.0 at 20°C, 1 mM EDTA.
11. NBT solution. Prepare by dissolving 75 mg/mL of nitroblue tetrazolium salt in 70% (v/v) dimethylformamide.
12. X-Phosphate solution. Prepare by dissolving 50 mg/mL of 5-bromo-4-chloro-3-indoyl phosphate, toluidinium salt in pure dimethylformamide.
13. Color substrate solution (freshly prepared): 45 μL of NBT solution and 35 μL of X-phosphate solution are added to 10 mL of Buffer 3.
14. Lumiphos (AMPPD) substrate solution: 10 mg/mL (w/v) Lumiphos (AMPPD). Store the solution at 4°C in the dark.
15. Luminescence solution (freshly prepared): 300 μL of Lumiphos (AMPPD) substrate solution is added to 30 mL of Buffer 3.

3. Methods

In the case of detecting nucleic acids by PCR labeling, the reaction can be started with as low as 3.5–50 ng of target DNA. If the PCR labeled nucleic acid is used as hybridization probe, the starting amount of DNA should be ≥2 ng, and the volume should not exceed 40 μL.

3.1. Labeling via PCR

1. Assemble a PCR (100 µL reaction vol) containing 10 µL of PCR buffer, 20 µL of dNTP labeling mixture, 200 ng of each primer, and 2 U of *Taq* DNA polymerase.
2. Mix the compounds in the reaction tube and add the template DNA. After the addition of the DNA, close each reaction tube before proceeding to the next one. Do not vortex. In a negative control reaction add sterile, autoclaved water instead of template DNA. Overlay reactions with 100 µL mineral oil to prevent evaporation during thermal cycling.
3. For amplification of DNA fragments >500 bp, a three-step PCR is recommended using the following cycle profile *(see* Note 3):
   30 main cycles   94°C, 0.5–2.0 min (denaturation)
                             40–60°C, 0.5–2.0 min (annealing)
                             72°C, 1–2 min (extension)
   Final extension   72°C, 5 min
4. For short sequences (<500 bp), a two-step PCR protocol is recommended using the following cycle profile:
   30 main cycles   92°C, 0.5–1.0 min (denaturation)
                             65–75°C, 0.5–1.0 min (annealing and extension)
   Final extension   72°C, 5 min
5. Perform agarose gel electrophoresis *(see* Chapter 1) with 10-µL aliquots of the PCR-amplified samples in order to verify the correct size of the amplified product.

3.2. Direct Detection of Labeled Material in Southern Blots

3.2.1. Agarose Gel Electrophoresis of the Labeled Products

1. Completely dissolve 1.5 g of agarose in 100 mL of 1X TAE buffer in a 250-mL Erlenmeyer flask by boiling for 5 min in a microwave oven (600 W), then cool the solution to 60°C in a water bath. Adjust the volume again to 100 mL with H_2O, add 5 µL of ethidium bromide stock solution to a final concentration of 0.5 µg/mL and mix thoroughly.
2. Seal the edges of the gel tray with autoclave tape and position the comb 0.5 mm above the plate. Pour the warm agarose solution in the mold. The gel thickness should be between 5 and 8 mm.

After the gel is completely set (30–40 min at room temperature), carefully remove the comb and the autoclave tape and mount the gel to the electrophoresis unit. Cover the gel with 1X TAE electrophoresis buffer to a depth of approx 1 mm.

3. Mix a 10-μL aliquot of the PCR reaction together with 8 μL of sterile H_2O and 2 μL of gel loading buffer.
4. Slowly load each sample into a slot of the submerged gel using a disposable micropipet. Do not produce bubbles and do not overfill slots. For the accurate mol-wt determination of the DNA to be examined, run a set of DNA mol-wt markers in parallel.
5. Close the lid of the electrophoresis unit and apply a voltage of 5 V/cm (distance between the electrodes). If the power supply has been attached correctly, the bromophenol blue should migrate from the wells into the body of the gel.
6. When the bromophenol blue dye is migrated approx 2/3 of the gel-length, examine the gel (without the gel-tray!) by UV light (wavelength 300 nm) with a UV transilluminator (**Caution**: Wear protective eyewear and handle the gel with gloves). Under these conditions, the pattern of the ethidium bromide stained DNA fragments are visualized and can be documented by photography.

3.2.2. Transfer of DNA from Agarose Gels to Nylon Membranes (Southern Blot)

1. Soak the agarose gel by constant shaking in a glass-tray for 30 min with 100 mL of denaturing solution. This alkali step is designed to denature and depurinate double-stranded DNA.
2. To return gel to a neutral pH, displace the denaturing solution with 100 mL of renaturating solution and soak the gel for another 10 min (**Caution**: NaOH causes irritation of the skin and eyes; avoid contact and wear gloves).
3. Two methods for transferring DNA from agarose gels to nylon membranes are recommended, the capillary transfer and the vacuum transfer:
   a. Capillary transfer: Wrap a piece of Whatman 3MM paper around a glass plate. Place the wrapped support inside a large baking dish. The support should be longer and wider than the gel. Fill the dish with 20X SSC almost to the top of the support and smooth out all bubbles in the 3MM paper using a smooth glass rod. Place the gel in a inverted position on the damp 3MM paper (make sure that there are no bubbles between the 3MM paper and the gel). Prewet a nylon membrane (11×13 cm) with renaturation buffer and place the wet membrane on top of the gel. Carefully remove all air bubbles that are trapped between the gel and the filter. Wet two pieces of 3MM paper, cut exactly to the same size as the gel, in 2X SSC and place them on top of the membrane. After removing all air bubbles, cut a stack of paper towels (just smaller than the 3MM paper) and place it on the top. A flow of 20X SSC is set up through the gel and the membrane, eluting the DNA fragments out of the gel and depositing them on the membrane.

Let the transfer proceed for 8–24 h, then remove the paper towels and 3MM papers and let the membrane air-dry.

b. Vacuum blotting: The same result is obtained with a vacuum blotting device. The vacuum transfer is more rapid than capillary transfer. Prewet a nylon membrane (approx 1–2 mm larger than the gel in both dimensions) with renaturation buffer and adjust the wet membrane at the blotting device. Carefully remove all air bubbles that are trapped between the membrane and the porous gel support. After the sealing gasket is positioned, place the renatured agarose gel on top of the membrane and apply the vacuum gently. Cover the gel with 20X SSC to a depth of approx 2 mm. Let the transfer proceed for 30–60 min, discard the buffer, and let the membrane air-dry.

4. Bind the transferred DNA to the nylon membrane by UV crosslinking with a transillumination device for 3 min.

See Notes 4 and 5.

3.2.3. Immunological Detection of the Labeled Amplification Products

1. Wash the membrane briefly in buffer 1, then incubate for 30 min in 100 mL of buffer 2.

2. Dilute <DIG>:AP-conjugate for the detection of digoxigenin-labeled probes in buffer 2 to a final concentration of 150 mU/mL. For the detection of biotin-labeled probes, dilute streptavidin:AP-conjugate to a final concentration of 100 mU/mL, in buffer 2. Soak the membrane for 30 min in 20 mL of the appropriate antibody-conjugate solution with periodic shaking.

3. Remove unbound antibody-conjugate by washing 2 × 15 min with 100 mL of buffer 1 and equilibrate the membrane for 2 min with 20 mL of buffer 3.

4. a. BCIP/NBT detection: Incubate the membrane in the dark with 10 mL freshly prepared color substrate solution in a plastic bag sealed air-bubble free (*see* Note 6). The color precipitate starts to form within a few minutes and the reaction is usually complete after 2–16 h. Do not shake while color is developing. The concentration of the labeled probe affects the sensitivity and time of the detection. When the desired bands are detected stop the reaction by washing the membranes for 5 min with 50 mL of buffer 4. The results can be documented by photography.

b. Lumiphos (AMPPD) detection: Alternatively, the labeled PCR products can be detected by chemiluminescent light emission from alkaline phosphatase activated Lumiphos (AMPPD). Rinse the membrane with 30 mL of freshly prepared Lumiphos (AMPPD) substrate solution for 5 min at room temperature in an appropriate glass tray. Place the wet membrane shortly on a Kleenex™ paper towel and then seal the semidry membrane in a plastic bag. The chemiluminescent light emission starts at room temperature within a few minutes and the results are recorded on X-ray films just like an autoradiograph. The diluted Lumiphos (AMPPD) substrate solution can be stored in the dark at 4°C and reused up to five times (*see* Note 7).

3.3. Hybridization of the Labeled Probe to Immobilized Target DNA/RNA and Detection of Probe-Target Hybrids

3.3.1. Agarose Gel Electrophoresis of the Target DNA and Subsequent Southern Transfer to Membrane (see Note 8)

1. Digest the target DNA with the appropriate restriction enzyme and separate the fragments on an agarose gel (described in Section 3.2.1.). In case of identifying recombinant molecules with this method, 0.3 µg of recombinant lambda phage DNA or 0.1 µg of a recombinant plasmid DNA are sufficient to allow inserted DNA sequences to be easily detected by hybridization with the PCR-labeled probes. Between 0.5 and 10 µg of total mammalian DNA are applied to a single gel slot in order to detect genomic sequences at a single copy level, depending on the nature of the gene to be detected.
2. After electrophoresis is completed, photograph the ethidium bromide stained gel for documentation. Denature and depurinate the target DNA in the gel by 30 min soaking with 100 mL of denaturing solution. Neutralize the gel by displacing the denaturation solution with 100 mL of renaturation solution and soak the gel for another 10 min.
3. A detailed description of the DNA transfer from agarose gels to nylon membranes (Southern blot) is given in Section 3.2.2. of this chapter.
4. Bind the transferred DNA to the nylon membrane by UV-crosslinking with a transillumination device for 3 min.
5. Prehybridize membranes in a sealed plastic hybridization bag with 25 mL of hybridization solution at 68°C (or at 42°C when using 50% formamide) for at least 60 min. Redistribute the solution over the membrane from time to time.
6. Mix 10–300 ng of PCR-labeled probe (*see* Note 9) with 3 mL of hybridization solution, denatured at 100°C for 10 min and chilled quickly on ice.
7. Displace the prehybridization solution in the bag with the hybridization solution containing the freshly denatured (2 min in boiling water bath) labeled probe DNA and incubate the membrane for at least 6 h at 68°C (or at 42°C when using formamide) with hybridization buffer. DNA/ RNA hybridization is performed at 50°C. If strong signals are expected,

it is possible to raise the concentration of labeled probe up to 500 ng and reduce hybridization time to approx 2 h. Redistribute the solution occasionally or use a shaking water bath.

8. Wash the membrane 2 × 5 min at room temperature with 50 mL of wash solution 1 and 2 × 15 min at 68°C with 50 mL of wash solution 2. Nylon membranes can be used directly for detection of hybridized DNA or stored air-dried for later detection (*see* Note 10).

3.3.2. Immunological Detection of Probe-Target Hybrids

The procedure of immunological detection of labeled probe-target hybrids is analogous to the procedure described in Section 3.2.3. of this chapter.

4. Notes

1. After synthesis and deblocking of the oligonucleotides, extract these with 5 vol of 1-butanol. During this step, organic by-products of the synthesis, which may inhibit the *Taq* DNA polymerase, are separated. This might enhance the performance of the reaction.
2. If the blocking reagent is poured into hot water, it dissolves more rapidly.
3. The temperature varies dependent on the primer length and their GC content. The optimal annealing and primer extension temperature has to be evaluated for every individual primer pair. Furthermore, the elongation period should be adapted to the length of the fragment to be amplified. A good approximation is 30 s for every 250 bp.
4. After blotting, mark the gel slots on the wet nylon membrane with a ballpoint pen. This is useful for exact determination of the gel lanes for detection of the labeled material.
5. Digoxigenin-labeled DNA mol-wt markers may be applied (*see* refs. *14,15*). After agarose gel electrophoresis, these labeled mol-wt markers can be directly transferred to the membrane by capillary blotting or with vacuum blotting devices. The labeled markers are detected according to the protocol described in Section 3.2.3.
6. To facilitate the air-bubble free sealing of the plastic bags, stretch the bag and tear it over the edge of the laboratory bench.
7. After the chemiluminescence detection, a color-detection reaction may be added. After 5 min soaking in buffer 3 to remove unbound Lumiphos (AMPPD), the labeled probe-target hybrids are capable to develop the color precipitate even after the chemiluminescence detection. Start the optical color detection by adding 30 mL of the color substrate solution to the membrane sealed in a plastic bag, as previously described in step 4a in Section 3.2.3.

8. If only the presence of a probe-complementary region in the target DNA needs to be detected, use a slot-blot manifold instead of agarose gel electrophoresis and subsequent Southern blotting for transferring the template-DNA to the membrane.
9. The concentration of the probe can be estimated by comparing serial dilutions of the PCR product with a standard DIG label (Boehringer Mannhein Digoxigenin Labeling Kit). The concentration of the labeled probe during hybridization affects the sensitivity and time of the detection reaction and depends on the amount of DNA to be detected on the membrane.
10. The hybridization solution containing labeled probes can be stored at –20°C and reused several times. Immediately before use, redenaturate the probe by heating the hybridization solution at 100°C for 10 min.
11. For further reading, *see* refs. *20–23*.

References

1. Saiki, R. K., Scharf, S., Faloona, F., Mullis, K. B., Horn, G. T., Erlich, H. A., and Arnheim, N. (1985) Enzymatic amplification of ß-globin sequences and restriction site analysis for diagnosis of sickle cell anemia. *Science* **230**, 1350–1354.
2. Murakawa, G. J., Wallace, B. R., Zaia, J. A., and Rossi, J. J. (1987) Method for amplification and detection of RNA sequences. European Patent Application 0272098.
3. Wu, D. Y. and Wallace, R. B. (1989) The ligation and amplification reaction (LAR)—amplification of specific DNA sequences using sequential rounds of template dependent ligation. *Genomics* **4**, 560–569.
4. Segev, D. (1990) Amplification and detection of target nucleic acid sequences— for in vitro diagnosis of infectious disease, genetic disorders or cellular disorders, e.g., cancer. Published under Patent Corporation Treaty (PCT) International Application WO 90/01069.
5. Kwoh, D. Y., Davis, G. R., Whitfield, K. M., Chapelle, H. L., DiMichelle, L. J., and Gingeras, T. R. (1989) Transcription-based amplification system and detection of amplified human immunodeficiency virus. *Proc. Natl. Acad. Sci. USA* **86**, 1173–1177.
6. Guatelli, J. C., Whitfield, K. M., Kwoh, D. Y., Barringer, K. J., Richman, D. D., and Gingeras, T. R. (1990) Isothermal, in vitro amplification of nucleic acids by a multienzyme reaction modeled after retroviral replication. *Proc. Natl. Acad. Sci. USA* **87**, 1874–1878.
7. Davey, C. and Malek, L. T. (1988) Nucleic acid amplification process. European Patent Application 0329098.
8. Schuster, D., Thornton, C., Buchmann, G., Berninger, M., and Rashtchian, A. (1990) A method for isothermal amplification of nucleic acid sequences. 5th San Diego Conference on Nucleic Acids, American Association of Clinical Chemistry (AACC), Abstract Poster 40.
9. Lizardi, P. M., Guerra, C. E., Lomeli, H., Tussie-Luna, I., and Kramer, F. R.

(1988) Exponential amplification of recombinant RNA hybridization probes. *Biotechnology* **6**, 1197–1202.

10. Saiki, R. K., Gelfand, D. H., Stoffel, S., Scharf, S. H., Higuchi, R., Horn, G. T., Mullis, K. B., and Erlich, H.A.(1988) Primer-directed enzymatic amplification of DNA with a thermostable DNA polymerase. *Science* **239**, 487–491.

11. Kessler, C. (1990) Detection of nucleic acids by enzyme-linked immuno-sorbent assay (ELISA) technique: An example for the development of a novel non-radioactive labeling and detection system with high sensitivity, in *Advances in Mutagenesis Research* (Obe, G., ed.), Springer-Verlag, Berlin, pp. 105–152.

12. Kessler, C., Höltke, H.-J., Seibl, R., Burg, J., and Mühlegger, K. (1990) Non-radioactive labeling and detection of nucleic acids: I. A novel DNA labeling and detection system based on digoxigenin:anti-digoxigenin ELISA principle (digoxigenin system). *Biol. Chem. Hoppe-Seyler* **371**, 917–927.

13. Höltke, H.-J., Seibl, R., Burg, J., Mühlegger, K., and Kessler, C. (1990) Non-radioactive labeling and detection of nucleic acids: II. Optimization of the digoxigenin system. *Biol. Chem. Hoppe-Seyler* **371**, 929–938.

14. Seibl. R., Höltke, H.-J., Rüger, R., Meindl, A., Zachau, H. G., Raßhofer, R., Roggendorf, M., Wolf, H., Arnold, N., Wienberg, J., and Kessler, C. (1991) Non-radioactive labeling and detection of nucleic acids: III. Applications of the digoxigenin system. *Biol. Chem. Hoppe-Seyler* **371**, 939–951.

15. Mühlegger, K., Huber, E., von der Eltz, H., Rüger, R., and Kessler, C. (1990) Non-radioactive labeling and detection of nucleic acids: IV. Synthesis and properties of the nucleotide compounds of the digoxigenin system and of photodigoxigenin. *Biol. Chem. Hoppe-Seyler* **371**, 953–965.

16. Rüger, R., Höltke, H.-J., Sagner, G., Seibl, R., and Kessler, C. (1991) Rapid labelling methods using the DIG-system: incorporation of digoxigenin in PC reactions and labelling of nucleic acids with photodigoxigenin. *Fresenius' Z. Anal. Chem.* **337**, 114.

17. Schaap, A. P., Sandison, M. D., and Handley, R. S. (1987) Chemical and enzymatic triggering of 1,2-dioxetanes. Alkaline phosphatase-catalyzed chemiluminescence from an aryl phosphate-substituted dioxetane. *Tetrahedron Lett.* **28**, 1159–1162.

18. Bronstein, I., Edwards, B., and Voyta, J. C. (1989) 1,2-Dioxetanes: novel chemiluminescent enzyme substrates. Applications to immunoassay. *L. Biolum. Chemolum.* **4**, 99–111.

19. Höltke, H. J., Sagner, G., Kessler, C., and Schmitz, G. G. (1992) Sensitive chemiluminescent detection of digoxigenin-labeled nucleic acids: A fast and simple protocol and its applications. *Biotechniques* **12**, 104–113.

20. Innis, M. A., Gelfand, D. H., Sninsky, J. J., and White, T. J. (1990) *PCR Protocols. A Guide to Methods and Applications.* Academic, New York.

21. Kessler, C. (1991) The digoxigenin:anti-digoxigenin (DIG) technology - a survey on the concept and realization of a novel bioanalytical indicator system. *Mol. Cell. Probes* **5**, 161–205.

22. Keller, G. H. and Manak, M. M. (1989) *DNA Probes.* Stockton, New York.

23. Kricka, L. J. (1992) *Nonisotopic DNA Probe Techniques.* Academic, San Diego, CA.

CHAPTER 6

Quantitation and Purification of Polymerase Chain Reaction Products by High-Performance Liquid Chromatography

Elena D. Katz

1. Introduction

The polymerase chain reaction (PCR) has rapidly become a standard laboratory technique. With the continuous development of PCR technology there is now a growing need for PCR product quantitation in areas such as therapeutic monitoring and quality control, disease diagnosis, and regulation of gene expression. One of the most common methods currently employed for post-PCR analysis is agarose or polyacrylamide gel electrophoresis. However, the method is time-consuming and only semiquantitative. In contrast, high-performance liquid chromatography (HPLC) is well accepted as a quantitative technique in many diverse applications areas such as pharmaceutical, biotechnology, food, and environmental, since the technique can provide reliable, precise, and sensitive sample detection, and wide dynamic range.

During the last two decades, column liquid chromatography has been employed for the separation, purification, and detection of nucleic acids. Different modes of chromatography of nucleic acids described previously have been the subject of a recent review *(1)*. Of the different chromatography methods, anion-exchange chromatography has been most commonly employed for the isolation and purification of not only oligonucleotides but also large double-stranded DNA. In

From: *Methods in Molecular Biology, Vol. 15: PCR Protocols: Current Methods and Applications*
Edited by: B. A. White Copyright © 1993 Humana Press Inc., Totowa, NJ

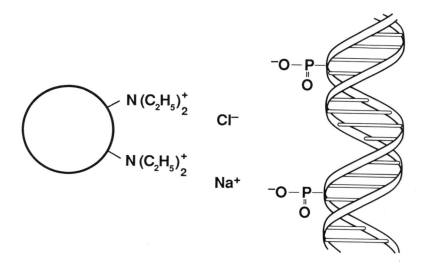

Fig. 1. Anion-exchange chromatography of DNA.

anion-exchange chromatography, DNA retention is dependent on electrostatic interactions between the negatively charged phosphate groups of DNA and cationic sites of the chromatographic matrix. This process is schematically depicted in Fig. 1. DNA can be eluted from the anion-exchange column by altering the ionic strength of the buffer solution. In the presence of a buffer of increasing ionic strength, DNA retention is generally a function of the number of negative charges associated with the phosphate groups.

One of the most common anion-exchange materials is diethylamino ethyl (DEAE) bonded support, and it has been extensively used to modify original soft-gel supports, porous microparticular silica, and polymer-based materials. Availability of porous microparticular silica and polymer-based supports has led to enhanced column resolution and faster analysis times so that HPLC has become one of the dominant analytical methods in bioresearch. However, the columns packed with porous particles have been used only with limited success in the analysis of large double-stranded DNA fragments. Recently, new HPLC columns based on nonporous, small particle resins have become available, and very fast and efficient separations of large double-stranded DNA have been demonstrated *(2)*. The success of these columns is based on the fact that nonporous, small-particle size material with

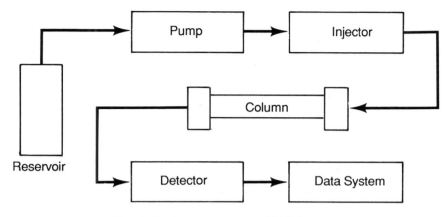

Fig. 2. Basic components of HPLC system.

which they are packed provides improved efficiency and short analysis time. This is owing to a much faster DNA transfer between the solid and liquid column phases than is possible for conventional porous support. In addition, these columns offer high DNA recoveries owing to the absence of pores.

It has been recently demonstrated that HPLC can be successfully employed for post-PCR analyses *(3,4)*. A typical HPLC system consists of an HPLC pump, sample injector, column, UV absorbance detector, and data-handling device, as is schematically represented in Fig. 2. A manual sample injector can be replaced by an autosampler, a fraction collector can also be added. Thus, the HPLC system can easily be automated and employed for the separation, quantitation, and purification of PCR products in a single step. In this chapter, the automated HPLC method for the rapid quantitation and purification of PCR products is discussed.

2. Materials

1. pBR322 DNA-*Hae* III digest.
2. 500-bp PCR product *(see* Note 1).
3. 115-bp HIV PCR product *(see* Note 2).
4. HPLC-grade water.
5. HPLC buffer A: 1*M* NaCl, 25 m*M* Tris-HCl, pH 9.0.
6. HPLC buffer B: 25 m*M* Tris-HCl, pH 9.0.
7. 0.2*N* NaOH.
8. 30% Acetonitrile in HPLC-grade water.

9. An automated biocompatible HPLC system: a binary pump, a UV-VIS detector, an autosampler, a data-handling device, an LC column oven (in this work, all instruments were from Perkin-Elmer, Norwalk, CT).

10. Anion-exchange analytical column: a PE TSK DEAE-NPR column (35-mm long and 4.6-mm id) packed with 2.5-μm particles of hydrophilic resin, bonded with DEAE groups.

11. Guard column: a PE guard column, 5-mm long and 4.6-mm id, packed with 5 μm of DEAE-NPR material.

12. UV spectrophotometer.

13. TE buffer: 10 mM Tris-HCl, pH 8, 1 mM EDTA.

3. Methods

3.1. HPLC Separation of DNA Fragments

1. Prepare an HPLC system for PCR analysis according to manufacturer's manual.

2. Make buffers A and B using HPLC-grade water.

3. Place a DEAE-NPR guard column followed by the analytical DEAE-NPR column between the injector valve and the UV detector.

4. Program an appropriate gradient profile according to the manufacturer's HPLC pump manual. A gradient protocol is recommended as follows:
   a. Step gradient from 44–55% A in B, for 0.1 min;
   b. Linear gradient from 55–61% A in B, for 3.5 min;
   c. Linear gradient from 61–100% A in B, for 0.5 min;
   d. Hold at 100% A for 1 min;
   e. Linear gradient from 100–44% A in B, for 0.1 min.

5. Select an appropriate starting buffer composition such as the one described in step 4 (44% A in B). The starting buffer composition can be changed, depending on the size of DNA fragments. Start with a lower molar concentration of NaCl (25–30% A) if the separation of PCR primers is desirable.

6. Equilibrate the column with the starting buffer composition for 15–30 min at 1 mL/min.

7. Obtain a blank chromatogram by injecting 10–20 μL of buffer A to establish a flat detector baseline. This will confirm that the total system is clean (potential sources of contamination, besides the column, could be the injector and connecting tubes).

8. Establish the performance of a new HPLC column by injecting an appropriate DNA standard. An example of the HPLC separation of an *Hae* III digest of pBR322 DNA employing the DEAE-NPR column is shown in Fig. 3.

9. Inject 10–50 μL of a PCR sample for PCR product separation and detection (*see* Note 3). Use the same operating conditions as in step 8 for the purpose of subsequent product identification.

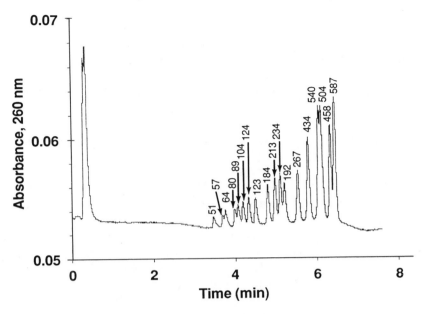

Fig. 3. Separation of *Hae* III digest of PBR322 DNA, operating conditions as described in the text. Sample concentration: 50 μg.mL in TE buffer. Injection size was 10 μL.

10. Identify a PCR product by matching its time of elution from the column with that of an appropriate DNA fragment in the chromatogram obtained in step 8.

3.2. HPLC Purification of PCR Products

1. Inject 100–200 μL of a given PCR sample onto the column.
2. Collect, manually or automatically, a fraction of the column buffer containing the PCR sample at the detector exit line during the chromatographic development at the appropriate time *(see* Note 4).
3. Desalt the purified PCR products using an established procedure *(5)*, if further PCR sample manipulations such as sequencing, cloning, or reamplification require a lower salt concentration *(see* Note 5).

3.3. HPLC Quantitation of PCR Products

1. Ensure that the performance of the HPLC system is reproducible *(see* Note 6).
2. Inject successively two 150-μL vols of the amplified 500-bp fragment.
3. Collect the eluted products at appropriate elution times *(see* Note 4).
4. Note their peak areas.

5. Concentrate the samples using a SpeedVac (Savant Instruments, Inc., Farmingdale, NY) and reconstitute in a known volume of TE buffer.
6. Take an absorbance reading of the PCR sample, obtained in step 18, using a UV spectrophotometer, set at 260 mm *(6)*.
7. Calculate the concentration of the PCR product (1 OD~50 µg/mL) and its amount in the buffer volume used in step 5.
8. Calculate the ratio of the amount of PCR product:its peak area.
9. Use this ratio to calculate the concentration of PCR product to be quantitated *(see* Note 7).

4. Notes

1. 500-bp product amplification. Amplifications were carried out using the Perkin-Elmer Cetus GeneAmp® PCR Reagent Kit (Norwalk, CT). A 500-bp segment of bacteriophage lambda DNA (nucleotides 7131–7630) was used as a target, the initial concentration of which was 10 pg/100 µL or $3 \times 10^{-13}M$. Using the Perkin-Elmer Cetus GeneAmp™ PCR System 9600, two-temperature step cycle PCR was carried out for 15 s at 95°C and 90 s at 68°C for 25 cycles, the final extension step was performed at 68°C for 6 min.
2. 115-bp product amplification. A 115-bp HIV product (PCR sample courtesy of Will Bloch of Cetus Corporation) was amplified using GeneAmplimer™ HIV-1 Control Reagents, the DNA Thermal Cycler (both Perkin-Elmer Cetus), and two-temperature step cycle PCR as follows: during the first two cycles, the target was denatured for 1 min at 98°C and annealed/extended for 2 min at 60°C. The subsequent cycles were carried out for 1 min at 94°C and for 1 min at 60°C for 38 cycles, the final extension step was performed for 10 min at 72°C. The initial concentration of this target was 10 copies in a 100-µL reaction vol. The 115-bp HIV product was amplified in the presence of a high genomic DNA background (10 µg crude human placental DNA). A hot start technique *(7)* was used to enhance the PCR specificity.
3. Sample volume. This volume is dependent on PCR product yield after amplification. Examples of PCR product separation are shown in Figs. 4A and 5A. A 10-µL injection vol was sufficient to clearly detect both amplified products. In some cases, injection of 50 µL may be necessary. For purification purposes, 100–200 µL can be injected without column overloading *(see* Figs. 4B and 5B).
4. Purification. Note in Fig. 4B that some nonspecific products, albeit at a very low level, were detected. However, these did not interfere with sample collection since nonspecific fragments were well separated from the 500-bp peak. During the development of this chromatogram, the

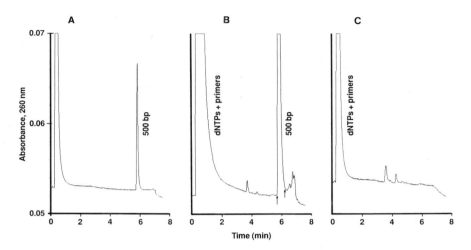

Fig. 4. (**A**) Separation of 500-bp product amplified as described in text. HPLC conditions as in Fig. 3. Injection size: 10 µL. (**B**) Separation of the same 500-bp product. Injection size: 100 µL. (**C**) Chromatogram of column eluent collected as described in text and amplified as in **A**. Injection size: 10 µL.

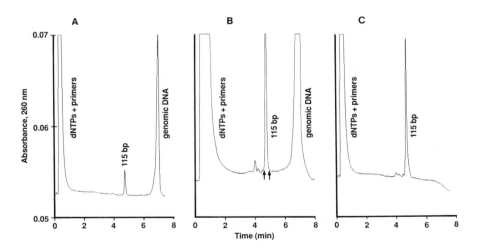

Fig. 5. (**A**) Separation of 115-bp HIV-1 product amplified as described in text. HPLC conditions as in Fig. 3. Injection size: 10 µL. (**B**) Separation of the same 115-bp product. Injection size: 100 µL. (**C**) Separation of 115-bp product reamplified as described in text. Injection size: 10 µL.

500-bp product was collected between 5.69 and 6.13 min (shown by the arrows in the figure). The 115-bp HIV fragment was collected between 4.64 and 4.90 min (corresponding arrows in Fig. 5B).

5. Reamplification. The HPLC-purified, 115-bp product was desalted using a Centricon R-30 microconcentrator (Amicon, Beverly, MA). A 10-µL sample vol was then amplified using the GeneAmplimer HIV-1 reagents and the GeneAmp PCR System 9600. The two-temperature step cycle PCR protocol consisted of denaturing for 15 s at 95°C and annealing/extending for 60 s at 60°C for 35 cycles. The final product was extended at 60°C for 9 min. The results of the HIV product reamplification are shown in Fig. 5C.

6. Reproducibility of the HPLC system. The HPLC performance can be monitored by periodically injecting an appropriate DNA standard such as an *Hae* III-digest of pBR322 DNA. The reproducibility data for the peak area and retention time for a 434-bp fragment of this digest are given in Table 1. Table 1 demonstrates that both the intraday and interday retention time reproducibility can be obtained within 1% relative SD. The interday peak area reproducibility can be obtained with 4% relative SD whereas the intraday precision can be within 2%.

7. Quantitation of PCR products. In the HPLC quantitation of PCR products, an area under a given PCR-amplified DNA peak is directly related to the DNA concentration. The PCR product quantitation can be performed using two different approaches. One approach is to generate an LC calibration curve employing an appropriate DNA standard quantitated by UV absorbance *(4)*. The LC calibration yields a straight line when the UV absorbance response is plotted against the mass or concentration of DNA injected onto the column. The second approach does not involve the generation of the calibration curve and is based on the measurement of UV absorbance of a PCR product collected during a chromatographic run. To quantitate the concentration of the collected PCR product, using a UV spectrophotometer, a reading is taken at 260 nm (1 OD-50 µg/mL of ds DNA), and the concentration values obtained are then related to the HPLC peak area corresponding to the PCR-product peak. This procedure can be recommended as follows. Two 150-µL portions of the amplified fragment are injected successively and the eluted products are collected in a volume of approx 400 µL. The samples are then concentrated using a SpeedVac and reconstituted in a known volume of 10 m*M* Tris-HCl, pH 9, 1 m*M* EDTA (TE) buffer. The amount of the PCR product is quantitated using a UV spectrophotometer. In this work, this protocol was employed to quantitate HPLC-purified, 500-bp product in 300-µL TE buffer. The same procedure was repeated for

Table 1
Peak Area and Retention Time Reproducibility

Day	Retention time (min)	Area (counts; arbitrary U)
1	5.900	276
	5.890	277
2	5.853	289
	5.903	286
	5.913	292
	5.910	281
	5.867	290
3	5.730	287
	5.730	291
4	5.810	293
	5.823	285
	5.853	299
	5.910	295
	5.833	292
	5.853	301
5	5.873	302
	5.813	305
	5.947	304
6	5.950	285
	5.930	290
	5.927	287
7	5.837	303
	5.830	306
	5.840	309
8	5.860	316
	5.863	315
	5.860	316
	5.880	313
	5.887	314
9	5.853	309
	5.887	307
	5.913	294
Mean	5.867	297
Relative SD (%)	0.9	3.9
4	5.810	293
	5.823	285
	5.853	299
	5.910	295
	5.853	301
Mean	5.847	294
Relative SD (%)	0.6	1.9

two 100-μL injections of the same PCR product, but the concentrated sample was reconstituted in 100-μL TE buffer. The results relating the DNA amount in the purified sample to the HPLC peak area are demonstrated in Table 2. The mean value of 0.139 ng/area, which was a constant as expected, can now be used to quantitate any separated PCR-amplified DNA fragment. For example, the area of the 500-bp peak (Fig. 4A) was equal to 855 counts (arbitrary units). Multiplying the value of 0.14 ng/area by the peak area of 855 counts, a value of 119 ng is obtained. If an injection size is 10 μL, the concentration of the 500-bp fragment is equal to 11.9 ng/μL or, if the sample was amplified in 100 μL, to 1190 ng in 100 μL.

8. HPLC sample cross-contamination. The exquisite sensitivity of PCR may lead to a serious problem of sample cross-contamination. To check the sample cross-contamination problem, immediately after the injection of a PCR fragment, the column should be washed by injecting 100 μL of 1*M* NaCl under the same gradient conditions. A volume of the eluting mobile phase can be collected during an appropriate time as indicated by the arrows of Fig. 4B. The resulting eluent should then be desalted using the Centricon-30 microconcentrator, and 10 μL of the concentrate amplified following an appropriate protocol. The sample cross-contamination was checked for the 500-bp product amplification. The results are shown in Fig. 4C, which clearly shows no detectable sample cross-contamination. It also shows that this amplification yielded nonspecific products, which was expected since no DNA template was present in the PCR mixture.

9. System contamination. Presence of additional peaks that cannot be attributed to PCR product being analyzed will indicate that the HPLC system has been contaminated. In general, sources of contamination could be DNA sample carryover from one injection to another and mobile phase impurities. To prevent sample carryover, the syringe (in the case of manual injections) and the injector loop should be rinsed with a strong solvent, e.g., 1*M* NaCl, prior to each sample injection. To prevent the system contamination by the mobile phase impurities, high-purity buffers and salts and HPLC-grade water must be used. To ensure that the HPLC system is clean, blank chromatograms demonstrating a flat detector baseline must be periodically obtained by injecting 10–50 μL of 1*M* NaCl.

10. High-inlet column pressure. Column inlet pressure is primarily dependent on column length, particle size, and flow rate. A normal operating pressure for DEAE-NPR columns employed in this work, should be in the range of 1000–1500 psi at a flow rate of 1 mL/min at room temperature. The column pressure can increase during the post-PCR analysis

Table 2
Purified Product Quantitation

Injection size, μL	Peak area, arbitrary U	Absorbance, OD	Concentration μg/mL or ng/μL	Amount, ng	Amount/Area, ng/area
			in 300 μL		
150	14,705	0.130	6.51	1953	0.133
150	14,831	0.132	6.62	1986	0.134
			in 100 μL		
100		0.321	16.06	1606	0.157
100		0.249	12.45	1245	0.133

owing to (1) partial clogging of connecting tubing that, in turn, can be caused by particulates from the mobile phase or injection devices, and (2) column frit clogging. To prevent the connecting tubing from clogging, in-line solvent filters should be installed in the pump outlet line before the injector. A column inlet filter and/or guard column will retain any particulate matter resulting from the rotor seal ware of the injector. HPLC columns packed with sub-5-μm particles contain low-porosity (1 μm) column frits that are more prone to clogging than the larger frits in conventional columns. Therefore, guard columns should be employed to protect the analytical columns.

11. Column maintenance. DEAE-NPR columns should be washed daily to maintain column performance during routine use. Using the injector valve, 1–2 mL of 0.2N NaOH (by repeat injections) should be injected. If PCR mixture that contains hydrophobic material such as mineral oil is repeatedly injected onto the column, increases in column pressure can be observed. Injecting 1–2 mL of pure organic solvent (e.g., acetonitrile, methanol, or ethanol) helps keep the column pressure stable. The column should be rinsed thoroughly with HPLC-grade water and filled with 20–30% organic solvent (e.g., acetonitrile) in water before removing from the system. Regular cleaning and thermostatting the column will ensure reproducible column performance during the post-PCR analysis.

References

1. McLaughlin, L. W. (1989) Mixed-mode chromatography of nucleic acids. *Chem. Rev.* **89**, 309–319.
2. Kato, Y., Yamasaki, Y., Onaka, A., Kitamura, T., Hashimoto, T., Murotsu, T., Fukushige, S., and Matsubara, K. (1989) Separation of DNA restriction frag-

ments by high-performance ion-exchange chromatography on a nonporous ion exchanger. *J. Chromatogr.* **478**, 264–268.

3. Katz, E. D., Eksteen, R., Haff, L. A. (1990) Rapid separation, quantitation and purification of products of polymerase chain reaction by liquid chromatography. *J. Chromatogr.* **512**, 433–444.

4. Katz, E. D. and Dong, M. W. (1990) Rapid analysis and purification of polymerase chain reaction products by high-performance liquid chromatography. *BioTechniques* **8**, 546–555.

5. Allard, M. W., Ellsworth, D. L., and Honeycutt, R. L. (1991) Uses for Centricon-30 microconcentrators. *Lab News. Amicon* **(Spring)**, 4,5.

6. Maniatis, T., Fritsch, E. F., and Sambrook, J. (1989) *Molecular Cloning. A Laboratory Manual* (2nd ed.), Cold Spring Harbor Laboratory, Cold Spring Harbor, New York.

7. Erlich, H. A., Gelfand, D., and Sninsky, J. J. (1991) Recent advances in the polymerase chain reaction. *Science* **252**, 1643–1651.

CHAPTER 7

Use of Polymerase Chain Reaction for Screening Transgenic Mice

Shizhong Chen and Glen A. Evans

1. Introduction

The production of transgenic mice by the direct microinjection of DNA fragments into isolated mouse embryos is now a standard technique for molecular and developmental analysis (1). Traditionally, the initial screening of litters of mice for those carrying the transgene has been carried out by the preparation of DNA from pieces of excised tail obtained at 2–3 wk of age ("tail blot"). The presence of the transgene is determined through Southern blot analysis, a procedure that usually takes a minimum of 5 d. Two important criteria for successful screening are that the tail DNA is sufficiently purified from connective tissues and proteins to enable restriction enzyme digestion, and that the amount of DNA is large enough to allow the detection by hybridization of radioisotope-labeled probes recognizing the transgene or endogenous gene sequences.

Polymerase chain reaction (PCR) allows amplification of very small amounts of DNA using specific oligonucleotide primers in a short time. The amounts of DNA necessary are small enough that adequate samples may be obtained from single drops of blood or small samples of tissue from "ear punches." It is also possible to carry out the PCR analysis on unpurified DNA so that time-consuming DNA preparation and purification are avoided. In the procedure described here, DNA is dissociated from the connective tissues and proteins by treatment with detergents and proteinase K digestion followed by dilution of the sample

From: *Methods in Molecular Biology, Vol. 15: PCR Protocols: Current Methods and Applications*
Edited by: B. A. White Copyright © 1993 Humana Press Inc., Totowa, NJ

in buffer. After dilution, the content of contaminants and other substances that might inhibit PCR amplification become very low and irrelevant to the reaction. PCR amplification may then be carried out directly without further DNA purification.

A number of rapid screening methods for transgenic mice have recently been developed based on PCR amplification *(2–8)*. All of these methods are similar in the conditions for PCR amplification, but differ in the amount of DNA preparation required, the source of the initial DNA (including samples of tail, blood, or ear punches tissue), and some other minor details. In our hands, the method that follows allows rapid screening of transgenic animals by PCR and requires a minimum of effort *(5)*. We also find this method to be extremely reproducible and reliable with a large variety of PCR amplification primers. Basically, small pieces of tissue obtained from marking infant mice with an ear punch are collected and digested in a one-step, detergent-proteinase K treatment, diluted 10-fold and boiled to denature and dissociate the proteins from DNA. One four-hundredth of the boiled sample is then used for PCR amplification and the resulting amplification products analyzed on an agarose gel. This method has a number of distinct advantages over alternate methods:

1. It is rapid in that results can be obtained in a matter of a few hours and a large number of nontransgenic animals discarded at an early stage in the study.
2. Ear punching is a routine procedure for marking the young animals and tissue may be used directly for PCR amplification.
3. Ear punching appears to be less traumatic to young mice and can be carried out at an earlier age than collecting blood or tail tissue. It also minimizes the risk of rejection of the pups by the mother.
4. Ear tissue contains less connective tissue than tail samples and can be completely digested in <1 h, compared to tail tissue, which must be digested overnight;
5. Ear tissue contains less contaminating factors that may inhibit PCR amplification than blood samples.

Although this method was originally developed for the screening of transgenic mice, it can equally well be applied to other research problems that require rapid screening of large number of samples. One example is the screening of transfected cell lines for integration of gene sequences. We and others have utilized a similar method *(9)* for that purpose.

2. Materials

2.1. DNA Extraction

1. Mouse ear puncher.
2. Digestion buffer: 50 mM Tris-HCl, pH 8.0, 20 mM NaCl, 1 mM EDTA, 1% SDS, stored at room temperature.
3. Proteinase K stock solution: prepared at 20 mg/mL in distilled water and stored at –20°C.

2.2. PCR Amplification

1. 10X PCR amplification buffer: 670 mM Tris-HCl, pH 8.8, 160 mM (NH$_4$)$_2$SO$_4$, 100 mM ß-mercaptoethanol, 15 mM MgCl$_2$ *(see* Note 1), stored at –20°C *(see* Note 2).
2. 100% DMSO stock solution.
3. 100X PCR primers stock solution: The sequences of PCR primers depend on the characteristics of the gene to be detected *(see* Chapter 2). In general, we use standard 30-bp primers selected to have about 60% GC content. Primers are synthesized on an ABI-PCR mate oligonucleotide synthesizer and used directly without further purification. Dilute each primer to a concentration of 25 µM in water and store at –20°C.
4. 100X dNTP stock solution: 20 mM of each dATP, dGTP, dCTP, and dTTP are prepared in water, aliquoted, and stored at –20°C. Repeated freezing and thawing are to be avoided.
5. *Taq* polymerase used at a stock concentration of 5 U/µL.

3. Methods

3.1. DNA Extraction

1. Prepare the digestion mixture fresh by adding 1 part of proteinase K stock solution to 20 parts of digestion buffer. Add 20 µL of the digestion mixture to each 0.5-mL Eppendorf tube.
2. Collect fresh tissue from ear punching and add one piece (about 50 mg) to each Eppendorf tube containing the digestion mixture *(see* Notes 3, 4, and 5). Close the caps of the tubes tightly and incubate for 15 min at 55°C.
3. Mix the samples vigorously, and incubate for an additional 15 min.
4. For screening transfected cell lines, add 10 µL of media containing cultured cells directly to the 20-µL digestion mixture and carry out digestion as previously described.
5. Add 180 µL of distilled water to the digestion mixture, cap the tubes and heat in a boiling water bath for 5 min. After heating, mix the samples vigorously.

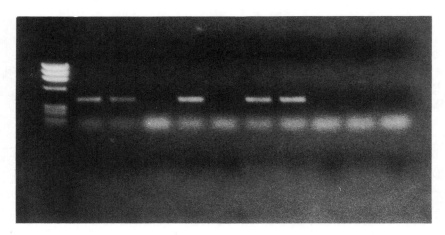

Fig. 1. PCR amplification-based screening of ear punch tissue from transgenic mice and nontransgenic littermates. PCR reaction products were analyzed on a 1.5% agarose gel. Size markers (lane M) are *Hae*III digestion products of ΦX174 DNA. Lanes 1, 2, 4, 6, and 7 are amplification products from transgenic mouse DNA. Lanes 3, 5, 8, 9, and 10 are amplification products from DNA of nontransgenic littermates.

3.2. PCR Reaction and Gel Electrophoresis

1. Prepare the PCR reactions in 0.5-mL Eppendorf tubes as follows:

DNA sample	0.5 μL
10X PCR buffer	5.0 μL
100% DMSO	5.0 μL
100X PCR primers	0.5 μL
100X each of four dNTPs	0.5 μL
Taq polymerase	1.0 U

   Add distilled water to a final volume of 50 μL.
2. Cover the reaction with one drop of mineral oil to prevent evaporation and cap the tube.
3. Amplify by PCR using the following cycle parameters *(see* Note 6):

30 main cycles	15 s at 93°C
	45 s at 65°C

4. Analyze 10 μL of the amplified sample using either an 8% polyacrylamide gel or 1.5% agarose gel *(see* Chapter 1). Stain the gels with 1 μg/mL of ethidium bromide and visualize the amplification products under UV transillumination. An example of typical results is shown in Fig. 1.

4. Notes

1. For efficient amplification using this technique, the concentration of magnesium ions in the final reaction is critical. Although most amplification primers we have used gave efficient amplification at a $MgCl_2$ concentration of 1.5 mM, other primers may amplify optimally at a different concentration. For best results, the optimal $MgCl_2$ concentrations should be determined for each new oligonucleotide primer set.

2. For PCR amplification, we routinely use a Tris-HCl/$(NH_4)_2SO_4$/DMSO/ß-mercaptoethanol/$MgCl_2$ buffer system, which differs from the more widely used Tris-HCl/KCl/gelatin/$MgCl_2$ buffer system. In our hands, this buffer system gives more consistent PCR amplification using nonpurified DNA samples as prepared here.

3. Cross-contamination of samples is a problem with all PCR-based methods. To avoid cross-contamination in the screening of transgenic mice, one can dip the ear puncher into 1% SDS after each clipping. However, we have not yet found any case of cross-contamination even without rinsing the ear puncher.

4. This procedure can also be used for screening tail DNA samples from transgenic mice. When tail tissue is used, 1 mm of tail is excised with a sterile razor blade. Since there is little or no bone in the last 1 mm of mouse tail, identical conditions of digestion to those previously described can be used for DNA preparation, rather than the overnight digestion of 5–10-mm tail pieces as described by Hanley and Merlie *(8)*. All other procedures can be carried out as for ear punch tissue.

5. This method also works quite well using single drops of blood for amplification. DNA is prepared from a blood droplet as previously described for ear punch tissue, and a final dilution of 1000-fold is carried out. We find that results may be less consistent with blood samples owing to inhibitors of PCR amplification present in blood. For efficient amplification, a dilution of at least 1000-fold must be carried out prior to amplification.

6. We obtain the best results when the size of the amplified product is approx 300 bp. Fragments of this size amplify well using the two-step temperature cycling used here rather than a three-step procedure, and allow reactions to be carried out faster. Amplification products that are smaller than 200–300 bp may not be resolved well on agarose gel electrophoresis.

References

1. Hogan, B., Costantini, F., and Lacy, E. (1987) *Manipulating the Mouse Embryo.* Cold Spring Harbor Laboratory, Cold Spring Harbor, NY.
2. Abbott, C., Povey, S., Vivian, N., and Lovell-Badge, R. (1988) PCR as a rapid screening method for transgenic mice. *Trends Genetics* **4**, 325.

3. Lin, C. S., Maguson, T., and Samols, D. (1989) A rapid procedure to identify newborn transgenic mice. *DNA* **8**, 297–299.
4. Walter, C. A., Masr-Schirf, D., and Luna, V. J. (1989) Identification of transgenic mice carrying the CAT gene with PCR amplification. *BioTechniques* **7**, 1065–1070.
5. Chen, S. and Evans, G. A. (1990) A simple screening method for transgenic mice using the polymerase chain reaction. *BioTechniques* **8**, 32–33.
6. Skalnick, D. G. and Orkin, S. (1990) A rapid method for characterizing transgenic mice. *BioTechniques* **8**, 34.
7. McHale, R. H., Stapleton, P. M., and Bergquist, P. L. (1991) Rapid preparation of blood and tissue samples for polymerase chain reaction. *BioTechniques* **10**, 20–23.
8. Hanley, T. and Merlie, J. P. (1991) Transgene detection in unpurified mouse tail DNA by polymerase chain reaction. *BioTechniques* **10**, 56–66.
9. Reddy, K. B., Yee, D., and Osborne, C. K. (1991) Rapid identification of transfected plasmid expression vectors in cells by PCR. *BioTechniques* **10**, 481–483.

CHAPTER 8

Polymerase Chain Reaction Analysis of DNA from Paraffin-Embedded Tissue

Matthias Volkenandt, Adam P. Dicker, Renato Fanin, Debabrata Banerjee, Anthony Albino, and Joseph R. Bertino

1. Introduction

One of the greatest potentials of polymerase chain reaction (PCR) lies in the fact that even minute amounts of target DNA or extensively damaged DNA can be successfully amplified in vitro and thus become amenable to further study. This enables a detailed molecular analysis of small amounts of DNA from tissue that has been damaged by fixation (e.g., in formalin) and long-term storage in paraffin. The applications of this methodology are nearly unlimited. For example, rare tumors that are stored as formalin-fixed, paraffin-embedded tissue in pathology departments throughout the world can be analyzed at the molecular level. Furthermore, tissue from small lesions (e.g., primary skin melanomas), which are only rarely available for molecular analysis since the entire specimen is usually needed for histopathological assessment, can be examined. For PCR analysis, only several sections from the paraffin block, which are usually dispensable, are sufficient. Even small amounts of very low quality DNA can be used, as the sensitivity of the detection of specific target DNA sequences is several orders of magnitude higher than that with any conventional method. For example, Southern blot analysis of DNA from paraffin-embedded tissue has been performed, but with limited success as only relatively

From: *Methods in Molecular Biology, Vol. 15: PCR Protocols: Current Methods and Applications*
Edited by: B. A. White Copyright © 1993 Humana Press Inc., Totowa, NJ

small amounts of degraded and irreversibly modified DNA can be obtained from embedded specimens *(1,2)*. Thus, PCR methodology creates an ideal link between traditional histology and modern molecular biology *(3)*.

In general, applications for PCR analysis of preserved tissue can be divided into two areas: detection of specific DNA sequences, which are normally not present in tissue (e.g., viral sequences), and amplification of specific DNA sequences, which are known to be present, but need to be expanded for further molecular analysis (e.g., for nucleotide sequence analysis of specific genes altered in certain diseases). The first purpose can usually be accomplished by a single PCR reaction and analysis of the product by hybridization techniques. For the second purpose, a large amount of one single discrete PCR product as detected by ethidium bromide staining of an agarose gel after electrophoresis is essential. This may require optimization of amplification conditions for an individual sample, which often is only achieved empirically by performing several reactions with various conditions. In general, the success of further analysis of the PCR product, especially sequence studies, greatly depends on the quality of the PCR product *(4)*.

2. Materials

2.1. DNA Extraction

1. Xylene.
2. Digestion buffer: 100 mM Tris-HCl, pH 8.5, 5 mM EDTA, 1% SDS, 500 µg/mL of proteinase K. Add proteinase K fresh from frozen aliquots of 20 mg/mL.
3. Storage buffer: 10 mM Tris-HCl, pH 8, 0.5 mM EDTA.
4. 100% Ethanol.
5. 95% Ethanol.
6. 70% Ethanol.
7. Phenol. Equilibrate phenol with 10 mM Tris-HCl, pH 7.5.
8. Chloroform-isoamylalcohol (24:1).
9. 3M Sodium acetate, pH 5.5

2.2. PCR

1. Standard PCR reagents *(see* Chapter 1).
2. Standard electrophoresis reagents *(see* Chapter 1).

3. Methods

3.1. DNA Extraction

1. Cut several sections (3–5 sections, 10–20-µm thick) and place into one 1.65-mL microcentrifuge tube using clean forceps *(see* Note 1*)*. Avoid cross contamination. Clean forceps and blade of the microtome between samples *(see* Note 2*)*.
2. To deparaffinize the sample, add approx 1 mL of xylene, invert several times or gently vortex and incubate at room temperature for 10 min, preferably on a rocking platform.
3. Centrifuge for 3 min at high speed in a microcentrifuge to pellet the tissue. Carefully remove the supernatant and discard. Do not disrupt the pellet.
4. Repeat steps 2 and 3.
5. If undissolved paraffin is still present, repeat steps 2 and 3 again.
6. Wash the pellet by adding approx 1 mL of 95% ethanol, invert the tube several times and centrifuge at high speed for 3 min. Remove the supernatant solution.
7. Repeat step 6.
8. Wash the pellet again with 1 mL of 70% ethanol, invert several times, and pellet tissue by centrifugation at high speed for 3 min.
9. Remove the supernatant as completely as possible *(see* Note 3*)*. The sample may be dried under vacuum centrifugation. To avoid cross-contamination, cover the tubes by stretching parafilm across the top and poking several holes in it *(5)*.
10. To digest the tissue *(see* Note 4*)*, resuspend the pellet in 150 µL of digestion buffer containing freshly thawed proteinase K. Use more buffer if more than five sections were cut. Be sure that the pellet is thoroughly resuspended.
11. Close the tubes tightly and put parafilm around the caps of the tube.
12. Incubate at 55°C overnight *(see* Note 5*)*.
13. To extract the DNA *(see* Note 6*)*, spin the tubes for 5 s and add 150 µL of buffered phenol and 150 µL of chloroform-isoamylalcohol, vortex vigorously for 15 s, and spin for 2 min at high speed.
14. Remove the upper phase and put into new tube. Do not touch the protein at the interface.
15. Repeat steps 13 and 14.
16. Add 150 µL of chloroform-isoamyl alcohol, vortex for 15 s and spin for 2 min at high speed. Remove the upper phase and place into a new tube.
17. Measure the final volume. Precipitate with ethanol by adding 1/10 vol of 3M sodium acetate and 2 vol of ice-cold, 100% ethanol. Invert several times or vortex briefly.

18. Place the sample in a freezer at the lowest temperature available (–20°C or –70°C) for about 1 h and then spin the sample at high speed in a microfuge for 20 min. Lower temperatures and longer precipitation and centrifugation times increase the yield.
19. After centrifugation, a pellet may or may not be visible. Carefully remove the supernatant without touching the pellet or the area where the pellet is expected to be.
20. Wash the pellet by adding approx 1 mL of 70% ethanol. Invert several times. Spin for 15 min at high speed and carefully remove the supernatant.
21. Repeat step 18 and carefully remove the supernatant as completely as possible.
22. Resuspend the pellet in 100 µL of storage buffer and store at 4°C.

3.2. PCR

1. Use approx 1–10 µL of the solution as a template for PCR. Perform conventional PCR as described in Chapter 1.
2. Examine PCR product by conventional agarose gel electrophoresis and ethidium bromide staining, or by a blotting procedure *(see* Notes 7 and 8 and Chapter 1).

4. Notes

1. To reduce the amount of paraffin to be dissolved by xylene, excess paraffin surrounding the tissue of interest should be trimmed away as much as possible before cutting slices from the paraffin block.
2. Even small amounts of target DNA can be detected by PCR. Therefore, careful precautions to avoid contamination are crucial *(6).* When cutting sections from paraffin blocks, the microtome's blade and the tweezer used must be cleaned between samples (e.g., with xylene). Also, the blade can be moved slightly before cutting a new block to insure that a different part of the blade is used each time. Plain paraffin blocks without any tissue can be cut between samples to further reduce contamination. Of special concern are "carry-over" contaminations of amplified products to nonamplified reaction mixtures. Therefore, "post-PCR mixtures" should be handled in a different area and with different instruments than "pre-PCR mixtures." Negative controls (without template DNA) should be performed in parallel with each PCR reaction.
3. It is desirable to add the digestion buffer to a dry pellet of tissue with the ethanol from the washing steps completely removed. However, this has to be balanced against the risk of cross-contamination during drying procedures (e.g., vacuum centrifugation, *see* step 9). We remove the ethanol from the last washing step by careful aspiration with a clean

Pasteur pipet as completely as possible and then immediately resuspend the pellet in digestion buffer.

4. The deparaffinized tissue pellet may be resuspended in 100 µL of distilled water. The mixture is then boiled for 30 min followed by centrifugation through a 1-mL Sephadex G-50 column and the eluate is used as template for PCR amplification *(16)*. Alternatively, if only one 5–10-µm section is cut from the block, the tissue pellet may be resuspended directly in 100 µL of a PCR mixture heated to 100°C for 10 min, immediately followed by amplification *(17)*.

5. After an overnight incubation of the tissue in digestion buffer (with 1% SDS and 500 µg/mL proteinase K) the concentration of SDS and proteinase K may be doubled and the incubation continued for another 12–48 h *(2)*. Briefly vortex samples in between.

6. The protocol for DNA extraction as presented here may be altered in many ways, and various simplifications as well as additional steps have been used. In our experience, purification steps, such as the described phenol-chloroform extraction and ethanol precipitation, lead most consistently to good results. However, simplifications may be feasible, depending on the individual situation (e.g., the tissue used for analysis, the gene to be amplified) and have to be determined empirically in each laboratory setting. One successful variation is that instead of SDS, a nonionic detergent can be used in the digestion buffer (50 mM Tris-HCl, pH 8.5, 1 mM EDTA, 200 µg/mL of proteinase K, 0.5% Tween 20), which is compatible with *Taq* DNA polymerase *(18)*. After proteinase K digestion of the tissue, the proteinase K is inactivated by incubation at 95°C for 8 min and the residual tissue is pelleted by centrifugation for 30 s. An aliquot of the supernatant is used directly as a template for PCR amplification *(5)*.

7. If a certain target gene cannot be amplified and detected in a sample, a positive control reaction should be performed on this sample before designating the sample "negative." Sequences of any "housekeeping gene" (e.g., ß2-microglobulin, ß-actin) should be amplfied to determine the quality of the DNA. As DNA from preserved tissue may be extensively degraded, amplification of products that are >400 bp is frequently unsuccessful. In general, segments <400 bp of the positive control gene, as well as the target gene, should be amplified.

8. When analyzing the PCR product by agarose gel electrophoresis and ethidium bromide staining, a specific product may not be clearly visible, because of the lower efficiency of PCR using DNA from preserved tissue as a template. One suggestion to increase the efficiency is to double the times used for denaturing, annealing, and extension and to increase

the number of cycles when analyzing DNA from paraffin-embedded tissue *(7)*. However, this will also increase nonspecific background amplification products. When the PCR product is transferred to a membrane (by Southern blot or dot blot techniques) and hybridized to a labeled probe, specific amplification products frequently become discernible. Transfer of PCR products to membrane is much less labor-intensive than transfer of endonuclease-digested genomic DNA in Southern blot analysis. Owing to the small size and abundance of amplification products, transfer to membrane efficiently occurs within few hours and depurination, denaturation, and neutralization steps are not essential. As an alternative to detection of PCR products by transfer to a membrane and hybridization to a labeled probe, one cycle of PCR with an internally nested, end-labeled oligonucleotide probe may be performed after conventional PCR *(8)*.

If multiple bands are still present, we excise the band of interest from the gel and elute the DNA by various methods (e.g., the GeneClean method, Bio 101, La Jolla, CA; or by elution in 100 µL of H_2O overnight). Occasionally, reamplification of the eluted DNA is necessary to generate more specific product. An aliquot of the eluate is then used as the template for a further PCR reaction. Alternatively, the band of interest on the agarose gel can be touched with a sterile tip. A new PCR reaction mixture without template DNA is then "contaminated" with this tip carrying specific template DNA. However, any alteration detected when analyzing this reamplified product (e.g., point mutations) has to be confirmed by repeating the PCR reactions using original DNA as the template.

9. The amount of template DNA can dramatically affect the yield of the PCR reaction. When analyzing DNA from preserved tissue, there may be an inverse relationship and dilution of template DNA may increase the yield. This may be a result of inhibiting effects of certain fixatives on the *Taq* DNA polymerase. When amplification reactions for a housekeeping gene are unsuccessful for a particular sample, we dilute the DNA (1:10, 1:100, 1:1000) and use 1 µL of these dilutions as the template. This is especially necessary when more tissue is used for DNA extraction. Successful amplification may depend on the fixative used. In general, DNA from samples fixed in ethanol *(9,10)* or in Carnoy's reagent *(11)* gives best results, followed by DNA from tissue fixed in phosphate-buffered formaldehyde (formalin). The use of mercury-based fixatives (B-5, Zenker's) frequently limits amplification to very short DNA sequences (i.e., 100 bp), as DNA degradation is extensive *(12)*. Furthermore, picric acid containing fixatives (Bouin's) were found to affect extensively the quality of the DNA *(1,13)*. In addition, the length

of fixation (e.g., in formalin) affects the quality of the DNA with longer times (exceeding 12 h) being more harmful to DNA *(1)*. However, another study found no decrease in the ability of PCR amplification when fixation times up to 48 h were used *(14)*.

10. A recent publication *(15)* indicated that the strategy used for extraction of DNA from paraffin-embedded tissue is also useful for isolation of RNA from preserved tissue, the length of incubation of the tissue in digestion buffer being the main difference. Reflecting a possible sequential lysis of the cytoplasmic and nuclear cell compartments, maximum extraction of RNA was completed after only 5 h of digestion, whereas maximum DNA yields required an incubation time of 24 h.

References

1. Dubeau, L., Chandler, L. A., Gralow, J. R., Nichols, P. W., and Jones, P. A. (1986) Southern blot analysis of DNA extracted from formalin-fixed pathology specimens. *Cancer Res.* **46,** 2964–2969.
2. Goelz, S. E., Hamilton, S. R., and Vogelstein, B. (1985) Purification of DNA from formaldehyde fixed and paraffin embedded human tissue. *Biochem. Biophys. Res. Commun.* **130,** 118–126.
3. Shibata, D., Martin, W. J., and Arnheim, N. (1988) Analysis of DNA sequences in forty-year-old paraffin-embedded thin-tissue sections: a bridge between molecular biology and classical histology. *Cancer Res.* **48,** 4564–4566.
4. Volkennandt, M., McNutt, N. S., and Albino, A. P. (1991) Sequence analysis of DNA from formalin-fixed paraffin-embedded human malignant melanoma. *J. Cutaneous Pathol.* **18,** 210–214.
5. Wright, D. K. and Manos, M. M. (1990) Sample preparation from paraffin-embedded tissues, in *PCR Protocols: A Guide to Methods and Applications* (Innis, M. A., Gelfand, D. H., Sninsky, J. J., and White, T. J., eds.), Academic, New York, pp. 153–158.
6. Kwok, S. and Higuchi, R. (1989) Avoiding false positives with PCR. *Nature* **339,** 237,238.
7. Ting, Y. and Manos, M. M. (1990) Detection and typing of genital human papillomaviruses, in *PCR Protocols: A Guide to Methods and Applications* (Innis, M. A., Gelfand, D. H., Sninsky, J. J. and White, T. J., eds.), Academic, New York, pp. 356–367.
8. Parker, J. D. and Burmer, G. C. (1991) The oligomer extension 'hot blot': A rapid alternative to Southern blots for analyzing polymerase chain reaction products. *BioTechniques* **10,** 94–101.
9. Wu, A. M., Ben Ezra, J., Winberg, C., Colombero, A. M., and Rappaport, H. (1990) Analysis of antigen receptor gene rearrangements in ethanol and formaldehyde-fixed, paraffin-embedded specimens. *Lab. Invest.* **63,** 107–114.
10. Bramwell, N. H. and Burns, B. F. (1988) The effects of fixative type and fixation time on the quantity and quality of extractable DNA for hybridization studies on lymphoid tissue. *Exp. Hematol.* **16,** 730–732.

Volkenandt et al.

11. Jackson, D. P., Lewis, F. A., Taylor, G. R., Boylston, A. W., and Quirke, P. (1990) Tissue extraction of DNA and RNA and analysis by the polymerase chain reaction. *J. Clin. Pathol.* **43,** 499–504.
12. Crisan, D. and Mattson, J. C. (1991) PCR amplification of intermediate size DNA sequences from formalin and B-5 fixed bone marrow specimens. *Proceedings of the 1991 Miami Bio/Technology Winter Symposium 92* (Abstract).
13. Herbert, D. J., Nishiyama, R. H., Bagwell, C. B., Munson, M. E., Hitchcox, S. A., and Lovett, E. J., III (1989) Effects of several commonly used fixatives on DNA and total nuclear protein analysis by flow cytometry. *Am. J. Clin. Pathol.* **91,** 535–541.
14. Rogers, B. B., Alpert, L. C., Hine, E. A., and Buffone, G. J. (1990) Analysis of DNA in fresh and fixed tissue by the polymerase chain reaction. *Am. J. Pathol.* **136,** 541–548.
15. Weizsacker, F., Labeit, S., Koch, H. K., Oehlert, W., Gerok, W., and Blum, H. E. (1991) A simple and rapid method for the detection of RNA in formalin-fixed, paraffin-embedded tissues by PCR amplification. *Bioch. Biophys. Res. Comm.* **174,** 176–180.
16. Lampertico, P., Malter, J. S., Colombo, M., and Gerber, M. A. (1990) Detection of hepatitis B virus DNA in formalin-fixed, paraffin-embedded liver tissue by the polymerase chain reaction. *Am. J. Pathol.* **137,** 253–258.
17. Shibata, D. K., Arnheim, N., and Martin, W. J. (1988) Detection of human papilloma virus in paraffin-embedded tissue using the polymerase chain reaction. *J. Exp. Med.* **167,** 225–230.
18. Higuchi, R. (1990) Rapid, efficient DNA extraction for PCR from cells of blood. *Amplifications. A Forum for PCR Users.* (May 1989), 1–3.

CHAPTER 9

The Use of Polymerase Chain Reaction for Chromosome Assignment

Michael Jaye and Craig A. Dionne

1. Introduction

The polymerase chain reaction (PCR) approach to chromosome assignment consists of analyzing the profile of specific amplification products of the gene of interest using genomic DNAs from a panel of somatic cell hybrids as templates *(see* Fig. 1). Each hybrid somatic cell line contains the normal complement of chromosomes of one species, along with one or more additional chromosomes from a second species *(1,2)*. The PCR approach has obvious advantages over the conventional Southern blotting method of chromosome assignment, including speed, sensitivity, and conservation of the genomic DNAs, which are difficult and labor intensive to generate.

Chromosome assignment by PCR consists of four steps, which, for the sake of clarity, are presented in terms of assigning human gene sequences with a panel of human/rodent somatic cell hybrids:

1. Establish conditions for specific detection of the human target gene sequence vs rodent sequences in test reactions employing human and rodent genomic DNAs as templates.
2. Run PCR reactions on the entire test panel.
3. Analyze PCR products by gel electrophoresis and, if necessary, verify by Southern hybridization.
4. Correlate the presence and absence of the human PCR product in each reaction with the human chromosome content of the corresponding hybrid cell lines.

From: *Methods in Molecular Biology, Vol. 15: PCR Protocols: Current Methods and Applications*
Edited by: B. A. White Copyright © 1993 Humana Press Inc., Totowa, NJ

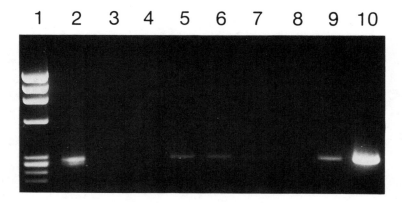

Fig. 1. The partial results of a typical chromosome assignment experiment are shown. A 287-bp DNA fragment of an FGF receptor gene, FLG, was amplified using the conditions described in Methods and the two oligonucleotides 5' AAACAGGAGTACCTGGACCTGTCC 3' and 5' CCAGCAGGAAAGGGGACA- GGGACG 3', which are located at the 3' end of the FLG coding region *(9)*. After PCR amplification, 15 μL of each reaction mixture was separated on a 1.5% agar- ose gel as described. A photograph of the stained gel is shown. The reactions con- tained DNA from: lane 2, human lymphocytes; lane 3, hamster E36 cells; lane 4, mouse RAG cells; lanes 5–9, human/hamster somatic cell hybrids 80H2-80H5 and 80P3; lane 10, 0.1 ng FLG cDNA. Lane 1 contains *Hae* III-digested φX174 DNA as mol-wt standards. These samples were chosen to exemplify the nature of discor- dant results. The results in lanes 7 and 9 are discordant with the karyotype analysis but when the PCR analysis is performed on all 45 hybrids in the panel *(10)*, FLG shows least discordancy with chromosome 8 as predicted from *in situ* hybridiza- tion studies *(9)*.

The most critical step in the analysis is step 1, in which the condi- tions for specificity determination are defined. If one has access to the human and rodent gene structures and sequences, then designing a PCR reaction to amplify specifically a portion of the human gene is quite straightforward. However, this is rarely the case for a gene that has not yet been assigned to a human chromosome. More generally, one has very little knowledge of gene structure or of interspecies dif- ferences in nucleotide sequence; therefore conditions for specific PCR amplification must be empirically determined. These include:

1. Placement of one or both priming oligonucleotides in the 3' or 5' untrans- lated region of the gene. Specificity for amplification of only the target human sequence can often be achieved by employing oligonucleotide prim- ers located within the untranslated regions of the gene. In general, these

regions are less conserved between species and contain fewer introns than the coding region, thus facilitating amplification of a PCR product of predictable size. We have used this approach successfully for the human aFGF and FGF5 genes *(3)* using one or both PCR oligonucleotides derived from the 3' untranslated regions of these genes, respectively. The assignment of the gene for human muscle-specific phosphoglycerate mutase, pGAM2, was obtained by PCR in the 5' flanking region of the gene *(4)*.

2. Placement of one or both priming oligonucleotides within an intron. This strategy takes advantage of the fact that sequences within introns are generally less conserved between species. In addition, intron length is variable between species. For these reasons, if nucleotide sequence information is available from an intervening sequence within the human gene, this strategy is quite attractive. The placement and specificity of the oligonucleotides will determine whether a PCR product will be generated from only the human gene, or whether PCR products of different sizes are generated from the human and rodent genes. Placement of priming oligonucleotides within two separate introns was used for the chromosome assignment of human complement component C9 by PCR across an exon *(5)*.

3. PCR across an intron. Two oligonucleotide primers may generate a PCR product that crosses an intron. Since the position but not size of introns is frequently conserved between species, it is possible to generate PCR products of different sizes that distinguish between the human and rodent genes. PCR across an intron has been used successfully for the chromosome assignment of the nuclear antigen p68 *(6)* and the FGF receptor, bek *(7)*.

If a clear distinction between human and rodent PCR products is not obtained by any of the aforementioned strategies, the PCR products can be digested with various restriction endonucleases to find one specific to the human sequence. The results of this approach are analyzed in terms of the presence or absence of the human-specific PCR product, which in this case is a restriction fragment of an empirically defined size.

2. Materials

1. A panel of somatic cell hybrid genomic DNAs, each adjusted to 0.1 µg/µL *(see* Note 1).
2. Genomic DNA samples from each species used to generate the somatic cell hybrids.
3. PCR oligomers 22–24 bases long, placed several hundred base pairs apart in the 3' untranslated region of a cDNA corresponding to the gene of interest *(see* Note 2).

4. Standard reagents for submarine agarose gel electrophoresis *(see* Chapter 1).
5. Standard reagents for PCR reactions *(see* Chapter 1).

3. Methods

1. Set up three tubes of PCR reactions to test specificity. Use standard PCR conditions (Chapter 1) in a final volume of 50 µL and containing 0.1 µg of genomic DNA from the species used to create the somatic cell hybrid panel (normally human, mouse, and hamster).
2. Amplify by PCR using the following cycle parameters:
   30 main cycles   94°C, 1 min (denaturation)
                         60°C, 1 min (annealing)
                         72°C, 2 min (extension)
   See Note 3.
3. Analyze the PCR reactions by submarine agarose gel electrophoresis *(8; see* Chapter 1). We generally run a 10 × 14 × 0.5-cm thick gel containing 1% agarose, 1X TAE and 5 µg/mL of ethidium bromide. To 15 µL of PCR reaction product, add 3 µL of 50% glycerol and load the samples into a well made by a 16-well comb. Perform electrophoresis at 75 mamp for approx 1 h or until the desired resolution is achieved.
4. Visualize the separated reaction products with a hand-held UV lamp or transilluminator. Under ideal circumstances, a single band corresponding to the amplified target sequence is found only in the reaction assembled with the human DNA *(see* Note 4).
5. If specificity is not achieved, perform the test PCR reactions again but with a higher annealing temperature, i.e., 65°C. If specificity is again not achieved, rerun the test reactions with a different pair of oligonucleotide primers. In the event that specificity cannot be obtained, restriction enzyme digestion of the PCR products may be attempted at this time.
6. Once specificity of amplification is achieved in test reactions, the same PCR conditions are used with each DNA sample from the panel of somatic cell hybrids. For these reactions, make a "master mix" containing all PCR reagents, oligonucleotide primers, and *Taq* polymerase and aliquot 49 µL into all the reaction tubes placed on ice. Add 1 µL of genomic DNA from the cell hybrids to each of the corresponding tubes, overlay with 50 µL of paraffin oil, and amplify the samples by PCR using the conditions previously shown to give specific amplification of the target sequence. After PCR amplification, store the samples at 4°C until analyzed.
7. Analyze the PCR reaction products of the amplified cell hybrid genomic DNA panel by agarose gel electrophoresis as described in step 3 above *(see* Note 4).
8. Correlate the presence or absence of the specific amplified target sequence in each PCR reaction with that of each human chromosome content of

each hybrids *(1,2)*. Comparison of the human chromosome content of each somatic cell hybrid with the distribution of the specifically amplified PCR product generally allows the gene to be assigned to the chromosome showing the least discordancy *(see* Notes 5–7).

4. Notes

1. Somatic cell hybrids are difficult and time-consuming to generate, characterize, and maintain. A panel of somatic cell hybrid genomic DNAs can best be obtained from an interested collaborator, although some panels of somatic cell hybrid DNA are commercially available.
2. Quite often, if a lab has been working with a particular sequence for a period of time, the oligonucleotides that have been synthesized for other purposes such as sequencing can be evaluated in different combinations with specificity being achieved by adjusting the annealing temperature in the PCR reaction.
3. As with all PCR techniques, special care must be taken to insure that cross-contamination between samples does not occur. This is especially critical in this application since the cloned cDNA or gene has very likely been in circulation within the laboratory for some time, and consequently there are potentially many sources of contamination. Likewise, a fingerprint contains much human genomic DNA and can render a somatic cell hybrid DNA sample completely useless.
4. It is often quite useful to verify the visual results of PCR analysis by Southern hybridization. In this case, the agarose gel containing the separated PCR reaction products is blotted to a nylon membrane and probed with a radiolabeled oligonucleotide or DNA fragment under standard conditions *(8; see* Chapters 1 and 5). This step can verify faint visual signals and can sometimes identify specific products in a reaction that was considered negative or uncertain by visual scoring.
5. There is usually a higher but acceptable rate of discordance associated with PCR methods of chromosome assignment when compared to the conventional Southern hybridization method *(3,7)*. This results from the much greater sensitivity of PCR techniques in detecting human gene sequences as compared to the methods of karyotype and isoenzyme analysis, which are standardly used to score the presence of human chromosomes in the hybrid cell panel.
6. One can strengthen the chromosome assignment of a particular gene by comparing its PCR amplification pattern in the somatic cell hybrid panel to the amplification pattern of a gene previously assigned to the same chromosome.
7. As has been demonstrated by the Southern hybridization technique, it is possible to get chromosome localization information by performing the

same analysis on a panel of somatic cell hybrids that have discrete, well characterized truncations or deletions in the chromosome of residence *(1)*.

Acknowledgments

The authors would like to acknowledge S. J. O'Brien, W. S. Modi, and H. Seaunez, who generously provided the hybrid cell DNA samples, which formed the basis of this PCR application, and Robin McCormick and Rosalie Ratkiewicz for preparation of the manuscript.

References

1. McKusick, V. A. and Ruddle, F. H. (1977) The status of the gene map of the human chromosomes. *Science* **196**, 390–405.
2. O'Brien, S. J., Simonson, J. M., and Eichelberger, M. (1982) Genetic analysis of hybrid cells using isozyme markers as monitors of chromosome segregation, in *Techniques in Somatic Cell Genetics* (Whay, J. W., ed.), Plenum, New York, pp. 513–524.
3. Dionne, C. A., Kaplan, R., Seuanez, H., O'Brien, S. J., and Jaye, M. (1990) Chromosome assignment by polymerase chain reaction techniques: Assignment of the oncogene FGF-5 to human chromosome 4. *Biotechniques* **8**, 190–194.
4. Edwards, Y. H., Sakoda, S., Schon, E., and Povey, S. (1989) The gene for human muscle-specific phosphoglycerate mutase, PGAM2, mapped to chromosome 7 by polymerase chain reaction. *Genomics* **5**, 948–951.
5. Abbot, C., West, L., Povey, S., Jeremiah, S., Murad, Z., DiScipio, R., and Fey, G. (1989) The gene for human complement component C9 mapped to chromosome 5 by polymerase chain reaction. *Genomics* **4**, 606–609.
6. Iggo, R., Gough, A., Xu, W., Lane, D. P., and Spurr, N. K. (1989) Chromosome mapping of the human gene encoding the 68-kDa nuclear antigen (p68) by using the polyemerase chain reaction. *Proc. Natl. Acad. Sci. USA* **86**, 6211–6214.
7. Dionne, C. A., Modi, W. S., Crumley, G., O'Brien, S. J., Schlessinger, J., and Jaye, M. (1992) BEK, a receptor for multiple members of the fibroblast growth factor (FGF) family, maps to human chromosome 10q25.3-10q26. *Cytogenet. Cell Genet.* **60,** 34–36.
8. Sambrook, J., Fritsch, E. F., and Maniatis, T. (1989) *Molecular Cloning: A Laboratory Manual* (2nd ed.), Cold Spring Harbor Laboratory, Cold Spring Harbor, NY.
9. Ruta, M., Howk, R., Ricca, G., Drohan, W., Zabelshansky, M., Laureys, G., Barton, D. E. Francke, U., Schlessinger, J., and Givol, D. (1988) A novel protein tyrosine kinase gene whose expression is modulated during endothelial cell differentiation. *Oncogene* **3**, 9–15.
10. O'Brien, S. J., Nash, W. G., Goodwing, J. L., Lowy, D. R., and Chang, E. H. (1983) Dispersion of the ras family of transforming genes to four different chromosomes in man. *Nature* **302**, 839–842.

CHAPTER 10

Mapping MHC Class II Genes and Disease-Susceptibility

Use of Polymerase Chain Reaction and Dot Hybridization for Human Leukocyte Antigen Allele Typing

Stephen C. Bain and John A. Todd

1. Introduction

The genes of the human leukocyte antigen (HLA) region control a variety of functions involved in the immune response and influence susceptibility to over 40 diseases. The region maps to the short arm of chromosome 6 and is divided into three regions, denoted class I, II, and III. The HLA class II gene complex is approx 1000 kb in length and is arranged into three main subregions (HLA-DP, -DQ, and -DR), each of which contains genes of two types, A and B. The A genes encode α-polypeptides and B genes encode β-polypeptides; together these form the functional class II $\alpha\beta$ dimer. Although some haplotypes contain up to 14 class II loci, not all are expressed, and limitations on functionally permissive heterodimer formation restricts the expressed repertoire to 4 class II molecules/haplotype: DP$\alpha\beta$, DQ$\alpha\beta$, DR$\alpha\beta_1$, and DR$\alpha\beta_{3,4,5}$ (the second DR molecule is either DR$\alpha\beta_3$, DR$\alpha\beta_4$, or DR$\alpha\beta_5$, depending on the haplotype). With the exception of the DRA locus, the genes encoding each of these polypeptides are highly polymorphic.

The class II heterodimers are transmembrane glycoproteins expressed on the surface of antigen-presenting cells, B lymphocytes and activated T lymphocytes. These molecules are involved in the presenta-

From: *Methods in Molecular Biology, Vol. 15: PCR Protocols: Current Methods and Applications*
Edited by: B. A. White Copyright © 1993 Humana Press Inc., Totowa, NJ

tion of antigen to the T-cell receptor and in the developmental selection of T-cell receptor specificities. Hence, they are intimately involved in the normal immune response. Population and family studies have demonstrated that polymorphisms within the class II region are associated with autoimmune diseases such as type I diabetes, rheumatoid arthritis, and coeliac disease. Previously, serological reagents directed at the products of the class II genes allowed the definition of the major HLA subtypes, with further subdivision by the mixed lymphocyte reaction. However, many allelic subtypes cannot be identified by these techniques and the products of the polymorphic DQA1 gene are not recognized at all.

The nucleotide sequences for most of the alleles of the class II HLA antigens are now available, allowing the design of synthetic oligonucleotides that distinguish between the allelic forms of DR, DQ, and DP. Together with the polymerase chain reaction (PCR), this has permitted the development of simple, highly specific procedures that identify the sequence variation responsible for the differences in the HLA class II proteins.

The methods to be described involve PCR amplification of the class II region of interest, dot-blotting of the PCR product, and probing using sequence-specific oligonucleotides that are end-labeled with [γ-^{32}P]ATP. The procedure for identifying HLA-DQ alleles is described first, followed by alterations in the method required for DR and DP alleles. The major differences relate to the PCR stage of the method.

Briefly, the second exons of the DQA1 and DQB1 genes are amplified in a two-stage procedure. The DQA1 gene is first amplified with primers GH26A and GH27A and then 2 μL of the primary amplification mixture is subjected to a second round using primers GH27A and GH27B (1). The DQB1 gene is initially amplified with primers GLPDQb3 and GAMPDQXb2 and then subjected to a second PCR using the primers GAMPDQXb2 and GLPDQb2 (2,3). These double amplifications improve the intensity of the hybridization signal. For HLA-DRB1 alleles, a single PCR is sufficient; primers GLPDRb and GAMPDRb amplify a 232-bp segment of the first domain of all known DRB alleles (2). Specific amplification of DR4-associated DRB1 alleles is achieved by using primers 336C and 337C (4). Primer 337C is identical to the sense strand of all DR4B1 sequences between amino acid positions 7 and 13 and contains multiple mismatches with DRB4 and all non-DR4, DRB1 genes. Primer 336C is complementary to the sense strand of all published HLA-DRB genes between amino acid positions 88 and 94. A similar strategy is used to define the DR3-specific alleles, DRw17 (DRB1*0301) and DRw18 (DRB1*0302) *(5)*. For HLA-DPB alleles, a single PCR is also adequate. Primers 5G and 3G amplify a 285-bp fragment of the second exon of DPB *(6)*.

For most of the HLA-DQA1, DQB1 alleles and many of the DRB1 alleles, the available sequence-specific probes are effectively allele-specific. Hence, a positive hybridization defines the allele under investigation. However, in the case of HLA-DPB1 the probes each recognize a number of alleles. It is therefore necessary to use a panel of probes and identify each allele according to the diagnostic pattern of probe hybridization *(7)*. Details of the alleles recognized by individual probes are given in Table 1.

2. Materials
2.1. Polymerase Chain Reaction

1. $1M$ KCl.
2. $1M$ $MgCl_2$ (1 mL, store at 4°C): From this stock, 1-mL aliquots of 20, 40, 60, 80, and 100 mM of $MgCl_2$ can be made (also store at 4°C).
3. "Tween 20": 2% Polyethylene sorbitan monolaurate (EIA purity).
4. 200 mM Tris buffer, pH 8.4 : Prepare 50 mL of this buffer from 6.5 mL of $1M$ Tris base, 3.5 mL of $1M$ Tris-HCl, 40 mL of double-distilled (dd) autoclaved water.
5. 4 mM Deoxyribonucleosides: deoxyadenine triphosphate, deoxycytidine triphosphate, deoxyguanine triphosphate, deoxythymidine triphosphate (dNTPs; 1-mL aliquots stored at –20°C).
6. *Taq* polymerase: 5 U/μL. Store at –20°C.
7. Mineral oil, without stabilizer.
8. Genomic DNA: 100 ng/μL in 10 mM Tris-HCl, pH 7.5, 1 mM EDTA. Store at 4°C. *See* Chapter 11 for DNA isolation.
9. Primers (*see* Table 2).

2.2. Dot Blotting

1. $6M$ NaOH.
2. $1M$ and $2M$ ammonium acetate: Prepare fresh 50-mL aliquots from $5M$ ammonium acetate stock using dd autoclaved water.
3. 6X Standard saline citrate (SSC): $0.9M$ NaCl, $0.09M$ trisodium citrate. Prepare 1 L from 800 mL of single distilled (sd) water, 52.5 g of NaCl and 26.46 g of trisodium citrate. Adjust volume to 1 L.

Table 1
Alleles Recognized by Probes

Probe	Alleles
HLA-DQ[a]	
A1.1	DQA1*0101 (DQA1.1)
A1.2a	DQA1*0102 (DQA1.2), DQA1*0103 (DQA1.3), DQA1*0501 (DQA4.1)
A1.2b	DQA1*0102 (DQA1.2), DQA1*0201 (DQA2), DQA1*0301 (DQA3)
A1.3	DQA1*0103 (DQA1.3)
A2	DQA1*0201 (DQA2)
A3	DQA1*0301 (DQA3)
A4.1	DQA1*0501 (DQA4.1), DQA1*0401 (DQA4.2)
A4.2	DQA1*0401 (DQA4.2)
B2	DQB1*0201 (DQw2)
B4	DQB1*0402 (DQw4)
B5a	DQB1*0501 (DQw5), DQB1*0604 (DQw1.19a, DQw1.19b)
B5b	DQB1*0501 (DQw5), DQB1*0502 (DQw1.AZH), DQB1*0503 (DQw1.9)
B6	DQB1*0602 (DQw6), DQB1*0603 (DQw1.18)
B7a	DQB1*0301 (DQw7), DQB1*0601 (DQw1.12)
B7b	DQB1*0301 (DQw7), DQB1*0303 (DQw9)
B8	DQB1*0302 (DQw8)
B9	DQB1*0602 (DQw6), DQB1*0302 (DQw8)
B1.9a	DQB1*0503 (DQw1.9), DQB1*0601 (DQw1.12)
B1.9b	DQB1*0503 (DQw1.9), DQB1*0502 (DQw1.AZH)
B1.12	DQB1*0601 (DQw1.12)
B1.18	DQB1*0603 (DQw1.18), DQB1*0604 (DQw1.19a)
B1.19b	DQB1*0604 (DQw1.19b)
B1.AZH	DQB1*0502 (DQw1.AZH)
HLA-DR[b]	
DR1	DRB1*0101(DR1,Dw1), DRB1*0102(DR1,Dw20), DRB1*0103 (DR'BR').
DR2	DRB5*0101(DRw15,Dw2), DRB5*0102(DRw15,Dw12), DRB5*0201(DRw16,Dw21), DRB5*0202(DRw16, Dw22).
Dw2	DRB5*0101(DRw15,Dw2).
Dw12	DRB5*0102(DRw15,Dw12), DRB5*0201(DRw16,Dw21), DRB5*0202(DRw16,Dw22).
DR3	DRB1*0301(DRw17,Dw3), DRB1*0302(DRw18,Dw'RSH').

Table 1 *(continued)*

Probe	Alleles
DR3/6/52	DRB1*0301(DRw17,Dw3), DRB1*1301(DRw13,Dw18), DRB1*1302(DRw13,Dw19), DRB1*1401(DRw14,Dw9), DRB3*0101(DRw52a,Dw24).
DR4	DRB1*0401-0408(DR4 subtypes, Dw4,10,13,14,15,'KTZ').
DR7	DRB1*0701(DR7,Dw17), DRB1*0702(DR7,Dw'DB1').
DRw8	DRB1*0801(DRw8,Dw8.1), DRB1*0802(DRw8,Dw8.2), DRB1*0803(DRw8,Dw8.3).
DR9	DRB1*0901(DR9,Dw23).
DRw10	DRB1*1001(DRw10).
DRw11	DRB1*1101-4(DRw11(5)).
DR1101/1104	DRB1*1101(DRw11,Dw5), DRB1*1104(DRw11,Dw'FS').
DRw12	DRB1*1201(DRw12(5)).
DRw13	DRB1*1301(DRw13,Dw18), DRB1*1302(DRw13,Dw19), DRB1*0402(DR4,Dw10), DRB1*1102(DRw11,Dw'JVM'), DRB1*1103(DRw11.3), DRB1*0103(DR'BR').
DRw14	DRB1*1401(DRw14,Dw9).
Dw14	DRB1*1402(DRw14,Dw16), DRB1*0101(DR1,Dw1), DRB1*0102(DR1,Dw20), DRB1*0404(DR4,Dw14), DRB1*0405(DR4,Dw15), DRB1*0408(DR4,Dw14).
Dw4	DRB1*0401(DR4,Dw4).

In combination with DR3-specific amplification:

153B	DRB1*0301(DRw17,Dw3)
GT86	DRB1*0302(DRw18,Dw'RSH')

With DR4-specific amplification:

152B	DRB1*0403(DR4,Dw13A),DRB1*0406(DR4,Dw'KT2'), DRB1*0407(DR4,Dw13B).
Dw15	DRB1*0405(DR4,Dw15).

HLA-DP[c]

A1(A/LFQG)	DPB1*0401(DPB4.1), DPB1*0402(DPB4.2), DPB1*0201(DPB2.1), DPB1*0202(DPB2.2), DPB1*0801(DPB8), DPB1*01601(DPB16), DPB1*0501(DPB5), DPB1*1901(DPB19), DPB7.
A2B(A/VYQL)	DPB1*0301(DPB3), DPB1*0601(DPB6), DPB1*1101(DPB11), DPB1*1301(DPB13).
A3(A/VHQL)	DPB1*1401(DPB14), DPB*1001(DPB10), DPB1*0901(DPB9), DPB12, DPB1*1701(DPB17).

Table 1 *(continued)*

Probe	Alleles
A4B(A/VYQG)	DPB1*0101(DPB1), DPB1*1801(DPB18), DPB1*1501(DPB15).
C1B(C/AAE)	DPB1*0401(DPB4.1), DPB7, DPB1*1101(DPB11), DPB1*1301(DPB13), DPB1*0101(DPB1), DPB1*1501(DPB15).
C2B(C/DED)	DPB1*0301(DPB3), DPB1*0601(DPB6), DPB1*1401(DPB14), DPB1*0901(DPB9), DPB12, DPB1*1701(DPB17).
C3/55(C/DEE)	DPB1*0402(DPB4.2), DPB1*0201(DPB2.1), DPB1*0801(DPB8), DPB1*1601(DPB16), DPB1*1801(DPB18), DPB*1001(DPB10).
DB16(C/EAE)	DPB1*0501(DPB5), DPB1*1901(DPB19), DPB1*0202(DPB2.2).
D1(D/LLEEK)	DPB1*0301(DPB3), DPB1*1401(DPB14).
D2(D/ILEEE)	DPB1*0201(DPB2.1), DPB1*0202(DPB2.2), DPB1*0801(DPB8), DPB1*1601(DPB16), DPB1*1901(DPB19), DPB1*1301(DPB13), DPB1*1001(DPB10), DPB1*0901(DPB9), DPB12, DPB1*1701(DPB17).
D3(D/ILEEK)	DPB1*0401(DPB4.1), DPB*0402(DPB4.2), DPB*0501(DPB5), DPB7, DPB*0101(DPB1), DPB1*1801(DPB18).
E1(E/M)	DPB1*0401(DPB4.1), DPB1*0402(DPB4.2) DPB1*0201(DPB2.1), DPB1*0202(DPB2.2), DPB1*1601(DPB16), DPB1*0501(DPB5), DPB1*0601(DPB6), DPB1*1101(DPB11), DPB1*1801(DPB18), DPB1*1501(DPB15), DPB1*1701(DPB17).
E2(E/V)	DPB1*0801(DPB8), DPB7, DPB1*0301(DPB3), DPB1*0101(DPB1), DPB1*1401(DPB14), DPB1*1001(DPB10), DPB1*0901(DPB9).
DB25(F/GGPM)	DPB1*0401(DPB4.1), DPB1*0402(DPB4.2), DPB1*0201(DPB2.1), DPB1*0202(DPB2.2).
DB26(F/DEAV)	DPB1*0801(DPB8), DPB1*1601(DPB16), DPB1*0501(DPB5), DPB1*1901(DPB19), DPB7, DPB1*0301(DPB3), DPB1*0601(DPB6), DPB1*1101(DPB11), DPB1*1301(DPB13), DPB1*0101(DPB1), DPB1*1401(DPB14), DPB1*1001(DPB10), DPB1*0901(DPB9), DPB12, DPB1*1701(DPB17).

[a]From ref. *8.* [b]From ref. *5.* [c]From ref. *9.*

Table 2
Oligonucleotide Primers

Primer	Sequence
HLA-DQA1	
GH26A	5'-GGT GTA AACTT GTA CCA G-3'
GH27A	5'-GGT AGC AGC GGT AGA GTT G-3'
GH27B	5'-GTA GAG TTG GAG CGT TA-3'
HLA-DQB1	
GAMPDDQXb2	5'-CCA CCT CGT AGT TGT GTC TGC A-3'
GLPDQb3	5'-GAT TTC GTG TAC CAG TTT AAG G-3'
GLPDQb2	5'-TGC TAC TTC ACC AAC GGG AC-3'
HLA-DRB1	
GLPDRb	5'-TTC TTC AAT GGG ACG GAG CG-3'
GAMPDRb	5'-CGC CGC TGC ACT GTG AAG CTC TC-3'
HLA-DR4B1 (*for DR4-specific subtyping*)	
337C	5'-TCT TGG AGC AGG TTA AAC A-3'
336C	5'-TCG CCG CTG CAC TGT GAA G-3'
HLA-DR3B1 (*for DR3-specific subtyping*)	
GLPDR3	5'-TTC CAT AAC CAG GAG GAG A-3'
GAMPDR3	5'-TAG TTG TGT CTG CAG TAG T-3'
HLA-DPB1	
5G	5'-CAG GGA TCC GCA GAG AAT TAC-3'
3G	5'-TCA CTC ACC TCG GCG CTG CAG-3'

4. 0.5*M* Disodium ethylenediamine tetracetic acid [EDTA], pH 8. Prepare 500 mL from 450 mL of dd water, 93.05 g of EDTA and 10 g of NaOH. Adjust pH with 1*M* NaOH. Adjust volume to 500 mL with dd water. Autoclave at 121°C in a 30-min cycle.

5. TE buffer: Prepare 50 mL from 45 mL of dd autoclaved water, 0.5 mL of 1*M* Tris-HCl, pH 7.5, and 0.01 mL of 0.5*M* EDTA, pH 8. Adjust volume to 50 mL with dd autoclaved water.

6. Denaturing solution. Prepare a 26-mL aliquot from 1.5 mL of 0.5*M* EDTA, pH 8, 2.0 mL of 6*M* NaOH and 22.5 mL of TE buffer. Prepare fresh as required.

7. Hybridot manifold (BRL, Gaithersburg, MD) and suction apparatus.

8. Zeta-Probe blotting membrane (Bio-Rad, Richmond, CA). *See* Note 1.

9. 3MM filter paper (Whatman, Maidstone, UK).

2.3. End-Labeling of Oligonucleotide by Kinase Reaction

1. Oligonucleotides. Store stocks at 500 µg/mL at –20°C.
2. 10X Kinase buffer: $0.5M$ Tris-HCl, pH 7.6, $0.1M$ MgCl$_2$. Store at –20°C.
3. 100 mM Dithiothreitol (DTT). Store at –20°C.
4. [γ-^{32}P]ATP in aqueous solution. The specific activity at reference date is >111 TBq/mmol (3000 Ci/mmol). Store at –20°C.
5. TE buffer. *See* Section 2.2.
6. 200 mM NaCl in TE buffer. Prepare as directed for TE buffer (Section 2.2.), with the addition of 2 mL of $5M$ NaCl before adjusting volume to 50 mL.
7. 500 mM NaCl in TE buffer. Prepare as directed for TE buffer (Section 2.2.), with the addition of 5 mL of $5M$ NaCl before adjusting volume to 50 mL.
8. Diethylamine ethyl cellulose (DE 52 Cellulose, Whatman) in 200 mM NaCl in TE buffer. Store at 4°C. Shake well immediately prior to use.
9. Heating block.
10. Disposable column (Bio-Rad Poly-Prep), supported by stand.
11. T4 polynucleotide kinase (PNK): 10,000 U/mL.

2.4. Hybridization and Washing

1. Denhardt's solution (100X): 2% bovine serum albumin (BSA), 2% poly-vinylpyrrolidone (PVP), 2% Ficoll: Prepare 50 mL from 45 mL of dd water, 1 g of BSA, 1 g of PVP, and 1 g of Ficoll. Adjust volume to 50 mL with dd water.
2. 10% Sodium dodecyl sulphate (SDS): Prepare 50 mL from 5 g of SDS and 50 mL of dd, autoclaved water. Store at 37°C and heat to 65°C to fully dissolve SDS.
3. 20X SSC: $3M$ NaCl, $0.3M$ trisodium citrate. Prepare 50 mL from 45 mL of sd water, 8.75 g of NaCl and 4.41 g of trisodium citrate. Adjust volume to 50 mL with sd water.
4. Prehybridization solution. Prepare 5 mL fresh from 3 mL of dd water, 0.25 mL of 100X Denhardt's solution, 1.5 mL of 20X SSC, 250 µL of 10% SDS, and 50 µL of yeast tRNA. Prepare yeast tRNA at 20 mg/mL in sterile dd water and store at –20°C.
5. Hybridization solution. Prepare 2.5 mL fresh from 1.75 mL of dd water, 0.75 mL of 20X SSC, and 25 mL of 10% SDS.
6. Oligonucleotide probes (Table 3). Oligonucleotide probes are end-labeled with [γ-^{32}P]ATP to a specific activity of approx 1×10^8 counts min^{-1}mg^{-1} prior to use (*see* Sections 2.3. and 3.3.).

Table 3
Oligonucleotide Probes

Probe	Sequence	Washing temp. (°C)
HLA-DQA1[a]		
A1 (group)	5'-TGA GTT CAG CAA ATT TG-3'	46
or		
A1.1	5'-ATG AGG AGT TCT ACG TG-3'	48
A1.2a	5'-AGA TGA GCA GTT CTA CG-3'	50
A1.2b	5'-CCT GGA GAG GAA GGA GA-3'	54
A1.3	5'-CCT GGA GAA GAA GGA GA-3'	52
A2	5'-TCT AAG TCT GTG GAA CA-3'	50
A3	5'-TTC CGC AGA TTT AGA AG-3'	48
A4.1	5'-GTT TGC CTG TTC TCA GA-3'	53–55
A4.2	5'-TGG AGA CGA GCA GTT CT-3'	52
HLA-DQB1[a]		
B2	5'-GCT GGG GCT GCC TGC CG-3'	62
B4	5'-TGG AGG AGG ACC GGG CG-3'	58–62
B5a	5'-GGC GGC CTG TTG CCG AG-3'	62
B5b	5'-CGT GCG GGG TGT GAC CA-3'	58
B6	5'-GGC GGC CTG ATG CCG AG-3'	62–64
B7a	5'-CGT GCG TTA TGT GAC CA-3'	52–56
B7b	5'-GGC CGC CTA ACG CCG AG-3'	68
B8	5'-GGC CGC CTG CCG CCG AG-3'	65–68
B9	5'-GCG TGC GTC TTG TGA CC-3'	50
B1.9a	5'-AGG GGC GGC CTG ACG CC-3'	62
B1.9b	5'-AGG AGT ACG TGC GCT TC-3'	54
B1.12	5'-GAG AGG AGG ACG TGC GC-3'	58
B1.18	5'-CGT CTT GTA ACC AGA CA-3'	50
B1.19b	5'-CTT GTA ACC AGA TAC ATC-3'	50
B1.AZH	5'-GCG GCC TAG CGC CGA GT-3'	60
HLA-DR[b]		
DR1	5'-GGA AAG ATG CAT CTA TA-3'	46
DR2 (DRB5 alleles)	5'-GCG GTT CCT GCA CAG AG-3'	57
DR3	5'-GGG TGG ACA ACT ACT GC-3'	57
DR3/DR6/DRw52	5'-GGA CAG ATA CTT CCA TA-3'	48
DR4	5'-ACT TCT ATC ACC AAG AG-3'	48
DR7	5'-GGA AAG ACT CTT CTA TA-3'	48
DRw8	5'-GGC GGG CCC TGG TGG AC-3'	64
DR9	5'-TGC GGT ATC TGC ACA GA-3'	52
DRw10	5'-GGA AAG ACG CGT CCA TA-3'	53

Table 3 (*continued*)

Probe	Sequence	Washing temp. (°C)
DRw11	5'-GGC CTG ATG AGG AGT AC-3'	56
DRw11(*1101,*1104)	5'-GAC AGG CGG GCC GCG GT-3'	66
DRw12	5'-GAG GAG CTC CTG CGC TT-3'	56
DRw13/DR4 (*0402)/ DRw11(*1102/*1103)	5'-TGG AAG ACG AGC GGG CC-3'	54
DRw14 (*1401)	5'-GCT GCG GAG CAC TGG AA-3'	50
Dw14	5'-GAG CAG AGG CGG GCC GCG-3'	67

Probes used for definition of subtypes of DR2, DR3, and DR4.

Probe	Sequence	Washing temp. (°C)
Dw4	5'-GGA GCA GAA GCG GGC CGC G-3'	70
Dw10	5'-GAA GAC GAG CGG GCC GCG-3'	64
Dw13.1,13.2, "DKT2"	5'-GGT GTC CAC CTC GGC CCG CC-3'	69
Dw14.1, 14.2, 15	5'-GAG CAG AGG CGG GCC GCG-3'	67
Dw15	5'-GCG GCC TAG CGC CGA GT-3'	62
DRw17, Dw10, Dw13.1, Dw14.1, "DKT2" (153B)	5'-GAA GCT CTC CAC AAC CCC GT-3'	64
DRw18, Dw4, Dw13.2, Dw14.2, Dw15 (GT86).	5'-AAG CTC TCA CCA ACC CCG TA-3'	62
Dw2 (DRB5*0101)	5'-GCA CAG AGA CAT CTA TA-3'	46
Dw12, Dw21, Dw22, (DRB5*0102, 0201, 0202)	5'-GCA CAG AGG CAT CTA TA-3'	48

HLA-DP[c]

Probe	Sequence	Washing temp. (°C)
A1(A/LFQ)	5'-CTT TTC CAG GGA-3'	36
A2B(A/VYQL)	5'-GTG TAC CAG TTA CGG CA-3'	52
A3(A/VHQL)	5'-GTG CAC CAG TTA-3'	36
A4B(A/VYQG)	5'-GTG TAC CAG GGA CGG CA-3'	56
C1B(C/AAE)	5'-GCC TGC TGC GGA GTA CT-3'	56
C2B(C/DED)	5'-GCC TGA TGA GGA CTA CT-3'	52
C3/55(C/DEE)	5'-GGC CTG ATG AGG AGT AC-3'	54
DB16(C/EAE)	5'-GAC TAC TCC GCC TCA GG-3'	56
D1(D/LLEEK)	5'-CTC CTG GAG GAG AAG-3'	48
D2(D/ILEEE)	5'-ATC CTG GAG GAG GAG-3'	48
D3(D/ILEEK)	5'-ATC CTG GAG GAG AAG-3'	46
E1(E/M)	5'-GAC AGG ATG TGC AGA CA-3'	52
E2(E/V)	5'-GAC AGG GTA TGC AGA CA-3'	52
DB25(F/GGPM)	5'-CTG CAG GGT CAT GGG CCC CCG-3'	74
DB26(F/DEAV)	5'-CTG CAG GGT CAC GGC CTC GTC-3'	72

[a]From ref. *8* and unpublished data. [b]From ref. *5*. [c]From W. M. C. Rosenberg, personal communication and ref. *9*.

7. Washing solution: 6X SSC, 0.1% SDS. Prepare from 400 mL of 6X SSC and 4 mL of 10% SDS.

2.5. Stripping the Membrane

1. 5X Stripping buffer: 250 mM NaOH, 0.5M NaCl, 2 mM EDTA. Prepare 1 L from 800 mL of dd water, 50 g of NaOH, 146.1 g of NaCl, and 20 mL of 0.5M EDTA. Adjust volume to 1 L with dd water.
2. Neutralization buffer: 1M Tris-HCl, pH 8, 1.5M NaCl. Prepare 1 L from 224 mL of 2M Tris base, 276 mL of 2M Tris-HCl and 500 mL of 3M NaCl.
3. 20X SSPE. Prepare 1 L from 800 mL dd water, 175.3 g NaCl, 27.6 g of sodium dihydrophosphate (NaH$_2$PO$_4$•H$_2$O) and 7.4 g of EDTA. Adjust to pH 7.4 by adding 6M NaOH dropwise. Adjust volume to 1 L with dd water.
4. 2X SSC. Prepare from 20X SSC (*see* Section 2.4.).

3. Methods
3.1. PCR
3.1.1. Amplification of HLA-DQ Alleles

The procedure for identifying HLA-DQ alleles is described in the following. Perform nested PCR using a 25-µL vol for the primary reaction, and a 50-µL vol for the secondary reaction.

1. Working on ice, pipet 16.2 µL of dd, autoclaved water into 0.5-mL microcentrifuge tubes.
2. Add 2 µL of genomic DNA (100 ng/µL), i.e., 200 ng/reaction.
3. Make up required volume of amplification mixture according to number of samples. Add *Taq* polymerase immediately prior to use.

	Vol/reaction, µL
4 mM dNTPs	1.25
1M KCl	1.25
40 mM MgCl$_2$	1.25
200 mM Tris-HCl, pH 8.4	1.25
2% Tween 20	1.25
Primer GH26A (0.5 mg/mL)	0.25
Primer GH27A (0.5 mg/mL)	0.25
Taq polymerase (0.5 U)	0.10

4. Aliquot 6.85 µL of amplification mixture into each tube. Vortex briefly, then pulse spin.
5. Overlay each sample with 1–2 drops of mineral oil and place tubes into thermal cycler.
6. Amplify by the primary PCR reaction using the following cycle parameters:

32 main cycles	94°C, 60 s (denaturation)
	55°C, 60 s (annealing)
	72°C, 30 s (extension)

7. After this and all subsequent PCR reactions, a small volume (e.g., 10%) of the reaction product may be separated by electrophoresis through 1.5% agarose and stained with ethidium bromide to confirm the expected amplification product (*see* Chapter 1 and Note 2).

8. Use PCR products immediately or store at −20°C.

9. Working on ice, pipet 35.3 µL of dd autoclaved water into microcentrifuge tubes.

10. Add 1 µL of primary PCR product.

11. Make up required volume of amplification mixture according to the number of samples. Add *Taq* polymerase immediately prior to use.

	Vol/reaction, µL
4 m*M* dNTPs	2.50
1*M* KCl	2.50
40 m*M* MgCl$_2$	2.50
200 m*M* Tris-HCl, pH 8.4	2.50
2% Tween	2.50
Primer GH26A	0.50
Primer GH27B	0.50
Taq polymerase	0.20

12. Aliquot 13.7 µL of PCR mix into each tube. Vortex briefly, then pulse spin (i.e., 30 s at max speed).

13. Overlay each sample with 1–2 drops of mineral oil and place tubes into thermal cycler.

14. Amplify in the secondary PCR reaction (*see* Note 3) using the following conditions:

32 main cycles	94°C, 60 s (denaturation)
	55°C, 60 s (annealing)
	72°C, 30 s (extension)

15. Analyze product as in step 7. *See* Note 2.

3.1.2. Amplification of HLA-DQB Alleles

1. Perform PCR as in Section 3.1., using primers GAMPDQXb2 and GLPDQb3 (*see* Table 2) for the primary PCR (25-µL vol) and primers GAMPDQXb2 and GLPDQb2 (*see* Table 2) for the second reaction (50-µL vol).

2. Amplify using the following cycle parameters:

Initial cycle	93°C, 5 min (denaturation)
	39°C, 3 min (annealing)
	64°C, 3 min (extension)

31 main cycles   93°C, 2 min (denaturation)
                        39°C, 2 min (annealing)
                        64°C, 2 min (extension)

3.1.3. Amplification of HLA-DRB Alleles

1. Use primers GLPDRb and GAMPDRb (*see* Table 2) and 1 µg of genomic DNA.
2. Perform a single PCR reaction in a 100-µL vol. The final concentrations of components are 50 mM KCl, 10 mM Tris-HCl, pH 8.3, 1.5 mM MgCl$_2$, 0.01% (w/v) gelatin, 200 µM of each dNTP, 1 µM of each primer, and 2.5 U of *Taq* polymerase). *See* Notes 4 and 5.
3. Amplify by PCR using the following cycle parameters:
   35 main cycles: 94°C, 1 min (denaturation)
                        39°C, 2 min (annealing)
                        64°C, 3 min (extension)
4. Examine PCR products as in step 7 in Section 3.1. *See* Note 2.

3.1.4. HLA-DR4-Specific Subtyping

1. Use primers 337C and 336C (*see* Table 2).
2. Perform a single PCR reaction in a 100-µL vol. Use the same amplification mixture as for HLA-DRB (*see* Section 3.3.), 1 µg of DNA and 2.5 U of *Taq* polymerase/reaction.
3. Amplify by PCR using the following cycle parameters:
   Initial denaturation step   95°C, 5 min
   30 main cycles   95°C, 1.3 min (denaturation)
                                  55°C, 2 min (annealing)
                                  72°C, 3 min (extension)
   Final extension step   72°C, 10 min

3.1.5. HLA-DR3-Specific Subtyping

1. Use primers GLPDR3 and GAMPDR3 (*see* Table 2).
2. Perform a single PCR in a 100-µL vol. Use the same amplification mixture as for HLA-DRB (*see* Section 3.3.), 1 µg of DNA and 2.5 U of *Taq* polymerase/reaction.
3. Amplify by PCR using the following cycle parameters:
   30 main cycles   94°C, 1.3 min (denaturation)
                        55°C, 2 min (annealing)
                        72°C, 3 min (extension)

3.1.6. Amplification of HLA-DPB1 Alleles

1. Use primers 5G and 3G (*see* Table 2).
2. Amplify 1 µg of genomic DNA in a single PCR reaction in a 100-µL vol. The final concentrations of components are 50 mM KCl, 10 mM Tris-HCl,

pH 8.3, 1.5 mM MgCl$_2$, 0.01% (w/v) gelatin, 200 µM of each dNTP, 0.02% Tween, 1 µM of each primer and 4 U of *Taq* polymerase (*see* Note 5).

3. Amplify by PCR using the following cycle parameters:

Initial cycle	94°C, 3 min (denaturation)
	60°C, 1 min (annealing)
	64°C, 2 min (extension)
30 main cycles	94°C, 1 min (denaturation)
	60°C, 1 min (annealing)
	64°C, 2 min (extension)

3.2. Dot Blotting

1. Transfer 10 µL of PCR product into 1.5-mL Eppendorf tube by placing pipet tip below level of mineral oil before fully expelling air, then withdrawing volume.
2. In groups of 12–24 tubes, add 104 µL of denaturing solution and place on ice for 10 min exactly. Neutralize with 114 µL of 2M ammonium acetate and leave on ice (samples can be frozen at –20°C at this stage).
3. Plan the sample map for 48 wells. Make duplicate maps of 24 samples each including controls (*see* Note 6).
4. Set up the dot-blot manifold by cutting the blotting membrane and two 3MM filter papers to size. Cut off top left-hand corner of the membrane for future orientation. Briefly wet the filter papers with dd autoclaved water and lay on middle section of the dot blot manifold, **avoiding bubbles**. Briefly wet the membrane and lay on top of the filter papers, **again avoiding bubbles**. Screw down the manifold and attach suction.
5. In groups of six samples, prewash the wells A1–6 and E1–6 with 200 µL of 1M ammonium acetate.
6. Mix the sample with pipet and load 110 µL to wells A1–E1, A2–E2, and so forth.
7. Wash the wells with 200 µL of 1M ammonium acetate.
8. Continue in groups of six samples until complete. Switch off suction and dismantle the manifold. Using forceps, remove the blotting membrane and briefly soak in 6X SSC. Allow to dry at room temperature for 30 min.
9. Label the membrane with pencil and bake for 120 min at 80°C. The blotting membrane can now be stored between filter papers at room temperature.

3.3. End-Labeling of Oligonucleotides (Kinase Reaction)

1. Work in a section of the laboratory reserved for radioactive materials. Use double gloves and protective eyewear and coat. Always keep radioactive solutions behind a screen.

2. Allow the $[\gamma\text{-}^{32}P]ATP$ to thaw in a perspex block. Meanwhile, prepare an "oligomix" in a 1.5-mL tube in wet ice using the following reagents:

dd H$_2$O	22.0 µL
100 mM DTT	1.5 µL
10X kinase buffer	3.0 µL
oligonucleotide	0.8 µL
PNK	1.0 µL

Vortex, then spin briefly in a microcentrifuge.

3. Add 2 µL of $[\gamma\text{-}^{32}P]ATP$ and gently "flick" the tube to mix.
4. Incubate tube in a heating block at 37°C for 30 min.
5. During incubation in step 4, set up the column. Allow sufficient room beneath the column for easy movement of tubes but not so much that solutions splash. Add 1 mL of DE 52 Sepharose solution, which should settle to a packed bed vol of approx 0.5 mL.
6. Increase heating block temperature to 70°C and incubate for another 15 min.
7. Add 80 µL of TE buffer to oligomix.
8. Break off the bottom of the column and allow the fluid to run out. **Do not allow the column to dry out at any point.**
9. Just before the meniscus reaches the top of the bed, carefully apply approx 5 mL of TE buffer. Avoid disturbing the packed bed.
10. Just before the TE buffer runs out, apply 110 µL of oligomix.
11. Wash with 1 mL of TE buffer. Discard this fluid into radioactive waste.
12. Wash with approx 5 mL of 200 mM NaCl in TE buffer and discard into radioactive waste.
13. Elute the oligonucleotide into three labeled tubes each with 0.5 mL of 500 mM NaCl in TE buffer. Tube 2 should be the most radioactive. Check counts in scintillation counter. Expect at least 50,000 cpm/µL in Tube 2. Store end-labeled oligonucleotide at –20°C.

3.4. Hybridization and Washing

1. Wash the blotting membrane briefly in 6X SSC.
2. Place the membrane between two layers of plastic and double heat-seal one end and both sides (the plastic should be approx 6 cm wider than the blotting membrane and 9 cm longer).
3. Pour in 5 mL of prehybridization solution, exclude air bubbles and double seal the forth side (plastic bag 1). Massage for 15 s and check for leaks. Place on a rocking platform for 30 min at 42°C.
4. Cut an end off the plastic bag and squeeze out the prehybridization solution.
5. Add 50 µL of $\gamma^{32}P$-labeled probe to the hybridization solution. Pour this solution into the plastic bag, exclude bubbles, and double seal the end.

6. Using two larger sheets of plastic, create plastic bag 2 around the first and double seal on all sides. Massage for 15 s.
7. Place on rocking platform at 42°C overnight or for a minimum of 4 h.
8. Cut the end off of each bag and discard radioactive solution appropriately.
9. Place blotting membrane in 300 mL of washing solution and gently agitate for 10 min at room temperature.
10. Remove the membrane and allow excess washing solution to run off. Place between layers of plastic wrap and expose to X-ray film for up to 1 h at room temperature with two intensifying screens in order to establish a baseline for the hybridizations prior to the high-stringency washing of the membrane to remove all but the completely matched probe. *See* Note 7.
11. Heat 300 mL of washing solution on heating block or in microwave to near desired washing temperature $[T_m]$ (*see* Table 3). Pour washing solution into a container in an agitating, heated water bath with the temperature set 2°C above T_m. Allow the solution to equilibrate at T_m (this is very important).
12. Place the membrane in washing solution and gently agitate for 30 min at the T_m. Meanwhile prepare the temperature of the next wash. Repeat this process with 300 mL of wash solution and wash for 30 min.
13. Using forceps, remove the blotting membrane and allow excess fluid to drain off. Wrap in Saran Wrap™ and check radioactivity using a Geiger counter. Expose to X-ray film for a variable length of time depending on radioactivity count at −70°C with two intensifying screens Alternatively, the membrane can be stored overnight at room temperature in 2X SSC.

3.5. Stripping the Blotting Membrane

The blotting membrane should be stripped as soon as possible after a satisfactory autoradiograph has been obtained.

1. Place membrane in 300 mL of 1X stripping buffer and agitate for 20 min at room temperature. Repeat.
2. Rinse the membrane twice in 50 mL of neutralization buffer and place the membrane in 300 mL of 5X SSPE.
3. Agitate for 20 min at room temperature. Repeat.
4. Expose the membrane for at least 4 h prior to probing with another oligonucleotide.
5. Keep the blotting membrane wet in 2X SSC in a bag, and do not allow the membrane to dry out. If membrane is to be stored for a long period, avoid exposure to light.

4. Notes

1. Alternative blotting membranes, such as Hybond C Extra (Amersham International plc, Amersham, UK) may be used.
2. After this and all subsequent PCRs, a small volume (e.g., 10%) of the reaction product may be separated by electrophoresis through 1.5% agarose and stained with ethidium bromide to confirm the expected amplification product. Once experience has been gained, this step is generally omitted.
3. We have found that, when DNA prepared from transformed lymphoblastoid cell lines is used, an adequate amplification can be obtained for the HLA-DQA1 alleles by using a single 25-µL vol PCR with primers GH26A and GH27B.
4. The HLA-DRB1 amplification mixture can be made in bulk and stored in aliquots for up to 3 mo at −20°C.
5. As a general guide, 1 µg of 20-mer primer in a volume of 100 µL gives a primer concentration of approx 1 μM.
6. Suitable controls include:
   a. A PCR product from reaction performed without DNA (replaced in amplification mix by dd, autoclaved water);
   b. Samples known to be positive and negative for the allele under investigation; and
   c. Previously successful samples, both positive and negative.
7. With experience this step can be omitted.

Acknowledgments

We gratefully acknowledge the help of Dr. W. M. C. Rosenberg in the preparation of this chapter.

References

1. Todd, J. A., Mijovic, C., Fletcher, J., Jenkins, D., Bradwell, A. R., and Barnett, A. H. (1989) Identification of susceptibility loci for insulin-dependent diabetes mellitus by trans-racial mapping. *Nature* **338,** 587–589.
2. Todd, J. A., Bell, J. I., and McDevitt, H. O. (1987) HLA-DQb gene contributes to susceptibility and resistance to insulin-dependent diabetes mellitus. *Nature* **329,** 599–604.
3. Todd, J. A., Fukui, Y., Kitagawa, T., and Sasazuki, T. (1990) The A3 allele of the HLA-DQA1 locus is associated with susceptibility to type 1 diabetes in Japanese. *Proc. Natl. Acad. Sci. USA* **87,** 1094–1098.
4. Lanchbury, J. S. S., Hall, M. A., Welsh, K. I., and Panayi, G. S. (1990) Sequence analysis of HLA-DR4B1 subtypes: Additional first domain variability is detected by oligonucleotide hybridisation and nucleotide sequencing. *Hum. Immunol.* **27,** 136–144.

5. Wordsworth, B. P., Allsopp, C. E. M., Young, R. P., and Bell, J. I. (1990) HLA-DR typing using DNA amplification by the polymerase chain reaction and sequential hybridization to sequence-specific oligonucleotide probes. *Immunogenetics* **32,** 413–418.
6. Rosenberg, W. M. C., Wordsworth, B. P., Jewell, D. P., and Bell, J. I. (1989) A locus telomeric to HLA-DPB encodes susceptibility to coeliac disease. *Immunogenetics* **30,** 307–310.
7. Bugawan, T. L., Horn, T. L., Long, C. M., Meckelson, E., Hansen, J. A., Ferrara, G. B., Angelini, G., and Erlich, H.A. (1988) Analysis of HLA-DP allelic sequence polymorphism using the in-vitro enzymatic DNA amplification of DP-a and DP-b loci. *J. Immunol.* **141,** 4024–4030.
8. Mijovic, C. H., Jenkins, D., Jacobs, K. H., Penny, M. A., Fletcher, J. A., and Barnett, A. H. (1991) HLA-DQA1 and -DQB1 alleles associated with genetic susceptibility to IDDM in a black population. *Diabetes* **40,** 748–753.
9. Bugawan, T. L., Angelini, G., Larrick, J., Auricchio, S., Ferrara, G. B., and Erlich, H. A. (1989) A combination of a particular HLA-DPb allele and an HLA-DQ heterodimer confers susceptibility to coeliac disease. *Nature* **339,** 470–473.

CHAPTER 11

The Use of the Polymerase Chain Reaction and the Detection of Amplified Products

Robert C. Allen and Bruce Budowle

1. Introduction

The most polymorphic genetic markers are DNA regions composed of variable number tandem repeats (VNTRs) *(1,2)*. Detection of the various VNTRs is possible by restriction fragment-length polymorphism (RFLP) analysis using the Southern blot procedure *(3)*. However, this procedure is time-consuming and requires an isotopic assay to achieve the sensitivity necessary to detect VNTR alleles in samples containing as little as 10–50 ng of human DNA *(4)*. The inability of RFLP technology to resolve discretely the alleles of VNTR loci also is a limitation. Where only limited quantities of DNA are available, or only single copy loci are to be analyzed, the DNA profile may be too weak to be detected with general RFLP methodology.

On the other hand, the polymerase chain reaction (PCR) *(5)* offers the ability to improve sensitivity of detection and specificity of analysis of discrete alleles of VNTR loci as well as specific regions of bacteria and viruses. The size of the individual segments of DNA from specific PCR amplification may be utilized in an analogous manner to RFLP analysis of VNTRs for detecting genetic and infectious diseases and identity testing with the advantage that a simple, horizontal surface-loaded system (*see* Section 3.5.) and silver staining, rather than blot hybridization with radioactive probes, may be used to separate and detect the nucleic acid fragments.

From: *Methods in Molecular Biology, Vol. 15: PCR Protocols: Current Methods and Applications*
Edited by: B. A. White Copyright © 1993 Humana Press Inc., Totowa, NJ

The use of multiple primers (multiplexing) in Duchennes muscular dystrophy (DMD) *(6,7)* has indicated that this procedure amplifies a spectrum of deficiencies marked by specific segments (Fig. 1). This approach might be extended to search for a variety of etiologic agents in diseases as indicated in Table 1. Multiplex PCR, under appropriate conditions, might provide a diagnostic approach by which confirmation can be based not only on the presence but also on the size of a specifically amplified DNA fragment or fragments.

PCR-amplified DNA fragments are normally in the size range of <100 up to 2000 bp. Fortunately, this size range is amenable to high resolution polyacrylamide gel electrophoresis on ultrathin-layer gels. Since repeats may differ by only 2 bp, electrophoretic separation systems should have this level of resolving capability. High resolution discontinuous buffer systems in ultrathin-layer polyacrylamide gels have demonstrated this resolving capability *(8,9)*. This capability is exemplified by the ability to resolve a 3-bp difference (94- and 97-bp fragments) of the delta F508 region deletion in cystic fibrosis *(9,10,11)* as shown in Fig. 2B. Continuous zone electrophoresis in agarose does not have sufficient resolving power to do this, in part, because of wide starting zones and diffusion during the long separation process. In addition, ethidium bromide stain sensitivity is less than silver and the small DNA fragments could migrate off the gel as there is no boundary to mark and limit their further migration.

Higher resolution of DNA on acrylamide gels may be obtained by initial zone sharpening by means of a discontinuity in ionic strength between the sample and the separation medium. This step may be followed by a moving ion boundary obtained by a discontinuous buffer system to sharpen further the starting zones. This procedure for maximizing the resolution of proteins and nucleic acids in polyacrylamide gel electrophoresis has been used in various forms for the separation of both native and denatured proteins *(12–14)* and RNA and DNA *(15)* for some 20 yr. Application of this technique in horizontal polyacrylamide gels to double-stranded DNA up to >5000 bp *(8)* and denatured, single-stranded DNA in vertical polyacrylamide sequencing gels *(16)* recently has been reported.

Improvements in resolution achieved by zone sharpening do not alone solve the problem of resolving a wide range of sizes of DNA on a single homogeneous gel. Gradient pore gels will resolve wide ranges of sizes of DNA. Some 112–21,000 bp can be resolved on a 3.5–7.5% gradient gel *(17)*. However, gradient gels are time-consuming to prepare and, even as commercially available gels, the gradient is fixed and may not provide the

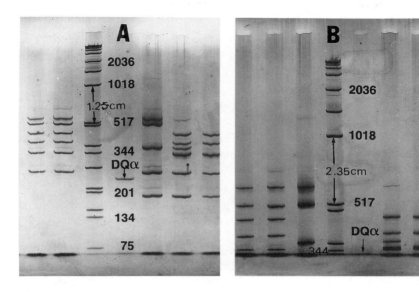

Fig. 1. Panels **A** and **B** are 5%T, 3.5%C rehydratable polyacrylamide gels. Gels were rehydrated in 35 m*M* Tris-sulfate (molarity with respect to sulfate ion). In panel **A**, 2.0*M* glycerol was added as a matrix modifier and in panel **B** no glycerol was used. Separations were carried out over a 9-cm distance and then stained with silver. In panel **A**, lanes 1, 2, and 5–7 are PCR-amplified products from multiplexing in analysis of Duchennes muscular dystrophy (DMD), lane 3 is the BRL 1-kb ladder and lane 4 is PCR-amplified DNA at the DQa locus. In panel **B**, lanes 1–3 and 6 and 7 are the same DMD samples as just mentioned but in reverse order. Lane 4 is the BRL 1-kb ladder and lane 5 is DQa-related fragment that has not unstacked from the boundary. Note that in the presence of the matrix modifier that the 344-bp fragment that is on the boundary in panel **B** has been retarded by almost 50% in the presence of the glycerol and that the 242-bp DQa fragment is now clearly resolved from the boundary. Amplified products of DMD were kindly supplied by Dr. C. T. Caskey, Baylor University.

Table 1
Size in Base Pairs of Some Disease PCR-Amplified DNAs

Use	Agent/allele	Fragments in bp	Ref.
AIDS	HIV-1 (gag region)	310	*(20)*
AIDS	HIV-2	294	*(20)*
Chlamydia	Genus specific	208	*(21)*
Chlamydia	Species specific region	517	*(21)*
Encephalitis	Herpes simplex	259	*(22)*
Encephalitis	H. influenzae Cap and noncap	273	*(22)*
Encephalitis	H. influenzae Capsule	343	*(23)*
DMD (multiplex)	9 alleles	196–547	*(6,7)*
Apolipo B 3'	14 alleles	560–970	*(24,25)*

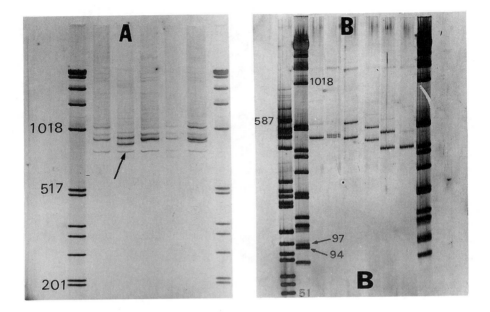

Fig. 2. Panel **A** shows a 4%T, 3%C ultrathin-layer polyacrylamide gel rehydrated with 30 m*M* Tris-formate (molarity with respect to formate ion) and with 0.25*M* ribose added as a matrix modifier. The separation distance is 9 cm and 0.5-μL samples were surface loaded with a parallel wire applicator (Gelman, Ann Arbor, MI). In lanes 1 and 7 are BRL 1-kb ladders, in lanes 2–6 are various 30-bp repeats of Apo B 3'. Panel **B** shows a similar gel with 1.0*M* ribose as a matrix modifier. The separation distance is 12 cm and the gel was surface-loaded using paper tabs with 2.0 μL of sample. Lane 1 is the Boehringer-Mannheim 587 ladder, lanes 2 and 7 are the BRL 1-kb ladder spiked with PCR-amplified DNA from a heterozygote for the delta 508 deletion of cystic fibrosis showing the 94- and 97-bp fragments. Lane 3 is 16-bp repeat D1S80-amplified DNA type 7–7, lane 4 is type 7–8, lane 5 is type 7–13 and lane 6 is type 6–10, lane 7 is type 1–9 and lane 8 is type 1–1. The DNA is stained with silver.

flexibility needed for a given separation. Until recently, multiple gel types have been required to assess a battery of different size ranges of amplified-fragment length polymorphisms (AMP-FLPs) from different loci to obtain sufficient resolution to identify each allele.

Matrix modification *(9)* offers another approach to tailor the resolution of DNA in polyacrylamide gels by altering the relative mobility (R_f) of each DNA component (see Fig. 3). Small polyols and various monosaccharides effect the R_f and resultant resolution of double-stranded DNA in

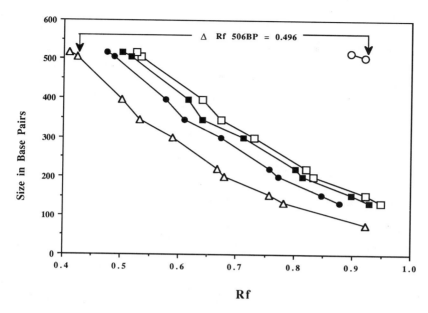

Fig. 3. Retardation in R_f of double-stranded DNA showing the portion of a 1-kb ladder from 517–75 bp by nonpolar compounds with a hydroyl group on the number 1 carbon atom and with zwitter ions from the Good buffer series used as trailing ions. The leading ion used was 30 m*M* formate. —O— , Borate; —●— , TAPS; —■— , TAPS + 1.0*M* Glycerol; —△— , MOPSO; —●— , Borate + 1.0*M* Ribose.

both rehydratable and conventionally cast polyacrylamide gels *(18)*. As mentioned in the Notes, members of the Good buffer series also have this capability when used as trailing ions *(19)*. Homogeneous gels with matrix modifiers, selected trailing ions, appropriate choice of pore size and or ionic strength of the separation medium may be used to optimize resolution of single-stranded RNA and single- and double-stranded DNA in discontinuous buffer system electrophoresis.

These procedures complement the use of the PCR technique and allow one now simply to select a dry gel and to rehydrate it with the buffer or matrix modifier of choice for the separation and resolution of a number of differently sized, amplified DNA products. These techniques, basically designed for forensic studies, also can be used to monitor a bone marrow transplant patient for the presence of donor markers, study molecular evolution, detect the presence of the AIDS virus or aid in the study of the sequence of the human genome. The rapidity of the new PCR instruments and the use of rapid, high-resolution electrophoresis with rapid silver staining of the amplified products has shortened the time for analysis from days with RFLPs to a matter of 2–3 h for AMP-FLPs.

2. Materials

2.1. DNA Extraction: Proteinase K/Phenol

1. EDTA vacutainer tubes.
2. 1X SSC. Prepare from 8.76 g of NaCl and 4.41 g of sodium citrate/L. Adjust pH to 7.0.
3. 200 mM Sodium acetate, pH 7.0.
4. 10% SDS.
5. Phenol:CHCl$_3$:isoamyl alchohol (25:24:10).
6. Absolute ethanol.
7. TE buffer: 10 mM Tris-HCl, pH 7.5, 0.1 mM EDTA.
8. 2.0M Sodium acetate, pH 7.0.
9. 70% Ethanol.

2.2. DNA Extraction: Chelex 100 (27,28)

1. Chelex 100 (Bio-Rad, Richmond, CA).
2. 1.5-mL Screw-cap tubes.

2.3. PCR

1. Sequence specific primers. For example: for the VNTR locus D1S80 *(26)*,
   5'-GAAACTGGCCTCCAAACACTGCCCGCCG-3' and
   5'-GTCTTGTTGGAGATGCACCGTGCCCCTTGC-3'
   for apoB *(24,25)*,
   5'ATGGAAACGGAGAAATTATG-3'
   5'-CCTTCTCACTTGGCAAATAC-3'
   Additional sets of primers can be included for multiplexing (*see* Fig. 2).
2. PCR reaction mixture (*see* Chapter 1).
3. Sample buffer (horizontal gels): 9 mM Tris-formate (or whatever leading ion buffer used for the gel), pH 9.0, 50% sucrose, and 0.1% bromophenol blue.
4. Sample buffer (vertical gels): 12 mM Tris-sulfate, pH 9.0, 50% sucrose, and 0.1% bromophenol blue.

2.4. Rehydratable Ultrathin-Layer Polyacrylamide Gel Reagents

1. Tris-formate buffer, 15–90 mM (we typically stay within the range of 30–60 mM), pH 9.0. Prepare from a stock buffer of 120 mM Tris-formate, pH 9.0. Prepare the stock buffer from 36.3 g of Trizma base and 40 mL of 1.0M formic acid. Adjust volume to 400 mL with pure, sterile

(e.g., MilliQ) water. Check pH after diluting stock buffer and correct pH with additional Trizma base.
2. Rehydratable gels (350-µm thick. 5%T, 3.5%C or 4%T, 3%C).
3. Fiberglass tabs for sample loading (LKB or Pharmacia, Rockville, MD).
4. LE Agarose (FMC Bioproducts, Rockland, ME).
5. 140 mM Tris borate buffer, pH 9.0. Prepare from 36.3 g of Trizma base and 4.37 g of boric acid. Adjust volume to 500 mL with pure sterile water.
6. 230 mM Tris-TAPS buffer, pH 9.0. Prepare from 36.3 g of Trizma base and 11.0 g of TAPS. Adjust volume to 200 mL with pure, sterile water.
7. 0.01% Bromophenol blue.
8. Saturated cupric sulfate.
9. E-C 1001 isothermically controlled electrophoresis apparatus.

2.5. Casting Utrathin-Layer Polyacrylamide Gel Reagents

1. Recrystallized acrylamide.
2. *N,N'*-methylene-*bis*-acrylamide.
3. Piperazine diacrylate.
4. 120 mM Tris-sulfate buffer, pH 9.0. Prepare from 36.3 g of Trizma base and 27 mL of 1.0M H_2SO_4. Dilute to 216 mL of pure, sterile H_2O.
5. TEMED (*N,N,N',N'*-tetramethylethylethylenediamine).
6. 0.05% Ammonium persulfate. Prepare 50-mL batches fresh.
7. GelBond-Pag (FMC, Bioproducts).
8. Gel-Fix® (Serva, Heidelberg, Germany).
9. Repel Silane® (LKB-Pharmacia).

2.6. Vertical Polyacrylamide Gel Reagents

1. Vertical gel electrophoresis apparatus.
2. *See* Section 2.5.

2.7. Silver Stain

1. 2% Nitric acid.
2. 3.4 mM Potassium dichromate in 3.2 mM nitric acid.
3. 140 mM Sodium carbonate containing 0.019% formalin.
4. 12 mM Silver nitrate.

3. Methods

3.1. DNA Extraction: Proteinase K/Phenol

1. Obtain whole blood in EDTA vacutainer tubes (*see* Note 1) and store at –20°C in 700-µL aliquots in 1.5-mL microfuge tubes.
2. Thaw sample, add 800 µL of 1X SSC, and mix.
3. Centrifuge sample for 1 min in a microcentrifuge.

4. Remove and discard 1 mL of supernatant.
5. Add 1 mL of 1X SSC, mix, and centrifuge for 1 min.
6. Discard supernatant.
7. Add 375 µL of 0.2*M* sodium acetate, pH 7.0, to the pellet and mix contents briefly.
8. Add 25 µL of 10% SDS and 5 µL of proteinase K and mix contents briefly.
9. Incubate sample at 56°C for 1 h and then cool to ambient temperature.
10. Add 120 µL of phenol/chloroform/isoamyl alcohol, vortex contents for 20 s and then centrifuge for 2 min.
11. Carefully remove the aqueous (top) layer and place in a 1.5-mL centrifuge tube.
12. Add 1 mL of cold (4°C) absolute alcohol to the aqueous layer, gently mix contents and centrifuge for 30 s.
13. Decant the supernatant and remove residual alcohol by pipeting.
14. Solubilize the pellet by vigorously mixing in 180 µL of TE buffer and incubate at 56°C for 10 min.
15. Add 20 µL of 2*M* sodium acetate, pH 7.0, and mix the contents gently.
16. Add 500 µL of cold absolute ethanol, mix the contents gently, centrifuge for 10 s and decant the supernatant.
17. Wash the pellet with 1 mL of 70% ethanol at 25°C, centrifuge for 10 s and decant the supernatant.
18. Place the tubes in a SpeedVac concentrator centrifuge (Savant, Farmingdale, NY) for 5 min to remove excess alcohol. Solubilize the dried DNA pellet in 200 µL of TE buffer overnight at 56°C.

3.2. DNA Extraction: Chelex 100 (27,28) (see Note 2)

1. Add 2–5 µL of whole blood to a 1.5-mL Sarstedt tube with a screw cap, then add 1 mL of sterile water and allow to remain at room temperature for 30 min.
2. Centrifuge the tube for 3 min in a microcentrifuge and remove and discard all but 30 µL of the supernatant.
3. Add Chelex 100 suspension (5% [w/v]) to the tube to a final volume of approx 200 µL.
4. Vortex the tube vigorously for 30 s, incubate for 30 min at 56°C, vortex again for 30 s, boil for 8 min, and finally vortex for 30 s.
5. Pellet the Chelex 100® by centrifugation in a microcentrifuge for 3 min, remove the supernatant and quantify (by A_{260}) and store the DNA at –20°C for subsequent PCR amplification.

3.3. PCR

1. Perform the PCR reactions in a 50-µL vol (*see* Chapter 1) using the appropriate specific primers (*see* Note 3).

2. Amplify D1S80 *(25)* by PCR using 100 ng of DNA and the following cycle parameters:
   25 main cycles   1 min, 95°C (denaturation)
                                     1 min, 65°C (annealing)
                                     8 min, 70°C (extension)
3. Amplify apolipoprotein B 3' *(15,16)* by PCR using 50 ng of DNA and the following cycle parameters:
   26 main cycles   1 min, 94°C (denaturation)
                                     6 min, 58°C (annealing and extension)
4. After amplification, transfer 45–50 µL of sample to a new microfuge tube and add an equivalent amount of sample buffer (according to the type of gel to be used; *see* Section 2.3.). Store samples at –20°C.

3.4. High Resolution Electrophoresis: Rehydratable Polyacrylamide Gels

1. Wear gloves. Place the dried, rehydratable gels in 30–60 mM Tris-formate, pH 9.0, buffer at 22°C for 45 min to 1 h (*see* Note 4). Gels may be held in rehydration buffer up to 1 wk at 4°C.
2. Following rehydration, carefully wipe the surface with the edge of a mylar sheet and blot the edges with a paper towel to remove all excess moisture.
3. Apply samples (0.3–5.0 µL) on fiberglass tabs laid directly on the rehydrated gel surface, or load samples directly with a micropipet in 0.3–5.0-µL amounts.
4. Place the gel on the cooling plate of an E-C 1001 isothermically controlled electrophoresis apparatus. Add a cathodal plug (1 cm × 1 cm × width of gel) of 1.5% LE agarose gel containing 140 mM Tris-borate buffer (or 230 mM Tris-TAPS buffer, pH 9.0) and 0.01% bromophenol blue as tracking dye across the entire width of the gel parallel to and 1-cm cathodal to the samples. Add another agarose plug (anodal) 9–15 cm away from the cathodal plug, depending on the desired length of the DNA separation (*see* Fig. 4).
5. Place two containers of saturated cupric sulfate in the space surrounding the cooling plate pedestal in order to maintain relative humidity (*see* Note 5).
6. Place electrodes directly on the gel plugs and perform electrophoresis. Use 6–7 mA constant current on 30–40 mM formate gels, and 20 mA on 90 mM formate gels that are 350–400-µm thick. Set the temperature to 15–20°C (*see* Note 6). Separation over a 9-cm distance with 30 mM formate as the leading ion should take 50 min.

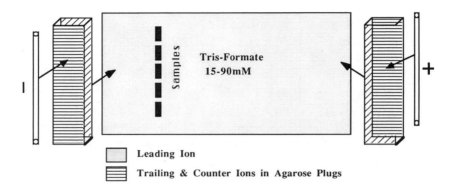

Fig. 4. Schematic set-up of horizontal rehydratable or freshly cast ultrathin-layer polyacrylamide gel bonded to GelBond-Pag® or GelFix®. The gel is rehydrated or cast with the leading ion (here Formate) and placed on an isothermically controlled plate I-C-E (E-C Apparatus, St. Petersburg, FL) . Samples are placed directly on the gel surface 1 cm from the cathodal plug edge. The agarose gel plugs containing the trailing and Tris counter ions then are placed on each end of the gel. The electrodes, mounted on bars in an electrode cover, are placed directly onto the surface of the gel plugs. Distances between plug edges varies from 9–15 cm depending on resolution requirements.

3.5. High Resolution Electrophoresis: Preparing Ultrathin-Layer Gels

1. Wear gloves. Place a glass plate, with dimensions 1/8 in. less on all sides than the GelBond-Pag or Gel-Fix® sheet, on a flat surface and wet the top surface.
2. Place a sheet of GelBond-Pag (hydrophobic side down) or Gel-Fix (either side down) on the glass plate and roll the plastic sheet flat with a photographic print roller. Pay particular attention to the surface of the film to assure that no bubbles or dirt are trapped beneath the sheet.
3. Dissolve 5 g of recrystallized acrylamide plus 150 mg of N,N'-methylene-*bis*-acrylamide or piperazine diacrylamide in a final volume of 25 mL.
4. Add 25 mL of acrylamide stock to 25 mL of a 120 mM leading ion buffer, preferably Tris-sulfate (*see* Note 7). Then add 50 mL of 0.05% ammonium persulfate and 90 µL of TEMED. This amount will produce seven ultrathin-layer, 11×25 cm gels that are 400-µm thick. Quickly degas the solution by drawing about 40 mL into a 60-mL syringe. Cover the barrel opening with a finger, draw down on the plunger to create a vacuum, and then tap the barrel of the syringe on the edge of the lab

bench. When all the bubbles have risen to the surface (20–30 s), remove finger from the barrel opening and cast the gel.

5. Place a 1/8-in. plastic or rubber frame 400-µm thick with outside dimensions 1/8 in. less than the GelBond or Gel-Fix sheet on the surface of the sheet. Pour approx 12–13 mL of the degassed acrylamide gel solution onto the surface at one end.

6. Carefully lay a cover glass treated with Repel Silane® on the frame edge at the liquid end and slowly lower the plate onto the frame allowing the solution to spread and to fill the frame without bubble formation. Let the gel polymerize for 1 h to overnight.

7. Remove the top cover and frame and load samples onto the gel surface. Carry out electrophoresis similarly to the instructions for rehydratable gels (*see* Section 3.4.).

3.6. High Resolution Electrophoresis: Vertical Systems

1. Wear gloves. Cut a sheet of GelBond-Pag or Gel-Fix to the same size as the larger casting plate. On the bottom of the back glass plate, wipe on a thin layer of stopcock grease and then lay the plastic sheet (hydrophobic side down with GelBond-Pag and either side with Gel-Fix™) and smooth the film to the back glass plate.

2. Place spacers on top of the plastic sheet and add a thin layer of stopcock grease to each side on the outside edge of the spacers.

3. Treat front inside surface of the cell with Repel Silane.

4. Dissolve 5 g of recrystallized acrylamide plus 150 mg of N,N'-methylene-*bis*-acrylamide or piperazine diacrylamide in a final volume of 25 mL. Add 25 mL of acrylamide stock to 25 mL of a 120 mM leading ion buffer, preferably Tris-sulfate (*see* Note 7). Then add 50 mL of 0.05% ammonium persulfate and 90 µL of TEMED. This amount will produce two 16×20 cm, 1.5-mm thick gels cast vertically in cassettes. Quickly degas the solution as described in Section 3.5.

5. Vertical gels should then be cast according to the directions of the manufacturer, since many casting devices have different methods of sealing the bottom.

6. Once the running gel has polymerized, rinse out the wells and fill wells with the cap gel solution. For the cap gel, prepare 1/5 as much gel solution. Thus, dissolve 1.0 g of acrylamide and 30 mg of N,N'-methylene-*bis*-acrylamide into 5 mL of water. Add 5 mL of 120 mM Tris-sulfate, pH 9.0. Then add 10 mL of 0.05% ammonium persulfate and 18 µL of TEMED.)

7. Immediately begin underlayering samples (usually 2–5 µL from the diluted PCR reaction) beneath nonpolymerized cap gel solution. Allow cap gel to polymerize.

8. Perform electrophoresis using conditions similar to those described for rehydratable gels (*see* Section 3.4.). Vertical gels will heat somewhat more, since they are considerably thicker. For "mini gels," 14 mA for 2 h with a divalent 30 mM sulfate buffer as the leading ion will give good results on a 6-cm separation distance. Carry out separation at 20 mA for 90 min on a 16×14-cm gel 1.5-mm thick in the presence of monovalent 30 mM formate. Run larger gels for 180 min using divalent 30 mM sulfate as the leading ion.

3.7. Rapid Staining of DNA with Silver (8) (see Note 8)

1. Following electrophoresis, place the gel in 2% nitric acid until the tracking dye boundary is completely yellow and any blue color below the cathode plug, in the case of horizontally run gels, is also yellow.

2. Place the gel in 3.4 mM dichromate, 3.2 mM nitrate solution for 5 min.

3. Rapidly wash the surface once and place in a 1.2 mM silver nitrate solution in the dark for 20 min. Alternatively, one may shorten the stain time by staining 10 min at 50°C achieved by placing the gel and stain in a microwave oven for 30 s. Use plastic or glass tray only.

4. Following the reaction with silver, wash the gel twice for 30 s with distilled water.

5. Add the cold (refrigerator temperature) sodium carbonate/formalin until the solution turns brown. Add new solution and continue to change every 5 min until the DNA bands show up well. During the development step, a cover should be placed over the staining dish to keep direct light off the gel to prevent a background from forming. If there is very little DNA in the electrophoretic tracks, then staining can be continued to a yellow background if more sensitivity is needed.

6. Stop the reaction in water for 5 min (*see* Note 9), then dry in a microwave oven for 1–2 min in short 1-min bursts. Wipe the excess silver from surface with a damp cotton pledget until no more blackening of cotton is apparent. This prevents further surface development during storage.

4. Notes

1. The use of ACD or 3.8% sodium citrate as an anticoagulant, rather than EDTA-coated vacutainer tubes, has been found to work well in whole blood that is to be stored at 4°C for extended times before centrifugation of the cells prior to DNA extraction.

2. The extraction procedures given in this presentation are only some of the many other procedures that will provide adequate genomic DNA for successful amplification. A number of these may be found in the references given in Table 1 and Chapter 1.

3. Amplification conditions can vary as to the tissue source, condition, and/or age of the genomic DNA. If samples are degraded significantly, one should not expect to amplify larger segments as may be done readily from fresh materials. Problems of unequal amplification of the larger of two alleles may often be overcome by varying $MgCl_2$ concentration, temperature of the cycles, increasing the *Taq* polymerase, and adding DMSO or formamide to the reaction mixture.

4. Matrix modifiers are small polyols and monosaccharides that retard the R_f of the nucleic acids in moving boundary systems and may be used to allow resolution of smaller DNA down to 50 bp on 4% T gels. In the absence of a modifier on this pore-size gel, 500 bp and smaller DNA do not unstack from the boundary. Ribose is the most effective one, although glycerol works well and is readily available in most laboratories. One-molar concentrations retard a 506-bp fragment of DNA by 43% and 47%, respectively in a 4%T, 3%C polyacrylamide gel. In rehydratable gels, incorporate the polyol into the rehydration buffer. Members of the Good buffer series, e.g., TAPS, Bicine, Tricine, used as trailing ions at 230–280 mM in place of borate, exert a similar mobility modifying effect and may be used also as matrix modifiers.

5. One of the major problems encountered in the separation of DNA over extended distances is the "smile effect," where the outside lanes migrate more slowly than the inner ones. This can be a particularly vexing problem on horizontal systems, where the edges of the gel are more subject to drying than is the center part. When drying occurs, the ionic strength at the outer edges of the gels is increased and the sample components move more slowly, causing the smile effect. This is readily prevented in moving boundary, or for that matter, on CZE systems, by simply adding 5-mm wide strips of rehydrated gel containing buffer at 1/2 the ionic strength of the separating gel to each edge of the gel. This effectively lowers the ionic strength at the edge. Thus, the mobility is increased in this region and prevents the smile effect. If wider strips are used, one can achieve a marked "frown effect." An alternative is to place 300–500 mL of saturated $CuSO_4$ in containers on each side of the gel in the buffer tray holder to maintain a high relative humidity of approx 92%.

6. Surface condensation is not a problem in gels run in closed systems. However, in open horizontal systems, overcooling should be avoided to prevent condensation. On the other hand, overheating from exces-

sive power application will cause drying. A separation temperature maintained within 3–10°C of ambient will prevent these effects and an isothermically controlled electrophoresis system allows proper conditions to be maintained at much higher power application than can be made in thicker, more poorly cooled gels.

7. For rapid separations, 30 mM of formate may be used as the leading ion to rehydrate the empty gels rehydratable gels. In gels cast horizontally on GelBond-Pag or those cast vertically in cassettes, sulfate is a better choice of leading ion, since the persulfate used as catalyst breaks down mainly to sulfate and one has a mix of leading ions if other anions are used.

8. The use of silver staining on polyacrylamide gels provides a much higher degree of sensitivity than does ethidium bromide staining of DNA in agarose or acrylamide gels. Therefore, unequal amplification of larger alleles in multiallelic systems is less likely to result in missing a larger-sized band and making the assumption that a sample is homozygotic, simply because the larger allele was less vigorously amplified. This is not a problem in VNTR loci of the size of D1S80 although it can manifest itself with apolipoprotein B 3'. Increasing the magnesium concentration to 2.5 mM of $MgCl_2$ will also improve amplification of larger alleles.

9. Stopping the silver stain reaction with acetic acid is often cited by authors where this stain technique is employed. In the presence of $0.28M$ sodium carbonate, this can lead to bubble formation in the ultrathin-layer gels. These later cause holes to appear on drying the gel and can interfere with photography and especially densitometry. It is better to wash extensively first in water.

References

1. Wyman, A. R. and White, R. (1980) A high polymorphic locus in human DNA. *Proc. Natl. Acad. Sci. USA* **77,** 6754–6758.
2. Nakamura, Y., Leppert, N., O'Connell, P., Wolff, R., Holm, T., Culver, M., Martin, C., Fujimoto, E., Hoff, M., Kumlin, E., and White, R. (1987) Variable number of tandem repeat markers for human gene mapping. *Science* **235,** 1616–1622.
3. Southern, E. M. (1975) Detection of specific sequences among DNA fragments separated by electrophoresis. *J. Mol. Biol.* **98,** 503–517.
4. Budowle, B. and Baectel, F. S. (1990) Modifications to improve the effectiveness of restriction length polymorphism typing. *Appl. Theor. Electrophoresis* **1,** 181–187.
5. Saiki, R. K., Scharf, S., Faloona, F., Mullis, K. B., Horn, G. T., Ehrlich, H. A., and Arnheim, N. (1985) Enzymatic amplification of beta globin genomic sequence and restriction analysis for diagnosis of sickle cell anemia. *Science* **230,** 1350–1354.

6. McCabe, E. R. B., Huang, Y., Descartes, M., Zahng, Y-H., and Fenwick, R. G. (1990) DNA from Guthrie spots for diagnosis of DMD by multiplex PCR. *Biochem. Med. Metabol. Biol.* **44,** 294–295.

7. Chamberlain, J. S., Gibbs, R. A., Ranier, J. E., Nguyen, P. N., and Caskey, C. T. (1988) Deletion screening of the Duchenne muscular dystrophy locus via multiplex DNA amplification. *Nucleic Acids Res.* **16,** 11,141–11,156.

8. Allen, R. C., Graves, G. M., and Budowle, B. (1989) Polymerase chain reaction amplification products separated on rehydratable polyacrylamide gels and stained with silver. *Biotechniques* **7,** 736–744.

9. Allen, R. C. and Graves, G. M. (1990) Rehydratable gels: A potential reference standard support for electrophoresing PCR-amplified DNA. *Bio/Technology* **8,** 1288–1290.

10. Kerem, B-s., Rommens, J. M., Buchanan, J. A., Markiewicz, D., Cox, T. K., Chakravarti, A., Buchwald, M., and Tsui, L-C. (1989) Identification of the cystic fibrosis gene: Genetic analysis. *Science* **245,** 1073–1080.

11. Riordan, J. R., Rommens, J. M., Kerem, B-s., Alon, N., Rozmahel, R., Grzelczak, Z., Zielenski, J., Lok, S., Plavsic, N., Chou, J-L., Drumm, M. L., Iannuzzi, M. C., Collins, F. S., and Tsui, L-C. (1989) Identification of the cystic fibrosis gene: Cloning and characterization of complementary DNA. *Science* **245,** 1066–1073.

12. Ornstein, L. (1964) Disc electrophoresis: Background and theory. *Ann. NY Acad. Sci.* **121,** 321–349.

13. Parrish, C. R. and Marchalonis, J. J. (1970) A simple and rapid method for estimating the molecular weights of proteins and protein subunits. *Anal. Biochem.* **34,** 436–450.

14. Laemmli, U. K. (1970) Cleavage of structural proteins during the assembly of the head of bacteriophage T4. *Science* **227,** 680–685

15. Maurer, H. R. and Allen, R. C. (1972) Some useful buffer systems for polyacrylamide electrophoresis. *Z. Klin. Chem. Klin. Biochem.* **10,** 220–225.

16. Carninci, P., Gustincich, S., Bottega, S., Patrosso, C., De Sal, G., Manfioletti, G., and Schneider, C. (1990) A simple discontinuous buffer system for increased resolution and speed in gel electrophoretic analysis of DNA sequences. *Nucleic Acids Res.* **18,** 204.

17. Jeppesen, P. G. N. (1974) A method for separating DNA fragments by electrophoresis in polyacrylamide concentration gradient slab gels. *Anal. Biochem.* **58,** 197–207.

18. Budowle, B., Charkraborty, R., Giusti, A. M., Eisenberg, A. J., and Allen, R. C. (1991) Analysis of the VNTR locus D1S80 by the PCR followed by high resolution PAGE *Am. J. Hum Genet.* **48,** 137–144.

19. Allen, R. C., Budowle, B., and Reeder, D. J. (1992) Resolution of DNA in the presence of mobility modifying polar and non-polar-compounds by Disc electrophoresis on rehydratable polyacrylamide gels. *Appl. Theor. Electrophoresis,* in press.

20. Rayfield, M., DeCock, K., Heyward, W. Goldstein, L., Krebs, J., Kwok, S., Lee, S., McCormick, J., Moreau, J. M., Odehouri, K., Schoctman, G., Sninsky,

J., and Ou, C-Y. (1988) Mixed human immunodeficiency virus (HIV) infection of an individual: demonstration of both HIV type 1 and type 2 proviral sequences by using polymerase chain reaction. *J. Infect. Dis.* **158,** 1170–1176.

21. Claas, H. C. J., Wagenvoort, J. H. T., Niesters, H. G. M., Tio, T. T., Van Rijsoort-Vos, J. H., and Quint, W. G. V. (1991) Diagnostic value of the polymerase chain reaction for *Chlamydia* detection as determined by a follow-up study. *J. Clin. Microbiol.* **29,** 42–45.

22. Aurelius, E. A., Johansson, B., Sköldenberg, B., Staland, A., and Forsgren, M. (1991) Rapid diagnosis of H*erpes simplex* encephalitis by nested polymerase chain reaction assay of cerebrospinal fluid. *Lancet* **337,** 189-192.

23. Van Ketal, R. J., De Wever, B., and Van Alphen, L. (1990) Detection of *Haemophilus influenzae* by polymerase chain reaction DNA amplification. *J. Med. Microbiol.* **33,** 271–276.

24. Boerwinkle, E., Xiong, W., Fourest, E., and Chan, L. (1989) Rapid typing of tandemly repeated hypervariable loci by the polymerase chain reaction: application to the apolipoprotein B 3' hypervariable region. *Proc. Natl. Acad. Sci. USA* **86,** 212–216.

25. Ludwig, E. H., Friedl, W., and McCarthy, B. J. (1989) High resolution of a hypervariable region in the human apolipoprotein B gene. *Am. J. Hum. Genet.* **45,** 458–464.

26. Kasai, K., Nakamura, Y., and White, R. (1989) Amplification of a VNTR locus by the polymerase chain reaction (PCR), in *Proceedings of an International Symposium on the Forensic Aspects of DNA Analysis.* US Government Printing Office, Washington, DC, pp. 279,280.

27. Hochmeister, M. N., Budowle, B., Jung, J., Boerer, U. V., Comey, C. T., and Dirndorfer, R. (1991) PCR-based typing of DNA extracted from cigarette butts. *Int. J. Legal Med.* in press.

28. Walch, P. S., Metzger, D. A., and Higuchi, R. (1991) Chelex 100 as a medium for simple extraction of DNA for PCR-based typing from forensic material. *BioTechniques* **10,** 506–513.

29. Allen, R. C. (1974) Polyacrylamide gel electrophoresis with discontinuous voltage gradients at a constant pH, in *Polyacrylamide Gel Electrophoresis and Isoelectric Focusing* (Allen, R. C. and Maurer, H. R., eds.), deGruyter, Berlin, pp. 105–113.

CHAPTER 12

Determination of Loss of Heterozygosity Using Polymerase Chain Reaction

Stephen J. Meltzer

1. Introduction

Loss of heterozygosity (LOH) is being detected with increasing frequency in a wide variety of human tumors *(1–10)*. Frequent LOH at a given chromosomal locus implies the existence of a tumor suppressor gene that is important in the pathogenesis of the particular cancer under study. LOH is believed to represent inactivation of one allele of a tumor suppressor gene by chromosomal or subchromosomal deletion. LOH studies have led to the isolation of new tumor suppressor genes *(11–13)*. They have also pinpointed known tumor suppressor genes that may be significant in specific types of cancer *(14,15)*.

One problem with LOH studies is that the currently accepted method, Southern blotting, requires large quantities of intact, high-mol-wt DNA. Such DNA is usually only obtainable from large freshly frozen tumor specimens, such as those available after surgical resection. However, surgical resection does not always occur, particularly in rare tumors and in certain advanced malignancies such as pancreatic cancer. Investigators conducting large studies might have to wait several years to obtain 50 specimens of certain tumors. Retrospective studies of archival tissues, valuable because the clinical significance of LOH can be determined by correlation with survival and disease-free interval, are

From: *Methods in Molecular Biology, Vol. 15: PCR Protocols: Current Methods and Applications*
Edited by: B. A. White Copyright © 1993 Humana Press Inc., Totowa, NJ

not possible unless frozen tumor banks are available. Analyses based on biopsies or aspirates are not feasible. Futhermore, the purity of small tumor specimens may have to be sacrificed in favor of DNA quantity, i.e., such tumor cell purification techniques as microdissection *(16)* and cell sorting *(17)* may be difficult to apply in this setting. The sensitivity of Southern blotting to detect LOH is poor when the malignant cell proportion in a specimen falls below 70% *(1,16)*.

We have developed a method of detecting LOH that utilizes polymerase chain reaction (PCR) *(18,19)*. Although this approach is fraught with several pitfalls that can cause artifact, with proper precautions and adequate repetition it is a dependable and rapid procedure for determining LOH. Small, very pure tumor cell preparations can be employed as sources of DNA. If the DNA is of high quality, only 50 ng are required; if degraded, 250 ng to 1 or 2 μg are needed. DNA from paraffin-embedded tissue can be studied, although it amplifies less reliably than DNA from frozen tissue. PCR-LOH carries the added advantage that methylation-sensitive enzymes can be used. Hypomethylation, which is common in human tumors, often makes LOH studies using Southerns and methylation-sensitive enzymes problematic. Hypomethylated genomic tumor DNA digests more completely than does fully methylated genomic normal DNA *(20)*. PCR-amplified DNA does not become methylated because the DNA polymerases are of bacterial origin; thus, DNA from normal and tumor digests equally. Two types of DNA polymorphisms can be used: the traditional restriction fragment length polymorphism (RFLP) or the newer variable number of tandem repeats (VNTR) polymorphism. Either one or two steps are involved after DNA has been extracted from tumor and matching normal tissue:

1. Amplification of DNA using PCR primers flanking the upstream and downstream sides of the polymorphism. For VNTRs, this is the only step.
2. For RFLPs, restriction enzyme digestion of PCR products.

The PCR products generated from normal and matching tumor DNA are then run side-by-side on polyacrylamide or agarose gels. Gels are stained with ethidium bromide and photographed under UV light. Photographs are analyzed visually by comparing normal to homologous tumor. LOH is manifested as a change in allele:allele ratio in the tumor relative to normal.

2. Materials

1. PCR reagents *(see* Chapter 1).
2. Oligonucleotide primers flanking the polymorphism. Examples: for p53, 5'-GATGCTGTCCGCGGACGATATT-3' (upstream) and 5'-CGTGCAAGTCACAGACTTGGC-3' (downstream; ref. *21).* One base in the upstream primer has been altered from the published sequence to create an artificial invariant *BstU* I site as an internal restriction enzyme digestion control. For retinoblastoma (Rb), 5'-CTCCTCCCTACTT-ACTTGT-3' (upstream) and 5'-AATTAACAAGGTTGTGGTGG-3' (downstream; ref. *22).* These primers flank a VNTR polymorphism.
3. Restriction enzyme and buffer (for example above, *BstU* I).
4. 50% Acrylamide:*bis*-acrylamide (19:1) stock solution.
5. TEMED.
6. 10% Ammonium persulfate stock solution.
7. Ethidium bromide stock solution (10 mg/mL).

3. Methods

1. Perform a standard PCR *(see* Chapter 1) using primers flanking desired RFLP or VNTR. Optimize the PCR so that the product is as abundant and pure as possible in order to minimize the need for further manipulations *(see* Notes 1 and 2).
2. For RFLPs, run an aliquot (e.g., 10 µL) of completed PCRs on a native polyacrylamide gel. For reactions using aforementioned primers, use 7% acrylamide *(see* Note 3 and Chapters 1 and 11). Stain with ethidium bromide and photograph. This test gel reveals which PCRs were successful and guides balancing of normal-tumor PCR pairs for subsequent restriction digestion. For VNTRs, proceed to step 4.
3. Digest the aliquots of each PCR product with an appropriate restriction endonuclease *(see* Note 4). For the p53-related RFLP indicated earlier, use *BstU* I. Select an aliquot amount based on the gel in step 2 so that normal and tumor products are balanced. In practice, this amount ranges from 5–40 µL of a typical 100-µL reaction. After digestion, dry down the entire digest using a vacuum concentrator (e.g., SpeedVac, Savant Instruments, Farmingdale, NY). Redissolve in a convenient amount (usually 20 µL) of H_2O for running on a polyacrylamide gel.
4. Run the products on a polyacrylamide gel. For RFLPs, make sure that the PCR product amounts from normal and tumor have been well balanced in step 3. Stain with ethidium bromide and photograph under UV light. For VNTRs, run 10–30 µL of each product on one gel. If the amounts of the products from normal and tumor tissues are not well bal-

anced, run a second gel using different amounts of each, reestimated in order to achieve balance.

5. Analyze photographs for LOH *(see* Note 5). This is manifested as partial or complete loss of one allele in the tumor DNA relative to homologous normal DNA *(see* Note 6). Stated alternatively, it is a decrease in allele:allele ratio in the tumor relative to normal tissue, which has an allele:allele ratio of 1:1. Examples are shown in Fig. 1.

4. Notes

1. The key parameters in optimization titrations are genomic template amount, annealing temperature, primer amount, and dNTP concentration. A good starting point is 100 ng of template, annealing temperature of 57°C, 400 nM of each primer, and 75 µM of each dNTP. The usual volume of reaction is 100 µL, but 50 µL or even 25 µL have been used by some authors. To diminish primer-dimer amplification, "hot start" PCR may be performed. This is done by withholding one essential component (e.g., dNTPs or enzyme) of PCR until the temperature has reached 80°C. An 80°C step may be programmed into the PCR machine for this purpose. This temperature should denature most primer-dimers. The optimal magnesium concentration is usually 2.5 mM, but this parameter can also be titrated. Some authors also add dimethyl sulfoxide (DMSO) to PCRs, although this is not necessary with the primers indicated above. PCRs using paraffin-extracted DNA require 250 ng to 2 µg of genomic template, whereas those employing frozen tissue-derived DNA can be performed with 50 or even 25 ng. Samples fixed in ethanol amplify better than those fixed in formaldehyde or picric acid. Keep total cycle number below 35, since higher numbers encourage the formation of heterodimer, which results from the annealing of mismatched PCR strands from different alleles in heterozygotes. This results in a double-stranded PCR product with one strand containing the restriction site and the opposite strand lacking it. This PCR product does not digest with the restriction enzyme.

2. To conserve limited quantities of tumor DNA, screening of normal DNA only may be performed first to determine constitutional heterozygosity.

3. Usual acrylamide percentage is 7%, but this may vary depending on size of PCR product. As low as 3% and as high as 15% can be used. The addition of 5–10% glycerol to gels improves resolution, especially in the analysis of VNTRs *(see* Chapter 11). Gels may be run at high voltage (e.g., 215 V for a 15-cm long gel). Stain with very low concentrations of EtBr, e.g., 1 µL of 10 mg/mL stock in 300 mL of H$_2$O. Stain briefly, 2–10 min, for low background.

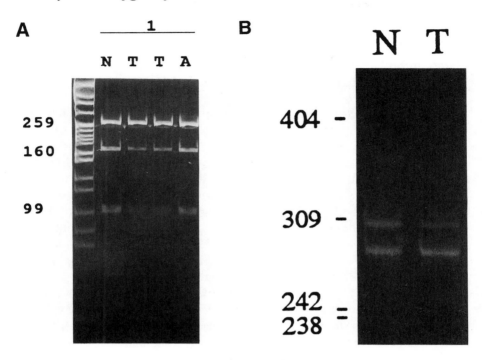

Fig. 1. Loss of heterozygosity demonstrated using PCR. **A,** LOH at p53 locus using primers indicated in text. N = normal tissue, T = two samples of a tumor from the same patient, A = histologically normal tissue adjacent to the tumor. LOH involving the "cut" allele is seen in both segments of tumor but not in neighboring normal tissue. Remaining signal at 160 and 99 bp probably represents contaminating nontumor cells within the tumor mass. **B,** LOH at Rb locus (VNTR) using primers in text. LOH involving the upper (larger) allele is seen in tumor. Some remaining signal in the upper band is again observed. This contaminating signal is common when relatively impure tumor preparations are used.

4. Performing digests in large volumes (100 µL) dilutes out salts from PCR and other potential inhibitors of restriction digestion. This is the most difficult step in the protocol. Use generous amounts of enzyme (20–50 U/digest) and continue digestion overnight. For p53 primers indicated earlier, make sure the temperature is 59°C for *BstU* I digestion. Repeat entire procedure (starting from fresh genomic DNA) at least twice to verify results. Artificial internal *BstU* I site in the aforementioned p53 primer will show incomplete digestion as blurring of the uncut allele. If possible, use an additional set of nonoverlapping primers to amplify same RFLP *(see* ref. *18)*. In addition, more than one

polymorphism at the same locus should be used if available. Two polymorphisms within the same gene should almost always agree if both are informative (heterozygous) for a given individual.

5. Analysis is usually by eye, but densitometry may be used. A good system is sold by Ambis (Olney, MD). A reasonable cutoff for positive LOH is a decrease in the allele:allele ratio in tumor to <50% of the ratio in normal tissue.

6. We have on occasion observed gain of heterozygosity (GOH) in the tumor relative to normal tissue. This is virtually always artifact. The most common cause is actual mix-up of tissue or genomic DNA samples. The next most common cause is incomplete restriction digestion of the normal PCR product. If the RFLP is not located in an evolutionarily conserved region of the tumor suppressor gene at which frequent mutations have been reported, interpret GOH results with great skepticism. Perform repeat assays using the same primers, nonoverlapping primers at the same RFLP, and primers surrounding a different, neighboring polymorphism. In addition, perform PCR-LOH studies at a locus on another chromosome: If sample mix-up is responsible for these findings, GOH will be seen elsewhere in the genome as well. Such results are more common when parraffin-derived DNA is used.

7. Other chromosomal loci may be evaluated for LOH as long as sequence information on both sides of RFLPs or VNTRs is available. In this way, "allelotypes" of specific cancers can be generated using PCR.

References

1. Vogelstein, B., Fearon, E. R., Hamilton, S. R., Kern, S. E., Preisinger, A. C., Leppert, M., Nakamura, Y., White, R., Smits, A. M. M., and Bos, J. L. (1988) Genetic alterations during colorectal-tumor development. *N. Engl. J. Med.* **319**, 525–532.

2. Wagata. T., Ishizaki, K., Imamura, M., Shimada, Y., Ikenaga, M., and Tobe, T. (1991) Deletion of 17p and amplification of the int-2 gene in esophageal carcinomas. *Cancer Res.* **51**, 2113–2117.

3. Meltzer, S. J., Ahnen, D. J., Battifora, H., Yokota, J., and Cline, M. J. (1987) Protooncogene abnormalities in colon cancers and adenomatous polyps. *Gastroenterology* **92**, 1174–1180.

4. Simon, D., Knowles, B. B., and Weith, A. (1991) Abnormalities of chromosome 1 and loss of heterozygosity on 1p in primary hepatomas. *Oncogene* **6**,765–770.

5. Khosla, S., Patel, V. M., Hay, I. D., Schaid, D. J., Grant, C. S., van Heeren, J. U. A., and Thibodeau, S. N. (1991) Loss of heterozygosity suggests multiple genetic alterations in pheochromocytomas and medullary thyroid carcinomas. *J. Clin. Invest.* **87**, 1691–1699.

6. Fey, M. F., Hesketh, C., Wainscoat, J. S., Gendler, S., and Thein, S. L. (1989) Clonal allele loss in gastrointestinal cancers. *Br. J. Cancer* **59**, 750–754.

7. Vogelstein, B., Fearon, E. R., Kern, S. E., Hamilton, S. R., Preisinger, A. C., Nakamura, Y., and White, R. (1989) Allelotype of colorectal carcinomas. *Science* **244**, 207–211.

8. Kovacs, G., Erlandsson, R., Boldog, F., Ingvarsson, S., Muller-Brechlin, R., Klein, G., and Sumegi, J. (1988) Consistent chromosome 3p deletion and loss of heterozygosity in renal cell carcinoma. *Proc. Natl. Acad. Sci. USA* **85**, 1571–1575.

9. Yokota, J., Tsukada, Y., Nakajima, T., Gotoh, M., Shimosato, Y., Mori, N., Tsunokawa, Y., Sugimura, T., and Terada, M. (1989) Loss of heterozygosity on the short arm of chromosome 3 in carcinoma of the uterine cervix. *Cancer Res.* **49**, 3598–3601.

10. Lee, J. H., Kavanagh, J. J., Wildrick, D. M., Wharton, J. T., and Blick, M. (1990) Frequent loss of heterozygosity on chromosomes 6q, 11, and 17 in human ovarian carcinomas. *Cancer Res.* **50**, 2724–2728.

11. Fearon, E. R., Cho, K. R., Nigro, J. M., Kern, S. E., Simons, J. W., Ruppert, J. M., Hamilton, S. R., Preisinger, A. C., Thomas, G., Kinzler, K. W., and Vogelstein, B. (1990) Identification of a chromosome 18q gene that is altered in colorectal cancers. *Science* **247**, 49–56.

12. Kinzler, K. W., Nilbert, M. C., Vogelstein, B., Bryan, T. M., Levy, D. B., Smith, K. J., Preisinger, A. C., Hamilton, S. R., Hedge, P., Markham, A., Carlson, M., Joslyn, G., Groden, J., White, R., Miki, Y., Miyoshi, Y., Nishisho, I., and Nakamura, Y. (1991) Identification of a gene located at chromosome 5q21 that is mutated in colorectal cancers. *Science* **251**, 1366–1370.

13. Kinzler, K. W., Nilbert, M. C., Su, L.-K., Vogelstein, B., Bryan, T. M., Levy, D. B., Smith, K. J., Preisenger, A. C., Hedge, P., McKechnie, D., Finniear, R., Markham, A., Groffen, J., Boguski, M. S., Altschul, S. F., Horii, A., Ando, H., Miyoshi, Y., Miki, Y., Nishosho, I., and Nakamura, Y. (1991) Identification of FAP locus genes from chromosome 5q21. *Science* **253**, 661–665.

14. Baker, S. J., Fearon, E. R., Nigro, J. M., Hamilton, S. R., Preisinger, A. C., Jessup, J. M., vanTuinen, P., Ledbetter, D. H., Barker, D. F., Nakamura, Y., White, R., and Vogelstein, B. (1989) Chromosome 17 deletions and p53 gene mutations in colorectal carcinomas. *Science* **244**, 217–221.

15. Baker, S. J., Preisinger, A. C., Jessup, J. M., Paraskeva, C., Markowitz, S., Willson, J. K. V., Hamilton, S., and Vogelstein, B. (1990) p53 mutations occur in combination with 17p allelic deletions as late events in colorectal tumorigenesis. *Cancer Res.* **50**, 7717–7722.

16. Meltzer, S. J., Mane, S. M., Wood, P. K., Resau, J. H., Newkirk, C., Terzakis, J. A., Korelitz, B. I., Weinstein, W. M., and Needleman, S. W. (1990) Activation of c-Ki-ras in human gastrointestinal dysplasias determined by direct sequencing of polymerase chain reaction products. *Cancer Res.* **50**, 3627–3630.

17. Reid, B. J., Haggitt, R. C., Rubin, C. E., and Rabinovitch, P. S. (1987) Barrett's esophagus: Correlation between flow cytometry and histology in detection of patients at risk for adenocarcinoma. *Gastroenterology* **93**, 1–11.

18. Meltzer, S. J., Yin, J., Huang, Y., McDaniel, T. K., Newkirk, C., Iseri, O., Vogelstein, B., and Resau, J. H. (1991) Reduction to homozygosity involving p53 in esophageal cancers demonstrated by the polymerase chain reaction. *Proc. Natl. Acad. Sci. USA* **88**, 4976–4980.
19. Boynton, R. F., Huang, Y., Blount, P. L., Reid, B. J., Rasking, W. H., Haggitt, R. C., Newkirk, C., Resau, J. H., Yin, J., McDaniel, T., and Meltzer, S. J. (1991) Frequent loss of heterozygosity at the retinoblastoma locus in human esophageal cancers. *Cancer Res.*, **51**, 5766–5769.
20. Feinberg, A. P. and Vogelstein, B. (1983) Hypomethylation distinguishes genes of some human cancers from their normal counterparts. *Nature* **301**, 89–92.
21. de la Calle-Martin, O., Fabregat, V., Romero, M., Soler, J., Vives, J. and Yague, J. (1990) Acc II polymorphism of the p53 gene. *Nucleic Acids Res.* **18**, 4963.
22. McGee, T. L., Yandell, D. W., and Dryja, T. P. (1989) Structure and partial genomic sequence of the human retinoblastoma susceptibility gene. *Gene* **80**, 119–128.

CHAPTER 13

Direct Sequencing of Polymerase Chain Reaction Products

Stephen J. Meltzer

1. Introduction

The advantages of direct sequencing of polymerase chain reaction (PCR) products over conventional sequencing of cloned, single-stranded DNA are manifold. Speed is perhaps the greatest asset. Time-consuming creation and screening of libraries, fragment purification, subcloning, bacterial transfection, and plasmid preparation steps are all eliminated. Cost is an advantage for similar reasons. The ability to sequence thousands or even millions of different templates at once, thereby obtaining pooled averages of mutated or polymorphic sequences, is another benefit. Moreover, short DNA sequences not obtainable by conventional cloning, such as DNA from paraffin-embedded tissues, can be sequenced using PCR sequencing.

Numerous refinements and additions to the PCR-sequencing armamentarium have been made over the past 3 years. However, most methods consist of two steps:

1. PCR amplification of the genomic DNA or cDNA template to be sequenced.
2. Dideoxynucleotide (Sanger) sequencing of the PCR product.

In particular, the ability to generate single-stranded sequencing template by PCR *(1)* and to sequence RNA by making a cDNA copy *(2)* have become standard applications of this technique. Some sequencing protocols include an incorporation or labeling reaction *(3)*, whereas

From: *Methods in Molecular Biology, Vol. 15: PCR Protocols: Current Methods and Applications*
Edited by: B. A. White Copyright © 1993 Humana Press Inc., Totowa, NJ

others contain only the termination reaction *(4)*. Nested sequencing primers internal to and not overlapping with the amplification primers can improve the specificity of sequenced product *(5)*.

The fewer the number of steps after amplification, the faster and more cheaply sequence data is obtained. In this chapter I describe a version of PCR sequencing that requires no purification of PCR product and only a sequencing termination reaction using end-labeled primer *(6)*. This protocol is particularly useful for sequencing short PCR products (<200 nucleotides long).

Perhaps the most important concern in direct PCR sequencing is the quality of PCR product used as template. If this PCR product is abundant and pure, essentially free of mispriming or nonspecific amplification products, clear sequencing ladders will usually result. Care should be taken to optimize the PCR by adjusting annealing temperature, original template amount, $MgCl_2$ concentration, and other parameters *(6)*. If pure product cannot be obtained by optimization experiments, gel purification prior to sequencing is preferred. This difficulty may be circumvented by using nested (internal, nonoverlapping) primers. Another pitfall relates to incomplete denaturation before sequencing: Some PCR products denature more easily than others owing to GC content, secondary structure, and other factors. This problem may be averted by using dITP or deaza-GTP rather than dGTP in the termination reaction, or by using an alternative denaturing procedure, such as alkaline denaturation *(7)*. There is also evidence that optimization of sequencing reaction time, postdenaturation temperature, and primer:template ratio may dramatically improve results *(8)*.

2. Materials

1. PCR kit reagents (Perkin-Elmer Cetus, Norwalk, CT).
2. Sequenase kit reagents (U.S. Biochemical, Cleveland, OH).
3. Genomic DNA template (1.0 µg).
4. Amplification primers (4 pmol).
5. Sequencing primer (5 pmol).
6. 100 µCi of [γ-^{32}P]ATP (5,000 Ci/mmol, 2 mCi/50 µL).
7. T4 polynucleotide kinase.
8. Acrylamide gel components: acrylamide and *bis*-acrylamide, urea, ammonium persulfate, TEMED *(see* Chapter 1).
9. Thermal cycling machine.
10. DNA sequencing apparatus.
11. 3000 V/150 mA power supply.

3. Methods

1. Amplify desired region of genomic template DNA in a 100-μL reaction *(see* Chapter 1) using 4 pmol of each PCR primer, 1.0 μg of genomic DNA, and 0.3 μL of *Taq* polymerase *(see* Note 1).
2. End-label sequencing primer as follows: Mix 5 pmol of primer and 100 μCi of [γ-^{32}P]ATP *(see* Note 2) with water to a final volume of 9 μL, boil for 5 min, incubate on ice for 10 min, add 1 μL of T4 polynucleotide kinase, and incubate at 37°C for 1.5 h.
3. Aliquot 2.5 μL of each Sequenase dideoxynuceotide termination mix into separate tubes *(see* Note 3). Make sure a separate set of ddG, ddA, ddC, and ddT has been aliquoted for each template to be sequenced. Keep on ice until ready for use.
4. Mix 10 μL of unpurified PCR product from step 1 with 2 μL of end-labeled primer from step 2; boil for 5 min, ice for 10 min *(see* Note 4). Meanwhile, prewarm ddNTPs to room temperature.
5. Add 4 μL of 5X Sequenase buffer, 1 μL of 0.1*M* dithiothreitol (from the Sequenase kit), and 2 μL of a freshly prepared 1:8 dilution of Sequenase enzyme *(see* Note 5) to the entire annealing reaction from step 4.
6. Mix the concoction from step 5 well but gently to avoid bubble formation, using a 20-μL pipettor set at 20 μL. Aliquot 4 μL of this mixture to each of the four termination mixes. Mix *(see* Note 6) and begin timing next step.
7. Incubate at room temperature for 3 min, then at 37°C for 2 min *(see* Note 7).
8. Stop the termination reaction by adding 4 μL of Sequenase stop buffer to each tube *(see* Note 8).
9. Denature the sequencing reactions at 90°C for 5 min, load onto a polyacrylamide-urea sequencing gel and electrophorese according to directions for gel apparatus *(see* Note 9).

4. Notes

1. When setting up multiple genomic templates using the same primers, prepare a master mix consisting of all components except genomic DNA and water. Aliquot this mix, then add water and genomic DNA last. In order to accurately pipet 0.3 μL of *Taq* polymerase, prepare a 1:10 dilution in water first, then pipet 3.0 μL.
2. 100 μCi of [γ-^{32}P]ATP is contained in 2.5 μL if radioisotope comes as a 2 mCi/50 μL stock and if specific activity is 5000 Ci/mmol. If stock is different, make calculations to achieve same molar ratio. Complete labeling is important, since unlabeled primer will compete with labeled primer in the sequencing reaction and lead to weak signal. Labeling for 1.5 h

rather than a shorter interval also maximizes completeness of the labeling reaction.

3. For better mononucleotide stability, ddNTP/dNTP termination mixes should be aliquoted into 2.5-µL lots immediately upon receipt from the company. Thereby, they will only be subjected to one freeze-thaw cycle. Repeated freeze-thawings speed breakage of the unstable phosphodiester bonds.

4. This annealing step is based on the "quick chill" or "mad rush" theory of hybridization: The low temperature should favor rapid motion by small primer molecules to their targets in genomic DNA. Again, sequencing primers may be located internal to PCR primers for added specificity.

5. Sequenase is extremely delicate. It tolerates temperatures > –20°C very poorly. Dilutions should be prepared immediately prior to this step, using either Sequenase dilution buffer (from the kit) or water.

6. If mixing is by a pipettor, separate tips must be used for each G,A,T,and C. A faster and more simultaneous method is to discharge the 4-µL aliquots onto the sides of each G,A,T,and C tube, then to spin all four briefly in a microfuge. Timing begins when the microfuge is turned on.

7. More than one sequencing reaction may be performed by staggering each termination reaction 1 min apart. Greater than three reactions at once are not recommended. For simplicity, one reaction at a time is easiest.

8. Again, mixing may be simultaneously accomplished as in Note 6.

9. Depending on the success of the labeling and sequencing reactions, between 1 and 8 µL of each termination reaction may be required for good signal. The smaller the volume loaded, the sharper the bands on the gel. The termination reactions may be stored at –20°C for up to 2 d. Longer storage times lead to fragmented termination products and high background across all lanes.

Acknowledgments

Supported by ACS Grant # PDT-419, the Crohn's and Colitis Foundation of America, and the Department of Veterans Affairs.

References

1. Gyllensten, U. B. and Erlich, H. A. (1988) Generation of single-stranded DNA by the polymerase chain reaction and its application to direct sequencing of the HLA-DQA locus. *Proc. Natl. Acad. Sci. USA* **85**, 7652–7656.

2. Sarkar, G. and Sommer, S. S. (1988) RNA amplification with transcript sequencing (RAWTS). *Nucleic Acids Res.* **16**, 5197.

3. Innis, M. A., Myambo, K. B., Gelfand, D. H., and Brow, M. A. D. (1988) DNA sequencing with Thermus aquaticus DNA polymerase and direct sequencing of polymerase chain reaction-amplified DNA. *Proc. Natl. Acad. Sci. USA* **85**, 9436–9440.

4. Engelke, D. R., Hoener, P. A., and Collins, F. S. (1988) Direct sequencing of enzymatically amplified human genomic DNA. *Proc. Natl. Acad. Sci. USA* **85**, 544–548.

5. Wrischnik, L. A., Higuchi, R. G., Stoneking, M., Erlich, H. A., Arnheim, N., and Wilson, A. C. (1987) Length mutations in human mitochondrial DNA: direct sequencing of enzymatically amplified DNA. *Nucleic Acids Res.* **15**, 529–541.

6. Meltzer, S. J., Mane, S. M., Wood, P. K., Johnson, L., and Needleman, S. W. (1990) Sequencing products of the polymerase chain reaction directly, without purification. *BioTechniques* **8**, 142–148.

7. Williams, J. F. (1989) Optimization strategies for the polymerase chain reaction. *Biotechniques* **7**, 762–770.

8. Wahlberg, J., Lundeberg, J., Hultman, T., and Uhlen, M. (1990) General colorimetric method for DNA diagnostics allowing direct solid-phase sequencing of the positive samples. *Proc. Natl. Acad. Sci. USA* **87**, 6569–6573.

9. Casanova, J.-L., Pannetier, C., Jaulin, C., and Kourilsky, P. (1990) Optimal conditions for directly sequencing double-stranded PCR products with Sequenase. *Nucleic Acids Res.* **18**, 4028.

CHAPTER 14

Manual and Automated Direct Sequencing of Product Generated by the Polymerase Chain Reaction

Adam P. Dicker, Matthias Volkenandt, and Joseph R. Bertino

1. Introduction

Identification of point mutations has been facilitated by a number of techniques, including transfection assays, oligonucleotide hybridization, electrophoretic migration of heteroduplexes, RNase mismatch analysis, direct sequencing, and DNA-polymerase catalyzed amplification. The large number of available techniques emphasizes the importance of developing rapid and reliable methods to identify molecular changes in genes. To date, we have concentrated on exploiting DNA-polymerase catalyzed amplification methods *(1,2)* in conjunction with direct manual and automated DNA sequencing to detect point mutations in the dihydrofolate reductase (DHFR) gene of methotrexate-resistant cells.

Acquired drug resistance is a major obstacle in the treatment of patients with drug sensitive tumors (e.g., leukemia and lymphoma). We are interested in determining whether a cause of methotrexate resistance in patients is owing to an altered DHFR gene giving rise to an altered enzyme with a decreased affinity for antifolates, e.g., methotrexate (MTX). We have used polymerase chain reaction (PCR) to study the problem of acquired drug resistance. This requires amplification of the human DHFR transcript and genes and a rapid and repro-

From: *Methods in Molecular Biology, Vol. 15: PCR Protocols: Current Methods and Applications*
Edited by: B. A. White Copyright © 1993 Humana Press Inc., Totowa, NJ

ducible method for directly sequencing the amplified DNA. We have developed and modified a series of methods that utilize [α-^{35}S]dATP without [γ-^{32}P]dATP end-labeled primers. We have improved on our initial method for directly sequencing PCR amplification products by allowing the universal and reverse primer to be incorporated at the 5' and 3' termini of the PCR products. Utilization of both the universal and reverse sequencing primers allows sequence information to be obtained from either strand of DNA. This method has also been successfully used for automated sequencing with the Applied Biosystems (Foster City, CA) Model 370A Sequencer using Sanger dideoxy-termination sequencing *(3)*. The incorporation of the universal and reverse primers for automated sequencing eliminates the need to covalently link specific primers to fluorescent dyes, because both primers are commercially available as dye-labeled primers from a number of companies.

The human DHFR gene will be used as an example of a target sequence to be amplified and directly sequenced. Using the numbering system of Morandi et al. *(4)*, primer 1 anneals to the 5' untranslated region of the first-strand cDNA between nucleotides 21 and 42 (Fig. 1). In addition, primer 1 contains 5' to the human DHFR sequence, unique, noncomplementary 17 nucleotides corresponding to the (−20) universal primer sequence. The rationale is to create an annealing site for the (−20) universal primer *(5)*. Primer 2 anneals to the coding strand, 77 bases 3' of the termination codon between nucleotides 680 and 650 (Fig. 1) and includes at the 5' terminus, 17 nucleotides corresponding to the (−24) reverse sequencing primer. These two primers define a region of 650 bp that include the entire coding region of the human DHFR without overlapping any translated region.

2. Materials

2.1. Asymmetric PCR

1. RNA isolated from 50×10^6 cells using the Total RNA Isolation kit (Invitrogen, San Rafael, CA), and mRNA isolated using the Fast Track kit (Invitrogen).
2. cDNA Cycle Kit for cDNA synthesis (Invitrogen).
3. RNase-free, siliconized tubes (National Scientific, San Diego, CA).
4. PCR reagents *(see* Chapter 1) including primer 1 and primer 2 (Fig. 1).
5. Agarose gel reagents *(see* Chapter 1).
6. Centricon 100 (Amicon, Danvers, MA).
7. GeneClean (Bio 101, La Jolla, CA).

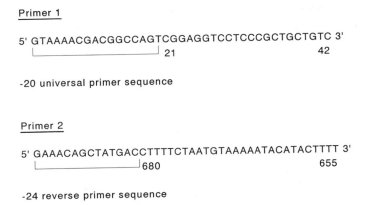

Fig. 1. Oligonucleotide primers used for cDNA synthesis and PCR of the human DHFR gene. Numbering is according to Masters and Attardi (1981).

2.3. Manual Direct DNA Sequencing

1. Centricon 100 (Amicon).
2. Sequenase 2.0 Kit (U.S. Biochemicals, Cleveland, OH).
3. Oligonucleotide primers: The (–20) universal primer and the (–24) reverse sequencing primer for manual sequencing were obtained from US Biochemicals.
4. [α-^{35}S]dATP (1000Ci/mmol).
5. Reagents for 6% polyacrylamide, 8M urea gel *(see* Chapter 4).

2.4. Automated Direct DNA Sequencing

1. The oligonucleotides used for automated sequencing contained either the (–20) universal sequencing primer (Applied Biosystems #400836) or the (–24) reverse sequencing (Applied Biosystems #400929) conjugated at the 5' terminus to the four dye-triethyl ammonium acetate esters TAMRA, FAM, ROX, JOE, reconstituted to 0.4 pmol/μL.
2. Applied Biosystems model 370A DNA sequencer and software version 1.30.
3. 95% Ethanol.
4. 3M Sodium acetate, pH 5.0.

3. Methods

3.1. Generation of Single-Stranded DNA via Asymmetric PCR

Owing to the uncertainty of how many cDNA molecules are created during first-strand cDNA synthesis, direct asymmetric PCR *(6)* of cDNAs did not provide an adequate template for dideoxy sequencing. We have therefore chosen to first amplify cDNAs with "symmetric"

PCR and use an aliquot of that reaction for "asymmetric" PCR. In addition, primer 1 contains the universal primer sequence and is present in a limited quantity, the result of the asymmetric PCR is the production of a single-stranded DNA that has at its 3' terminus sequences complementary to those of the *universal* primer sequence *(5)*. If one changes the ratios of the primers such that primer 1 is in excess and primer 2 is limiting, then the result of the asymmetric PCR is the production of a strand that has at its 3' terminus sequences complementary to those of the *reverse* primer sequence *(see* Figs. 2 and 3). Generally, our experience has been that as long as the symmetric PCR product is relatively distinct on an agarose gel, the asymmetric PCR and subsequent reactions will work. The combination of methylmercury hydroxide in the cDNA synthesis reaction and the use of high annealing temperatures for the amplification reaction probably contributes to the generation of a specific PCR product.

1. Synthesize cDNA in a 50-µL reaction on 100 ng of mRNA using the cDNA Cycle kit (Invitrogen) according to manufacturer's directions.
2. Assemble symmetric PCR reactions using 10–20% of the cDNA reaction, and 300 ng each of primer 1 and primer 2 and standard PCR reagents *(see* Chapter 1) in a 50-µL volume.
3. Amplify by PCR using the following cycle parameters *(see* Note 1):
   30 main cycles   95°C, 1 min (denaturation)
                        53°C, 1 min (annealing)
                        73°C, 1 min (extension)
4. Extract with an equal volume of CHCl$_3$/H$_2$O. Analyze 10% of the PCR reaction by nondenaturing agarose gel electrophoresis *(see* Chapter 1) to verify that the intended product was amplified *(see* Note 2).
5. If the majority of the PCR product was distinct (one band, no smear, total yield 100 ng to 20 µg), wash the sample three times with 2 mL of H$_2$O in a Centricon 100. To wash, spin the Centricon at approx 2500g for 10 min.
6. Recover the sample by a "recovery spin" at 2000g for 3 min. The reservoir should be empty (i.e., dry) after this step.
7. Use 5–10% of the eluate as a template for asymmetric PCR with a 1:50 ratio of primer 1 to primer 2 (6:300 ng respectively; *see* Note 3). Reverse the ratio to synthesize the opposite strand. The cycle number and parameters of asymmetric PCR are identical to those of symmetric PCR.
8. If multiple bands from the initial PCR reaction are observed, excise the band of interest from the agarose gel and elute in 200 µL of H$_2$O overnight at 4°C or alternatively, isolate the band by GeneClean. Use 10% of recovered DNA in asymmetric PCR.

primer 1   primer 2

                                             ·ȵ·············· CGACAA

GTAAAA -------->

universal primer sequence   reverse primer sequence

| *+ Taq polymerase*
↓

GTAAAA -- GCTGTT

CATTTT .... CGACAA

Incorporation of primer sequences into PCR product

Fig. 2. Strategy used to amplify the human DHFR cDNA and incorporate exogenous sequences that can be utilized for direct manual or automated sequencing of the PCR product.

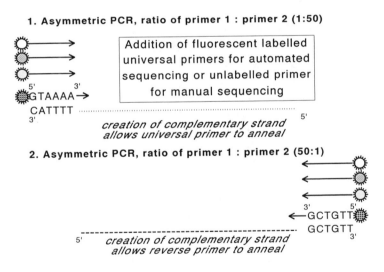

1. Asymmetric PCR, ratio of primer 1 : primer 2 (1:50)

Addition of fluorescent labelled universal primers for automated sequencing or unlabelled primer for manual sequencing

5'   3'

GTAAAA→

CATTTT ....

3'   5'

*creation of complementary strand
allows universal primer to anneal*

2. Asymmetric PCR, ratio of primer 1 : primer 2 (50:1)

3'   5'

←GCTGTT

-- GCTGTT

5'   *creation of complementary strand*   3'
allows reverse primer to anneal

Fig. 3. Strategy used to generate single-stranded DNA via asymmetric PCR for manual and automated direct sequencing. The ratio of primers used for asymmetric PCR determines whether the coding or noncoding strand is produced.

3.2. Manual Direct DNA Sequencing

1. Wash the asymmetric PCR product three times with 2 mL of distilled H2O in a Centricon 100 as described in Section 3.1. *(see* Note 4).
2. Use approx 20% of the eluate for manual sequencing by the chain-termination sequencing *(3)* using a modified T7 DNA polymerase *(7)* using the Sequenase 2.0 kit *(see* Note 5).

3. Add 1 µL (10 pmol) of the universal or reverse primer to 7 µL of single-stranded template in the presence of 2 µL of a 5X Sequenase buffer.
4. Anneal by heating the tubes to 80°C for 2 min and allowing them to cool down slowly to 35°C.
5. To the annealed primer/template mixture, add 1 µL 0.1M DDT, 2.0 µL of labeling mix (1.5 µM dGTP, dCTP, dTTP), 1 µL 10 µM [α-^{35}S]dATP *(see* Note 6) and 1.5 U of Sequenase 2.0 was added.
6. After 3–5 min, aliquot the mixture into four tubes to which 2.5 µL of the respective deoxy/dideoxy termination mixture was added. Each mixture contained 80 µM dATP, 80 µM dCTP, 80 µM dGTP, 80 µM dTTP, and 8 µM of the respective dideoxy nucleotide triphosphate.
7. Incubate the tubes at 37°C for 2–5 min and add stop buffer as per kit directions. Heat tubes to 94°C for 2 min and then load the reactions on a 6% polyacrylamide, 8M urea gel. Run sequencing gel and analyze according to standard procedures. *See* Note 7.

3.3. Automated Direct DNA Sequencing

1. Wash the asymmetric PCR product three times with 2 mL of H$_2$O in a Centricon 100 and use 50–100% of the eluate for automated sequencing.
2. Assemble the following reactions:
   A: 3 µL template, 1 µL JOE, 1 µL 5X sequencing buffer
   C: 3 µL template, 1 µL FAM, 1 µL 5X sequencing buffer
   G: 9 µL template, 3 µL TAMRA, 3 µL 5X sequencing buffer
   T: 9 µL tempalte, 3 µL ROX, 3 µL 5X sequencing buffer
3. Anneal by heating the tubes to 80°C for 2 min and allowing them to cool slowly to room temperature.
4. Perform extension and termination reactions as follows:
   A and C reactions contain 1.5 µL of the appropriate deoxy/dideoxy mixture, 5 µL of the annealed template/primer mixture and 1 µL of diluted Sequenase™ (1 µL of Sequenase, 4.5 µL of 0.1M DTT, 3.5 µL of 5X Sequenase buffer). The G and T reactions contain 4.5 µL of the appropriate deoxy/dideoxy mixture, 15 µL of the annealed template/primer mixture, and 3 µL of diluted Sequenase.
5. Incubate the tubes at 37°C for 5 min, followed by heat inactivation of the enzyme at 68°C for 10 min.
6. Ethanol precipitate by 1.4 µL of 3M NaAcetate, pH 5.0, to the A and C reactions and 4.1 µL of 3M NaAcetate, pH 5.0, to the G and T reactions. Mix by flicking and then add 35 µL of 95% ethanol to the A and C reactions, and 105 µL of 95% ethanol to the G and T reactions. Incubate on powdered dry ice for 30 min, then spin for 10 min at maximum speed in a microcentrifuge.

7. Decant the supernatant and dry in a dessicator container attached to a lyophilizer (cover tube with parafilm and poke holes with a 21-g needle, or dry in a SpeedVac (Savant).
8. Resuspend the sequencing reactions in 1 μL of 50 m*M* EDTA, pH 8.0, and 5 μL of deionized formamide and heat to 94°C for 2 min before loading a 6% polyacrylamide, 8*M* urea gel. Run sequencing gel according to standard procedure.
9. Perform the fluorescence analyses and base-calling with the Applied Biosystems software version 1.30. *See* Note 7.

4. Notes

1. All techniques involved in the sequence analysis of DNA amplified via PCR have common intrinsic problems associated with the technique of PCR and should be pointed out. Contamination from unintended sources with only some molecules of the DNA fragment to be detected may give false positive results. In the case of amplification of possibly mutated genes, minute amounts of DNA contamination may dramatically alter the results. This problem is of great concern especially when starting with very small amounts of DNA (e.g., DNA from paraffin-embedded tissues), because the number of contaminating DNA molecules may be in vast excess over the number of DNA molecules from the intended source. Further analysis of the PCR product may falsely show heterozygosity or even only the wrong sequence at the analyzed locus. An article *(8)* summarized precautions to be taken to avoid contamination from the unintended sources. Samples should be prepared in an area of the laboratory or under a hood different from any area where PCR reactions are performed or amplified material is processed. Water, buffer solutions, as well as pipet tips and microcentrifuge tubes should be autoclaved prior to use. Reagents such as buffer and primers should be divided into aliquots to limit any contamination. Disposable gloves should be used and changed frequently. Adding DNA to the tubes last may reduce the risk of cross-contamination. Lastly, negative controls (without DNA) should also be performed.
2. The power of PCR is its sensitivity, and, thus, its ability to amplify small amounts of DNA. The sensitivity of the symmetric/asymmetric technique is such that a PCR product is produced from a 1:100 dilution of a cDNA synthesis from 0.8 ng of mRNA (1 ng equivalent to 840 cells) using an altered DHFR cell line, which provides a signal that is clearly visible by ethidium bromide staining of a nondenaturing agarose gel *(9)*. Lower levels are detectable by the use of Southern blotting. The desired PCR product is very distinct, with little nonspecific hybrid-

ization. These characteristics may be owing to the combined effects of adding methylmercury hydroxide to the cDNA synthesis to reduce mRNA secondary structure and utilizing very high annealing temperatures for the amplification reaction *(10)*.

3. The use of two primers for asymmetric PCR in a 1:50 ratio yielded better results than the use of only one.

4. Purifying the asymmetric product by washing is crucial for manual sequencing, presumably because the unlabeled dATP is in excess of [α-^{35}S]dATP and will compete effectively during the extension reaction. Such washing is not as important for automated sequencing because of the use of end-labeled sequencing primers, although we have found that washing greatly improves the results.

5. The advantage of using *Taq* 1 polymerase for DNA sequencing instead of Klenow or T7 DNA polymerase is that *Taq* 1 polymerase is a heat-stable enzyme with very little exonuclease activity, and thus allows an incorporation of dNTPs and ddNTPs at a reaction temperature of 70–85°C, a temperature at which GC-rich, palindromic sequences and other secondary structures will melt out and provide unambiguous sequence data *(11)*. Our experience has been that although both modified T7 and *Taq* 1 polymerase work well for manual sequencing, the use of the modified T7 DNA polymerase usually results in a decreased background when compared to sequencing with *Taq* 1 polymerase. Our laboratory prefers to use the modified T7 DNA polymerase unless significant DNA secondary structure is present.

6. The advantage of ^{35}S labeling is that all bands are of equal intensity, making the determination of heterozygosity unambiquous. Further, because ^{35}S has a much longer half-life than ^{32}P, there is an extended period of time for analysis during which reloading of sequencing reactions onto gels or a reexposure of the gel to film is possible. However, if ^{32}P-labeled sequencing reactions are analyzed the day they are prepared, sequence data should be as definitive as that acheived with ^{35}S, although the bands will be broader.

7. One of the earliest concerns with PCR was the rate of misincorporation during DNA polymerization. Three groups have estimated the error rate of *Taq* 1 DNA polymerase to be in the range of 1.1×10^{-4} to 2.8×10^{-4} substitutions/nucleotide replicated *(12–14)* and have observed that the predominant mutations are AT to GC transitions. Krawczack et al. *(15)* estimated that replication errors can be neglected if a large number of staring templates (at least 100,000) are used. When small amounts of material are used, the probability of generating a polymerase-induced error is about 1%. The misincorporation rate may be minimized by starting with more than a limiting number of copies of target DNA sequences

and analyzing multiple copies of amplified PCR product to confirm true changes or by direct sequencing of PCR product DNA. In our studies, sequencing directly from the PCR product has only shown the expected single-base mutation. In addition, we have cloned this PCR product into a T7 sequencing vector and obtained analogous sequence data (data not shown). Discrepancies with automated sequencing have been very infrequent (99% agreement). The most common errors are those of single-base deletions. This is most likely owing to an event that ocurred during the automated sequencing reaction or the base-calling sequence analysis. It is interesting to note that others have found PCR using *Taq* 1 polymerase to be the preferable method of cloning DNA from mummified tissues for faithful amplification and to avoid cloning artifacts *(16,17)*.

References

1. Saiki, R. K., Scharf, S., and Faloona, F. (1985) Enzymatic amplification of beta-globin genomic sequences and restriction site analysis for diagnosis of sickle cell anemia. *Science* **230,** 1350–1354.
2. Mullis, K. B. and Faloona, F. A. (1987) Specific synthesis of DNA in vitro via a polymerase-catalyzed chain reaction. *Methods Enzymol.* **155,** 335.
3. Sanger, F., Nicklen, S., and Coulson, A. R. (1977) DNA sequencing with chain-terminating inhibitors. *Proc. Natl. Acad. Sci. USA* **74,** 5463–5467.
4. Morandi, C., Masters, J., Mottes, M., and Attardi, G. (1982) Multiple forms of human dihydrofolate reductase messenger RNA. Cloning and expression in Escherichia coli of their DNA coding sequence. *J. Mol. Biol.* **156,** 583–607.
5. McBride, L. J., Koepf, S. M., Gibbs, R. A., Salser, W., Mayrand, P. E., Hunkapiller, M. W., and Kronick, M. N. (1989) Automated DNA sequencing methods involving polymerase chain reaction. *Clin. Chem.* **35,** 2196–2201.
6. Gyllensten, U. B. and Erlich, H. A. (1988) Generation of single-stranded DNA by the polymerase chain reaction and its application to direct sequencing of the HLS-DQA locus. *Proc. Natl. Acad. Sci. USA* **85,** 7652–6765.
7. Tabor, S. and Richardson, C. C. (1987) DNA sequence analysis with a modified bacteriophage T7 DNA polymerase. *Proc. Natl. Acad. Sci. USA* **84,** 4767–4771.
8. Kwok, S. and Higuchi, R. (1989) Avoiding false positives with PCR. *Nature* **339,** 237–238.
9. Dicker, A. P., Volkenandt, M., and Bertino, J. R. (1989) Detection of a single base mutation in the human dihydrofolate reductase (DHFR) gene from a methotrexate resistant cell line using the polymerase chain reaction (PCR). *Cancer Commun.* **1,** 7–12.
10. Dicker, A. P., Volkenandt, M., Adamo, A., Barreda, C., and Bertino, J. R. (1989) Sequence analysis of a human gene responsible for drug resistance. A rapid method for manual and automated direct sequencing of products generated by the polymerase chain reaction. *Bio Techniques* **7,** 830–838.
11. Heiner, C., Mallick, S., Wan, A., and Hunkapiller, M. (1988) Taq polymerase: increased enzyme versatility in DNA sequencing. Applied Biosystems, Inc. Sequencer Model 370 User Bulletin.

12. Tindall, K. R. and Kunkel, T. A. (1988) Fidelity of DNA synthesis by the Thermus aquaticus DNA polymerase. *Biochemistry* **27**, 6008–6013.
13. Cooper, D. N. and Krawczak, M. (1990) The mutational spectrum of single base-pair substitutions causing human genetic disease: patterns and predictions. *Hum. Genet.* **85**, 55–74.
14. Keohavong, P. and Thilly, W. G. (1989) Fidelity of DNA polymerases in DNA amplification. *Proc. Natl. Acad. Sci. USA* **86**, 9253–9257.
15. Krawczak, M., Reiss, J., Schmidtke, J., and Rosler, U. (1989) Polymerase chain reaction: replication errors and reliability of gene diagnosis. *Nucleic Acids Res.* **17**, 2197–2201.
16. Rollo, F., Amici, A., Salvi, R., and Garbuglia, A. (1988) Short but faithful pieces of ancient DNA. *Nature* **335**, 774.
17. Paabo, S. and Wilson, A. C. (1988) Polymerase chain reaction reveals cloning artifacts. *Nature* **334**, 387,388.

CHAPTER 15

Genomic Footprinting by Ligation Mediated Polymerase Chain Reaction

Gerd P. Pfeifer and Arthur D. Riggs

1. Introduction

Chromatin structure analysis at single-nucleotide resolution can be done by genomic sequencing, and, recently, several techniques have been developed *(1)* that give improved specificity and sensitivity over the original method of Church and Gilbert *(2)*. The most sensitive method uses ligation-mediated polymerase chain reaction (LM-PCR) to amplify all fragments of the genomic sequence ladder *(3,4)*. The unique aspect of LM-PCR is the ligation of an oligonucleotide linker onto the 5' end of each DNA molecule. This provides a common sequence on the 5' end, and in conjunction with a gene-specific primer, allows conventional, exponential PCR to be used for signal amplification. Thus by taking advantage of the specificity and sensitivity of PCR, one needs only 1 μg of mammalian DNA per lane to obtain good quality DNA sequence ladders, with retention of methylation, DNA structure, and protein footprint information. The LM-PCR procedure is outlined in Fig. 1. Briefly, the first step is cleavage of DNA, generating 5' phosphorylated molecules. This is achieved, for example, by chemical DNA sequencing (β-elimination) or by cutting with the enzyme DNase I. Next, primer extension of a gene-specific oligonucleotide (primer 1) generates molecules that have a blunt end on one side. Linkers are ligated to the blunt ends, and then an exponential PCR amplification of the linker-ligated fragments is done using the longer oligonucle-

From: *Methods in Molecular Biology, Vol. 15: PCR Protocols: Current Methods and Applications*
Edited by: B. A. White Copyright © 1993 Humana Press Inc., Totowa, NJ

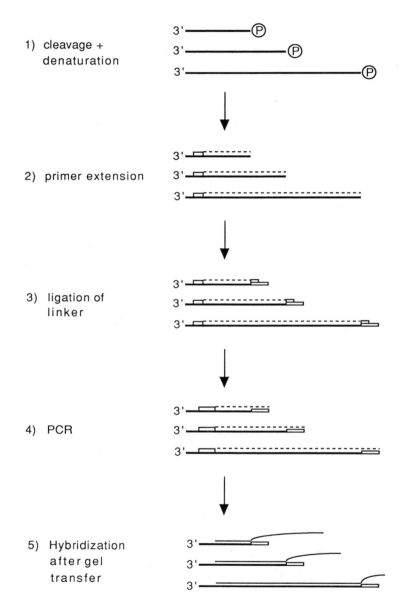

Fig. 1. Outline of the ligation-mediated PCR procedure. Step 1, cleavage and denaturation of genomic DNA; step 2, annealing and extension of primer 1; step 3, ligation of the linker; step 4, amplification of gene-specific fragments with primer 2 and the linker-primer; step 5, detection of the sequence ladder by hybridization with a single-stranded probe that does not overlap primers 1 and 2.

otide of the linker (linker-primer) and a second gene-specific primer (primer 2). After 15–20 amplification cycles, the DNA fragments are separated on a sequencing gel, electroblotted onto nylon membranes, and hybridized with a gene-specific probe to visualize the sequence ladder. By rehybridization, several gene-specific ladders can be sequentially visualized from one sequencing gel *(4)*.

LM-PCR is generally suitable for detection of specific DNA strand breaks *(5)*. The method has been used for sequencing genomic DNA and for determination of DNA cytosine methylation patterns *(4–8)*. It provides adequate sensitivity to map rare DNA adducts, like those formed after UV irradiation *(9)*. To obtain information about protein binding and other aspects of DNA structure, in vivo footprinting experiments can be done on intact cells using dimethylsulfate (DMS), a small molecule that penetrates the cell membrane *(3,6)*. DMS does not, however, reveal most contacts. Nucleosomes, for example, are transparent to DMS. Reasoning that more bulky agents such as enzymes would give more chromatin structural information, we adapted DNase I footprinting for use with LM-PCR and obtained very informative DNase I footprint data using cells that have been permeabilized with lysolecithin *(10)*.

2. Materials

2.1. Preparation of Genomic DNA

1. Phenol, equilibrated with 0.1M Tris-HCl, pH 8.
2. Chloroform.
3. Ethanol.
4. 3M Sodium acetate, pH 5.2.
5. Appropriate restriction enzyme and buffer.

2.2. G-Reaction

1. DMS buffer: 50 mM sodium cacodylate, 1 mM EDTA, pH 8.
2. DMS (dimethylsulfate, 99+%) from Aldrich (Milwaukee, WI). DMS is a highly toxic chemical and should be handled in a well ventilated hood. Detoxify DMS waste (including plastic material) in 5M NaOH. DMS is stored under nitrogen at 4°C.
3. DMS stop: 1.5M sodium acetate, pH 7, 1M β-mercaptoethanol.
4. Absolute ethanol, precooled to –70°C.
5. 3M Sodium acetate, pH 5.2.
6. 75% Ethanol.
7. Piperidine (99+%, Fluka, Ronkonkoma, NY), stored under nitrogen at –20°C.

2.3. Genomic Footprinting with DMS

1. DMS.
2. Cell culture medium.
3. Phosphate buffered saline (PBS).
4. 1*M* Piperidine, freshly diluted.

2.4. Genomic Footprinting with DNase I

1. Lysolecithin, type I (Sigma, St. Louis, MO).
2. Solution I: 150 m*M* sucrose, 80 m*M* KCl, 35 m*M* HEPES, 5 m*M* K_2HPO_4, 5 m*M* $MgCl_2$, 0.5 m*M* $CaCl_2$, pH 7.4.
3. Solution II: 150 m*M* sucrose, 80 m*M* KCl, 35 m*M* HEPES, 5 m*M* K_2HPO_4, 5 m*M* $MgCl_2$, 2 m*M* $CaCl_2$, pH 7.4.
4. DNase I (grade I, Boehringer Mannheim, Indianapolis, IN).
5. Stop solution: 20 m*M* Tris-HCl, pH 8.0, 20 m*M* NaCl, 20 m*M* EDTA, 1% SDS, 600 µg/mL proteinase K.
6. 150 m*M* NaCl, 5 m*M* EDTA, pH 7.8.
7. RNase A (DNase-free).
8. Sequenase 2.0 (USB).
9. 5 µ*M* ddNTPs in 40 m*M* Tris-HCl, pH 7.7, 25 m*M* NaCl, 6.7 m*M* $MgCl_2$ (a 10X buffer of this can be prepared).
10. Terminal transferase (BRL).
11. 2*M* Sodium cacodylate, pH 7.4, 10 m*M* β-mercaptoethanol.
12. 7.5*M* Ammonium acetate.

2.5. First Primer Extension

1. 5X Sequenase buffer: 250 m*M* NaCl, 200 m*M* Tris-HCl, pH 7.7.
2. The primers we have used as primer 1 (Sequenase primer) are 17- to 19-mer oligonucleotides with a calculated T_m of 50–56°C. Calculation of T_m is done with a computer program *(11)*. Gel-purify the primers. This is especially recommended for the amplification primers (primers 2, *see* below). Purify the primers on a 6.7*M* urea, 20% polyacrylamide gel (19:1 ratio of acrylamide:*bis*-acrylamide) made in 100 m*M* TBE buffer (*see* Section 2.8.), using a gel 0.8-mm thick and 50-cm long. Load <5 OD_{260} U of primer per lane. The gel should be run in 100 m*M* TBE at a voltage that maintains a gel temperature of about 50°C until the bromophenol blue runs to the bottom of the gel. Crush gel slices containing the desired oligonucleotide and soak in water on a shaker at 37°C overnight. Pass the solution through a Spin-X column (Costar, Cambridge, MA) and evaporate to dryness in a SpeedVac. Redissolve the oligonucleotide in 100 µL of 0.5*M* ammonium acetate, pH 7.0, and precipitated by addition of 900 µL of ethanol. After incubation on dry ice for 20 min and centrifugation at about 14,000*g* for 20 min, redis-

solve the oligonucleotide in water, determine the A_{260}, and adjust the concentration to 50 pmol/µL (stock solution).

3. Mg-DTT-dNTP mix: 20 mM MgCl$_2$, 20 mM DTT, 0.25 mM of each dNTP.
4. Sequenase™ 2.0 (USB), 13 U/µL.
5. 300 mM Tris-HCl, pH 7.7.

2.6. Ligation

1. 2M Tris-HCl, pH 7.7.
2. Linkers. Prepare in a total volume of 500 µL, containing 250 mM Tris-HCl, pH 7.7, and 10 nmol of each primer. Anneal a 25-mer (5'-GCGGTGACCC GGGAGATCTGAATTC-3',) to an 11-mer (5'-GAATTCAGATC-3') by heating to 95°C for 3 min and gradually cooling to 4°C over a time period of 3 h. Store linkers at –20°C for at least 3 mo. Always thaw and keep linkers on ice.
3. Ligation mix: 13.33 mM MgCl$_2$, 30 mM DTT, 1.66 mM ATP, 83 µg/ mL of BSA, 3 U/reaction of T4 DNA ligase (Promega, Madison, WI), and 100 pmol of linker/reaction (= 5 µL linker).
4. 3M Sodium acetate, pH 5.2.
5. *E. coli* tRNA.

2.7. Polymerase Chain Reaction

1. 2X *Taq* polymerase mix: 20 mM Tris-HCl, pH 8.9, 80 mM NaCl, 0.02% gelatin, 4 mM MgCl$_2$, and dNTPs at 0.4 mM each.
2. Primers. Use 10 pmol of the gene-specific primer (primer 2) and 10 pmol of the 25-mer linker-primer (5'-GCGGTGACCCGGGAGATCT-GAATTC) per reaction along with 3 U *Taq* polymerase, and include in the 2X *Taq* polymerase mix.

    The primers used in the amplification step (primer 2) are 23- to 26-mers (calculated T_m between 65 and 70°C). They are designed to extend 3' to primer 1. Primer 2 can overlap several bases with primer 1, but we have also had good results with a second primer that overlapped only two bases with the first.
3. Mineral oil.
4. 3M Sodium acetate, pH 5.2.
5. 400 mM EDTA, pH 7.7.
6. *E. coli* tRNA.

2.8. Gel Electrophoresis and Electroblotting

1. Formamide loading dye: 94% formamide, 2 mM EDTA , pH 7.7, 0.05% xylene cyanol, 0.05% bromophenol blue.
2. 1M TBE: 1M Tris, 0.83M boric acid, 10M EDTA, pH ~8.3. Use this stock to prepare 100, 90, and 50 mM TBE buffers.

3. 40% acrylamide stock. Prepare from 380 g/L of acrylamide and 20 g/L of *bis*-acrylamide. Dissolve in H_2O and adjust volume to 1 L. Filter through Whatman 3MM paper and store at 4°C in dark.
4. 10% Ammonium sulfate. Store at 4°C, stable for 1 wk.
5. Urea.
6. TEMED.
7. Electroblotting apparatus. Use stainless steel plates from a Bio-Rad gel drier as electrodes (obtainable as spare parts from Hoefer Scientific, San Francisco, CA). Connect electrodes directly with a platinum wire and a cord to the power supply. Use a homemade plastic tank for the buffer chamber. The size of the chamber is 50 cm × 40 cm × 14 cm (length × width × height). Opening the lid interrupts the current (safety precaution). A Bio-Rad 200/2.0 power supply is sufficient.
8. Whatman 3MM and Whatman 17 paper.
9. GeneScreen™ nylon membranes (New England Nuclear).

2.9. Hybridization

1. Hybridization buffer: 250 m*M* sodium phosphate, pH 7.2, 1 m*M* EDTA, 7% SDS, 1% BSA.
2. Washing buffer: 20 m*M* sodium phosphate, pH 7.2, 1 m*M* EDTA, 1% SDS.
3. UV irradiation apparatus. UV irradiation is performed by mounting six 254-nm germicidal UV tubes (15 W) into an inverted transilluminator from which the upper lid is removed. The distance between membrane and UV bulbs is 20 cm. Note that commercially available UV cross-linkers (e.g., Stratalinker, Stratagene, La Jolla, CA) may be used.

3. Methods

In Sections 3.1.–3.4., we describe two different procedures that can be used for genomic footprinting: DMS footprinting of intact cells and DNase I footprinting of permeabilized cells. Sections 3.5.–3.7. contain a detailed protocol of the ligation-mediated PCR procedure that is used to amplify gene-specific fragments. Sections 3.8.–3.10. comprise the separation of amplified fragments and their detection.

3.1. Preparation of Genomic DNA

1. Isolate DNA by standard procedures (*see* vols. 2 and 4 of this series, and Chapter 1) using phenol/chloroform extraction and ethanol precipitation. The DNA preparation should be mostly free of RNA.
2. Digest 50 µg of the DNA in a 100-µL vol with a restriction enzyme that does not cut within the region to be sequenced. This is done to reduce

the viscosity of the solution. After digestion, extract the DNA once with an equal volume of phenol/chloroform and once with chloroform and then ethanol precipitate by addition of sodium acetate, pH 5.2, to 0.3M and 2.5 vol of ethanol. Dissolve the DNA in water, determine the amount of DNA by absorbance measurement or agarose gel electrophoresis, and adjust the concentration to 2–10 µg/µL.

3.2. G-Reaction

Genomic DNA is chemically cleaved according to the Church-Gilbert procedures *(2,12)*. The conditions described in this section work well for 10–50 µg DNA. This DNA is used as an in vitro control for genomic footprinting with DMS (*see* Note 1).

1. Mix, with caution, on ice: 5 µL of genomic DNA (10–50 µg), 200 µL of DMS buffer and 1 µL of DMS.
2. Incubate at 20°C for 2 min.
3. Add 50 µL of DMS stop.
4. Add 750 µL of precooled ethanol (–70°C).
5. Keep the samples in a dry ice/ethanol bath for 20 min.
6. Spin 15 min at maximum speed in an Eppendorf centrifuge (or more than 10,000g in a microfuge) at 0–4°C.
7. Remove the supernatant and respin the pellet.
8. Resuspend pellet in 225 µL of water.
9. Add 25 µL of 3M sodium acetate, pH 5.2, then add 750 µL of precooled ethanol (–70°C).
10. Put on dry ice, 15 min.
11. Spin 10 min at maximum speed in an Eppendorf centrifuge at 0–4°C.
12. Remove the supernatant and respin the pellet.
13. Wash with 1 mL of 75% ethanol, spin at maximum speed for 5 min in Eppendorf centrifuge, and dry the pellet in SpeedVac.
14. Dissolve the pellet in 100 µL of 1M piperidine (freshly diluted).
15. Secure the caps with Teflon™ tape and heat at 90°C for 30 min in a heat block (lead weight on top).
16. Transfer the sample to a new tube.
17. Add 1/10 vol of 3M sodium acetate, pH 5.2, 2.5 vol of ethanol and put on dry ice for 20 min.
18. Spin for 15 min at maximum speed in Eppendorf centrifuge at 0–4°C.
19. Wash twice with 75% ethanol.
20. Remove traces of remaining piperidine by drying the sample overnight in a SpeedVac concentrator. Dissolve DNA in water to a concentration of approx 1 µg/µL.
21. Proceed with first primer extension (Section 3.5.).

3.3. Genomic Footprinting with DMS

1. Replace the cell culture medium with complete medium containing 0.2% of DMS (freshly prepared) and incubate at room temperature for 5–10 min. For cells cultured in a T75 flask, use 20 mL.
2. Remove the DMS-containing medium (centrifuge and wash at 4°C for cells in suspension).
3. Quickly wash the cells with 10–20 mL of medium without DMS.
4. Remove cells by trypsinization (or whatever procedure is used for a particular cell line), dilute with ice-cold PBS, and centrifuge.
5. Wash cell pellet with 20 mL of ice-cold PBS.
6. Isolate and digest the DNA with a restriction enzyme (1–2 U/µg DNA) to reduce viscosity if necessary.
7. Redissolve the DNA pellet in 100 µL of 1M piperidine, heat to 90°C for 30 min and continue as described in Section 3.2. (G-reaction), step 16.
8. Check the size of DNA fragments on an alkaline agarose gel. Prepare alkaline agarose gels (1.5%) in 50 mM NaCl, 4 mM EDTA. The running buffer is 30 mM NaOH, 2 mM EDTA. Denature samples prior to running by addition of 1 vol of 2X loading buffer (50% glycerol, 1M NaOH, 0.05% bromocresol green) and incubation at room temperature for 15 min. Run the gel for about 2 h at 40 V. After the run, neutralize the gels in 0.1M Tris-HCl, pH 8.0, and stain in ethidium bromide (1 µg/ mL). There should be DNA fragments smaller than 100 bp.
9. Proceed with first primer extension (Section 3.5.).

3.4. Genomic Footprinting with DNase I

A cell permeabilization system (*see* Note 2) is used for *in situ* digestion of chromatin with DNase I. DNase I footprints of single-copy genes are then obtained by LM-PCR *(10)*. Cell permeabilization can be used for both monolayer and suspension culture cells *(13)*.

1. Grow the cells as monolayers to about 80% confluency (we use a T75 flask).
2. Permeabilize the cells by treating cell monolayers (approx 4×10^6 cells) with 4 mL of 0.05% lysolecithin in solution I (prewarmed) for 1 min at 37°C *(14)*.
3. Remove the lysolecithin and wash with 10 mL of solution I.
4. Treat the cells with DNase I (10–50 µg/mL) in solution II at room temperature for 5 min. DNase I concentration and incubation times may have to be adjusted for different cell types. During this incubation, <10% of the cells should become detached from the plastic surface.
5. Stop the reaction and lyse the cells by removal of the DNase I solution

and addition of 2.5 mL of stop solution. Add 2.5 mL of 150 m*M* NaCl, 5 m*M* EDTA, pH 7.8 and incubate for 3 h at 37°C.

6. Purify the DNA by phenol/chloroform extraction and ethanol precipitation. Remove RNA by digestion with RNase A (50 µg/mL in TE buffer, 1 h at 37°C) followed by a second phenol/chloroform extraction and ethanol precipitation.

7. To obtain naked DNA controls (*see* Note 1), digest 40 µg of purified DNA in solution II with 0.8 or 1.6 µg/mL DNase I for 10 min at room temperature.

8. The 3' OH groups of DNase I-treated genomic DNA fragments, which may contribute to nonspecific priming, are blocked by addition of a dideoxynucleotide. This step, which theoretically should leave only the primers with 3' hydroxyls, significantly increases the signal-to-noise ratio of the procedure. To block 3' OH groups after DNase I treatment, denature 10 µg of DNA in 50 µL of 5 µ*M* of ddNTPs, 40 m*M* Tris-HCl, pH 7.7, 25 m*M* NaCl, 6.7 m*M* MgCl$_2$ by incubation at 95°C for 3 min and then add 5 U of Sequenase 2.0 and incubate for 20 min at 45°C.

9. Denature the DNA by incubation at 95°C for 3 min and incubate the DNA with 30 U of terminal transferase at 37°C for 30 min in the same reaction mixture (step 8, this section) supplemented to 200 m*M* of sodium cacodylate, pH 7.4, and 1 m*M* of β-mercaptoethanol. After phenol/chloroform extraction, selectively precipitate the DNA fragments at room temperature for 10 min by addition of ammonium acetate to a final concentration of 2*M* and 2 vol of ethanol. Pellet the DNA by centrifugation at room temperature. Wash the pellets in 75% ethanol and dissolve the DNA in water.

10. Check the size of the fragments obtained after chemical or enzymatic cleavage on alkaline 1.5% agarose gels (*see* step 8 in Section 3.3.). For optimum molecule usage, the average fragment size should be between 100 and 250 nt. DNase I digestion often results in a broader distribution of fragment sizes (50–1500 nucleotide), but make sure there are fragments smaller than 100 nt. The amount of DNA used in the Sequenase reaction (*see* Section 3.5.) is estimated from the relative amount of DNA fragments within the lower size range. This estimation is important and allows one to obtain similar band intensities on the sequencing gel in all lanes.

11. Proceed with first primer extension (Section 3.5.).

3.5. First Primer Extension

1. Mix in a siliconized tube: 0.5–5 µg of cleaved DNA, 0.6 pmol of primer 1, and 3 µL of 5X Sequenase buffer in a final volume of 15 µL.

2. Incubate at 95°C for 3 min, then at 45°C for 30 min.

3. Cool on ice, quick spin.
4. Add 7.5 µL of cold, freshly prepared Mg-DTT-dNTP mix and then 1.5 µL of Sequenase, diluted 1:4 in cold 10 m*M* Tris-HCl, pH 7.7.
5. Incubate at 48°C, 15 min, then cool on ice.
6. Add 6 µL of 300 m*M* Tris-HCl, pH 7.7.
7. Incubate at 67°C for 15 min (heat inactivation), cool on ice, and quick spin.

3.6. Ligation

The primer-extended molecules that have a 5' phosphate as a remnant from the chemical sequencing or DNase I cleavage are ligated to an unphosphorylated synthetic double-stranded linker.

1. Add 45 µL of freshly prepared ligation mix.
2. Incubate overnight at 18°C.
3. Incubate 10 min at 70°C (heat inactivation).
4. Add 8.4 µL of 3*M* sodium acetate, pH 5.2, 10 µg of *E. coli* tRNA, and 220 µL of 95% ethanol.
5. Put samples on dry ice for 20 min and then centrifuge 15 min at 4°C in an Eppendorf centrifuge.
6. Wash pellets with 950 µL of 75% ethanol and remove ethanol residues in a SpeedVac.
7. Dissolve pellets in 50 µL of H_2O and transfer to 0.5-mL siliconized tubes. Proceed with PCR (Section 3.7.).

3.7. Polymerase Chain Reaction

1. Add 50 µL of freshly prepared 2X *Taq* polymerase mix containing the primers (*see* Note 3) and the enzyme and mix by pipetting.
2. Cover the samples with 50 µL of mineral oil and spin briefly.
3. Amplify by PCR using the following cycle profile:
   16–20 main cycles: 95°C, 1 min (denaturation)
                                       66°C, 2 min (annealing)
                                       76°C, 3 min (extension)
4. To extend completely all DNA fragments, an additional *Taq* polymerase step is performed. Add 1 U of fresh *Taq* polymerase per sample is added together with 10 µL reaction buffer. Incubate for 10 min at 74°C.
5. Stop the reaction by adding 11 µL of 3*M* sodium acetate and 2.8 µL of 400 m*M* EDTA, and add 10 µg of tRNA.
6. Extract with 70 µL of phenol and 120 µL of chloroform (premixed).
7. Add 2.5 vol of 95% ethanol and put on dry ice for 20 min.
8. Centrifuge samples 15 min in an Eppendorf centrifuge at 4°C, wash pellets in 1 mL of 75% ethanol and dry pellets in SpeedVac.

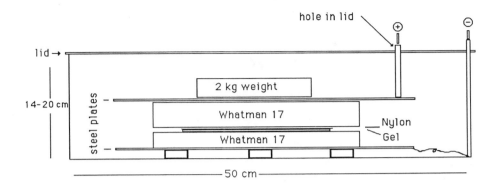

Fig. 2. Schematic drawing of the electroblotting apparatus.

3.8. Gel Electrophoresis and Electroblotting (see Notes 4 and 5)

1. Dissolve pellets in 1.5 µL of water and add 3 µL of formamide dye.
2. Heat samples to 95°C for 2 min prior to loading. Perform loading with a very thin flat tip and load only one half of the sample.
3. Use a gel that is 0.4-mm thick, 50–95-cm long, consisting of 8% polyacrylamide and $7M$ urea in 100 mM TBE. Prepare gel by mixing 20 mL of acrylamide stock, 50 g of urea, 10 mL of $1M$ TBE and 30 mL of H_2O. After urea dissolves, degas gel and add 1 mL of 10% ammonium persulfate and 10 µL of TEMED. Prerun and run the gel at a voltage or power setting that maintains gel at about 50°C (*see* manufacturer's suggestions for particular apparatus) until the xylene cyanol marker reaches the bottom. Fragments below the xylene cyanol dye hybridize only very weakly.
4. Electroblotting is performed with a simple home-made apparatus (Fig. 2; *see* Section 2.8.). After the run, cut the gel into one or two pieces, transfer to Whatman 3MM paper, and cover with Saran Wrap™.
5. Perform electroblotting essentially as described by Saluz and Jost (*12*). On the lower electrode, which is resting on three plastic incubation racks (height approx 3 cm), pile ten layers of Whatman 17 paper, 47 × 17 cm, presoaked in 90 mM TBE, and squeeze with a rolling bottle to avoid air bubbles between the paper layers.
6. Place the gel pieces covered with Saran Wrap on the paper pile and remove the air bubbles between gel and paper by wiping over the Saran Wrap using a soft tissue. When all air bubbles are squeezed out, remove the Saran Wrap and cover the gel with a nylon membrane cut somewhat larger than the gel and presoaked in 90 mM TBE.

7. Prepare the upper paper pile of 12 layers of Whatman 17 paper presoaked and cut as previously mentioned. Place the upper electrode on the upper pile by putting 2-kg lead weights on top of it.

8. Fill the electroblotting apparatus with 90 mM TBE until the buffer level is about 5 layers of paper below the gel, and electroblot at 1.6 amps and 30 V. After 1 h, remove the nylon membrane after marking the DNA side.

3.9. Hybridization

1. After electroblotting, briefly dry the membrane at room temperature or at 37°C and subsequently bake at 80°C for 20 min in a vacuum oven.

2. Crosslink the DNA by UV irradiation for 30 s. Calibrating the UV irradiation time for different batches of nylon membranes was found to be unnecessary.

3. Prehybridize and hybridize in a rotating 250-mL plastic or glass cylinders in a hybridization oven. Soak the nylon membranes in 50 mM TBE and roll into the cylinders so that the membranes stick completely to the walls of the cylinders without air pockets between wall and membrane. This can be easily done by rolling the membrane first onto a 25-mL pipet and then unspooling it into the cylinder.

4. Prehybridize the membrane in 15 mL of hybridization buffer for 10 min at 68°C. Remove this 15 mL of hybridization buffer immediately before adding probe.

5. Hybridize with the labeled probe (*see* Note 6) by diluting the probe into 5 mL of hybridization buffer and adding it to the bottle. Hybridize at 68°C for probes with a G+C content of 60–75%.

6. After hybridization, wash each nylon membrane in 2 L of washing buffer at 60°C. Several washing steps are performed at room temperature with prewarmed buffer. After washing, dry the membranes at room temperature, wrap in Saran Wrap, and expose to Kodak XAR-5 X-ray film. If the procedure has been done without error, a result can be seen after 0.5–8 h of exposure with intensifying screens at –70°C.

7. Nylon membranes can be used for rehybridization if several sets of primers have been included in the primer extension and amplification reactions. Probes can be stripped from the nylon membranes by soaking in 0.2M NaOH for 30 min at 45°C.

4. Notes

1. Identification of in vivo protein-DNA contacts requires a comparison of the in vivo treated sample with a purified DNA control. We emphasize that purified DNA samples might be different depending on cell type. This is owing to heterogenous cytosine methylation patterns of genomic DNA samples, which can give altered DNase I cleavage pat-

terns *(10)*. Therefore, it is imperative to compare all in vivo samples with an in vitro control from the same cell type.

2. Genomic footprinting of permeabilized cells with DNase I has given very clear footprints at the PGK-1 promoter *(10)*. An example is shown in Fig. 3. DNase I digestion of isolated nuclei has consistently resulted in less clear footprints with the occasional complete absence of transcription factors (no footprints).

3. When new combinations of primers are initially tested, it may be necessary to adjust certain conditions (Mg^{2+} concentration and temperatures in the PCR). This initial testing can be done with restriction enzyme cut DNA *(5)* instead of sequenced genomic DNA. This is technically simpler and should give information on the specificity of the primers and the conditions resulting in the best signal-to-noise ratio for the amplified band.

4. We are currently using electroblotting and hybridization instead of directly extending a ^{32}P-labeled primer followed by gel electrophoresis *(3)*. The reasons are:
   a. Labeling the amplification primer does not provide sufficient specificity, so an additional labeled primer (third primer) has to be made that will compete with the unlabeled primer from the amplification reaction.
   b. Hybridization introduces an additional level of specificity if a probe is used that does not overlap with the amplification primer.
   c. The hybridization approach as described here results in significantly less exposure of the worker to radioactivity.
   d. Longer single-stranded probes provide a higher specific activity than kinased oligonucleotides.
   e. Nylon membranes can easily be rehybridized after inclusion of multiple primer sets in Sequenase reaction and PCR (multiplexing).

5. Transfer to a nylon membrane can also be accomplished by vacuum blotting using a gel drier, as described *(15)*. Transfer by vacuum blotting was approx 50% efficient for longer fragments, and we are not currently using it. However, many may find it adequate, since less efficient transfer can be easily compensated for by longer exposures.

6. We have been using single-stranded DNA probes made as described by Weih et al. *(16)*. Other hybridization probes can be used and might be much easier to produce. We have successfully tested oligonucleotides labeled with T4 polynucleotide kinase. Longer exposure times and/or more PCR cycles are required and the hybridization and washing conditions are more difficult to control with these probes, but their preparation is convenient. Another very useful method to prepare labeled single-stranded probes is to use repeated primer extension (approx 35 cycles) by *Taq* polymerase with a single primer (primer 3) on a double-

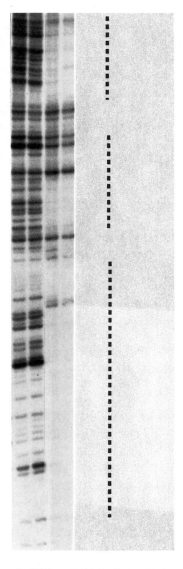

Fig. 3. An example of genomic DNase I footprints obtained by ligation-mediated PCR. The region shown contains promoter sequences from the human PGK-1 gene *(10)*. Three footprints (indicated by dashed lines) are seen at the promoter of the active gene in permeabilized cells. Purified genomic DNA cut with DNase I is shown in the control lanes.

stranded template of the respective cloned DNA *(17)*. The length of these probes can be easily controlled by an appropriate restriction cut. The primer that is used to make a probe (primer 3) should be on the same strand just 3' to the amplification primer (primer 2). It should not overlap more than a few bases. Even if a cloned DNA is not available, single-stranded probes can be made from PCR products. *See also* Espelund et al. for preparation of single-stranded probes *(18)*.

Acknowledgments

This work was supported by National Institute of Aging grant (AG08196) to A.D.R. and a fellowship from the Deutsche Forschungs-gemeinschaft (Pf212/1-1) to G.P.P.

References

1. Saluz, H., Wiebauer, K., and Wallace, A. (1991) Studying DNA modifications and DNA-protein interactions in vivo. *Trends Genetics* **7**, 207–211.
2. Church, G. M. and Gilbert, W. (1984) Genomic sequencing. *Proc. Natl. Acad. Sci. USA* **81**, 1991–1995.
3. Mueller, P. R. and Wold, B. (1989) In vivo footprinting of a muscle specific enhancer by ligation mediated PCR. *Science* **246**, 780–786.
4. Pfeifer, G. P., Steigerwald, S. D., Mueller, P. R., Wold, B., and Riggs, A. D. (1989) Genomic sequencing and methylation analysis by ligation mediated PCR. *Science* **246**, 810–813.
5. Steigerwald, S. D., Pfeifer, G. P., and Riggs, A. D. (1990) Ligation-mediated PCR improves the sensitivity of methylation analysis by restriction enzymes and detection of specific DNA strand breaks. *Nucleic Acids Res.* **18**, 1435–1439.
6. Pfeifer, G. P., Tanguay, R. L., Steigerwald, S. D., and Riggs, A. D. (1990) In vivo footprint and methylation analysis by PCR-aided genomic sequencing: comparison of active and inactive X chromosomal DNA at the CpG island and promoter of human PGK-1. *Genes Dev.* **4**, 1277–1287.
7. Pfeifer, G. P., Steigerwald, S. D., Hansen, R. S., Gartler, S. M., and Riggs, A. D. (1990) Polymerase chain reaction aided genomic sequencing of an X chromosome-linked CpG island: Methylation patterns suggest clonal inheritance, CpG site autonomy, and an explanation of activity state stability. *Proc. Natl. Acad. Sci. USA* **87**, 8252–8256.
8. Rideout III, W. M., Coetzee, G. A., Olumi, A. F., and Jones, P. A. (1990) 5-Methylcytosine as an endogenous mutagen in the human LDL receptor and p53 genes. *Science* **249**, 1288–1290.
9. Pfeifer, G. P., Drouin, R., Riggs, A. D., and Holmquist, G.P. (1991) In vivo mapping of a DNA adduct at nucleotide resolution: detection of pyrimidine (6-4) pyrimidone photoproducts by ligation-mediated polymerase chain reaction. *Proc. Natl. Acad. Sci. USA* **88**, 1374–1378.
10. Pfeifer, G. P. and Riggs, A.D. (1991) Chromatin differences between active and inactive X chromosomes revealed by genomic footprinting of permeabilized cells using DNase I and ligation-mediated PCR. *Genes Dev.* **5**, 1102–1113.

11. Rychlik, W. and Rhoads, R. E. (1989) A computer program for choosing optimal oligonucleotides for filter hybridization, sequencing and *in vitro* amplification of DNA. *Nucleic Acids Res.* **17,** 8543–8551.
12. Saluz, H. P. and Jost, J. P. (1987) *A Laboratory Guide to Genomic Sequencing.* Birkhauser, Boston.
13. Contreras, R. and Fiers, W. (1981) Initiation of transcription by RNA polymerase II in permeable SV40-infected CV-1 cells; evidence of multiple promoters for SV40 late transcription. *Nucleic Acids Res.* **9,** 215–236.
14. Zhang, L. and Gralla, J. D. (1989) In situ nucleoprotein structure at the SV40 major late promoter: melted and wrapped DNA flank the start site. *Genes Dev.* **3,** 1814–1822.
15. Gross, D. S., Collins, K. W., Hernandez, E. M., and Garrard, W. T. (1988) Vacuum blotting: a simple method for transfering DNA from sequencing gels to nylon membranes. *Gene* **74,** 347–356.
16. Weih, F., Stewart, A. F., and Schütz, G. (1988) A novel and rapid method to generate single stranded DNA probes for genomic footprinting. *Nucleic Acids Res.* **16,** 1628.
17. Stürzl, M. and Roth, W. K. (1990) "Run-off" synthesis and application of defined single-stranded DNA hybridization probes. *Anal. Biochem.* **185,** 164–169.
18. Espelund, M., Prentice-Stacy, R. A., and Jakobsen, K. S. (1990) A simple method for generating single-stranded DNA probes labeled to high activities. *Nucleic Acids Res.* **18,** 6157–6158.

CHAPTER 16

RNA Template-Specific Polymerase Chain Reaction (RS-PCR)

A Modification of RNA-PCR that Dramatically Reduces the Frequency of False Positives

Alan R. Shuldiner, Riccardo Perfetti, and Jesse Roth

1. Introduction

Reverse transcription of RNA followed by the polymerase chain reaction (RT-PCR or RNA-PCR) is an extraordinarily sensitive method to detect as few as 1–100 copies of a specific RNA *(1–3)*. However, we and others have found that false positives caused by contamination with minute quantities of DNA (i.e., cDNAs, plasmid DNAs, genomic DNA, or PCR carryover) is a major shortcoming of the method even when meticulous laboratory technique is employed *(4–6)*. RNA template-specific PCR (RS-PCR) is a modification of conventional RNA-PCR in which RNA is reverse-transcribed with a primer that contains at its 5' end a unique nucleotide sequence that may then be exploited in the PCR to amplify preferentially the RNA-derived sequence. RS-PCR retains full sensitivity, but reduces dramatically the frequency of false positives *(7–9)*.

From: *Methods in Molecular Biology, Vol. 15: PCR Protocols: Current Methods and Applications*
Edited by: B. A. White Copyright © 1993 Humana Press Inc., Totowa, NJ

1.1. Theory of the Method

1.1.1. Step 1: Reverse Transcription

First, total RNA is reverse-transcribed using an oligonucleotide primer 47 bases in length (designated primer $d_{17}t_{30}$) whose sequence contains 17 bases at its 3' end (segment d_{17}) that is complementary to a region of the target mRNA, and 30 bases at its 5' end (segment t_{30}) that is unique in sequence (Fig. 1). Thus, reverse transcription yields single-stranded DNA that contains a unique 30 base "tag" (segment t_{30}) at its 5' end.

1.1.2. Steps 2 and 3: Second-Strand Synthesis and Amplification

The second-strand of DNA is synthesized during the first cycle of PCR with primers u_{30} and t_{30} (Fig. 1). Upstream (sense) primer u_{30} is a 30-mer whose sequence corresponds to the target RNA a predetermined distance (optimally 200–500 bases) upstream from reverse transcription primer $d_{17}t_{30}$, whereas downstream (antisense) primer t_{30} is a 30-mer whose sequence is identical to segment t_{30} of reverse transcription primer $d_{17}t_{30}$. With these primers, sequences derived from RNA that had been tagged with unique sequence t_{30} during reverse transcription are amplified preferentially, whereas contaminating DNA that lack the unique tag are not amplified.

The lengths of the segments of reverse transcription primer $d_{17}t_{30}$ were derived empirically. The 17-base d_{17} segment hybridizes efficiently with its RNA during reverse transcription at 37°C, but does not hybridize to contaminating DNA that lacks the unique tag (t_{30}) during the PCR in which an elevated annealing temperature (70°C) is used. PCR primers of about 30 bases in length (primers u_{30} and t_{30}) retain efficient hybridization at the higher PCR annealing temperature.

2. Materials

1. 10X GeneAmp PCR buffer : 500 mM KCl, 100 mM Tris-HCl, pH 8.3 at 25°C, 15 mM MgCl$_2$, 0.1 mg/mL of gelatin. GeneAmp is the trademark of Perkin-Elmer/Cetus (Norwalk, CT).
2. dNTPs: 10 mM each of dATP, dGTP, dCTP, and dTTP.
3. AMV-reverse transcriptase.
4. *Taq* polymerase.
5. Oligonucleotides of the appropriate sequences should be synthesized chemically *(see* Note 1). We recommend leaving the last dimethoxytrityl (DMT) group on, and purifying with NENsorb Prep columns (New

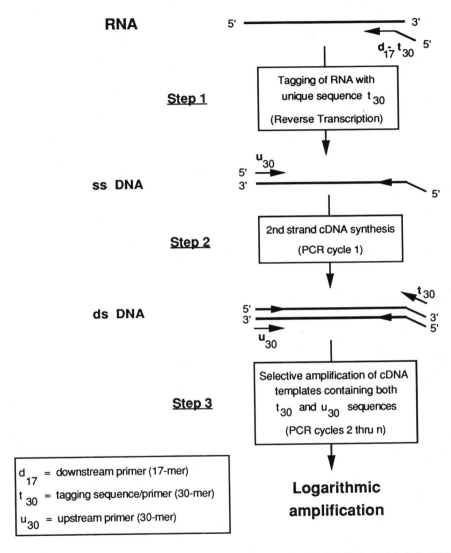

Fig. 1. Schematic of RNA template-specific PCR (RS-PCR). With RS-PCR, sequences derived from RNA that are tagged with the unique sequence (t_{30}) during reverse transcription (step 1) are amplified preferentially during PCR with primers t_{30} and u_{30} (steps 2 and 3). The 17-base d_{17} segment of reverse transcription primer $d_{17}t_{30}$ anneals efficiently to RNA during reverse transcription at 37°C, but does not hybridize to contaminating DNA that lacks the unique tag at the higher PCR annealing temperature (70°C). d_{17} = downstream primer (17-mer). t_{30} = tagging sequence/primer (30-mer). u_{30} = upstream primer (30-mer).

England Nuclear, Boston, MA) according to the manufacturer's directions. Alternatively, the last DMT group may be removed, and the oligonucleotides may be desalted on a Sephadex G-25 spin column (Boehringer Mannheim Biochemicals, Indianapolis, IN) and used without further purification. Primers should be diluted in water to a concentration of 5 μM, aliquoted, and stored at –20°C.

6. Autoclaved, distilled, deionized water.
7. RNAse A: 0.1 mg/mL in water. Boil for 10 min to remove any residual DNAse activity, and allow to cool to room temperature over approx 30 min. Aliquot and store at –20°C.
8. Paraffin oil.
9. Sterile 1.5-mL and 0.5-mL Eppendorf tubes.
10. Horizontal agarose gel electrophoresis apparatus with power supply.
11. 10X loading buffer: 30% glycerol, 0.25% bromophenol blue, and 0.25% xylene cyanol.
12. Tris-borate-EDTA (TBE) electrophoresis running buffer: 89 mM Tris - borate, 89 mM boric acid, 2 mM EDTA, pH 8.
13. Agarose.
14. Nuseive GTG (FMC Bioproducts, Rockland, ME).

3. Methods

3.1. Reverse Transcription

Reverse transcription is accomplished in a final volume of 20 μL. We recommend that a 2X reverse transcription master mix be prepared, and that 10 μL of the master mix be added to 10 μL of total RNA in water *(see* Note 2).

1. Place total RNA (1–10 μg) dissolved in 10 μL of sterile water in an autoclaved 0.5-mL Eppendorf tube. Keep at 4°C.
2. Make a 2X reverse transcription master mix. The following recipe is sufficient for 10 reactions:

GeneAmp PCR buffer	20 μL
dATP (10 mM)	4 μL
dCTP (10 mM)	4 μL
dGTP (10 mM)	4 μL
dTTP (10 mM)	4 μL
Primer d$_{17}$t$_{30}$ (5 μM)	20 μL
Sterile water	24 μL
RNAsin (40 U/μL)	10 μL
AMV-reverse transcriptase (7–10 U/μL)	10 μL
Total	100 μL

The final concentrations of components after RNA is added are 1X of

GeneAmp buffer, 200 μM of each dNTP, 0.5 μM of primer $d_{17}t_{30}$, 2 U/ μL of RNasin, and 0.35–0.5 U/μL of reverse transcriptase.

3. Add 10 μL of the 2X reverse transcription master mix to 10 μL of the RNA sample in water. Pipet up and down several times to mix.
4. Incubate for 1 h at 37°C.
5. Spin the tubes briefly in a microfuge to remove condensation of liquid from the inside walls of the tubes and place at 4°C *(see* Note 3).

3.2. Second-Strand Synthesis and PCR

PCR is performed in a final volume of 50 μL. We recommend that a PCR master mix be prepared, and that the PCR reaction mixture contain 45 μL of PCR master mix and 5 μL of the reverse transcription reaction mixture.

1. Make a PCR master mix. (The following recipe is sufficient for 10 reactions.)

10X GeneAmp PCR buffer	50 μL
dATP (10 mM)	10 μL
dCTP (10 mM)	10 μL
dGTP (10 mM)	10 μL
dTTP (10 mM)	10 μL
Primer u_{30} (5 μM)	50 μL
Primer t_{30} (5 μM)	50 μL
Sterile water	257 μL
Taq polymerase (5 U/μL)	3 μL
Total	450 μL

Final concentrations of components after an aliquot of the reverse transcription reaction mixture is added are 1X GeneAmp buffer, 200 μM of each dNTP, 0.5 μM of primers and 1.5 U/tube of *Taq* polymerase.

2. Add 45 μL of the PCR master mix to sterile 0.5-mL Eppendorf tubes.
3. Cover the PCR master mix with 50 μL of paraffin oil.
4. Add 5 μL of the reverse transcription reaction mixture to the PCR master mix. This may be accomplished easily by placing the pipet tip through the oil, and pipeting the reverse transcription reaction mixture into the lower aqueous phase. Pipet up and down several times to mix.
5. Centrifuge briefly to obtain a clean oil/aqueous interface and to remove air bubbles.
6. Place tubes in a thermocycler. We recommend the following cycle parameters:

Initial denaturation	94°C, 5 min
25–45 main cycles	70°C, 1 min (annealing/extension)
	94°C, 1 min (denaturation)
Final extension	70°C, 10 min

(Note that 70°C is used for both annealing and extension. *See* Note 4.)

7. After thermocycling is completed, transfer 18 µL of the PCR reaction mixture from each tube to individual Eppendorf tubes.
8. Add 2 µL of 10X loading buffer to each tube. Vortex and centrifuge briefly.
9. Load samples onto a gel consisting of 1% agarose and 2% Nuseive GTG in 1X TBE, and perform electrophoresis.
10. DNA may be visualized by staining with ethidium bromide followed by UV transillumination, or by Southern blot analysis using established procedures (*see* Chapter 1).

4. Notes

1. Selecting a unique oligonucleotide sequence (t_{30}). The nucleotide sequence of the unique "tag" may be selected using several methods. We use a computer to generate a random 30-mer. The sequence is inspected visually to be sure that it contains approximately equal numbers of each of the four nucleotides, has a low probability of containing appreciable secondary structure, and does not contain complementarity at its 3' end with the 3' end of upstream primer (u_{30}). A restriction endonuclease recognition sequence may be introduced into the unique tag for future subcloning. Examples of unique tag sequences that we have used successfully include 5'-GAACATCGATGACAAGCTTAGGTATCGATA-3' and 5'-CTTATACGGATATCCTGGCAATTCGGACTT-3'.
2. RNA isolation. Total RNA may be prepared using virtually any method. We use the RNAzol method (*10*) since it is easily adapted for small quantities of RNA, and can be done in a single tube thereby minimizing opportunity for cross-contamination of samples (*9*). For RS-PCR, 1–10 µg of total RNA may be used. We do not find it necessary to select for poly-A RNA.
3. Reverse transcription. After reverse transcription of RNA, the reaction mixture may be used immediately, or may be stored indefinitely at –20°C.
4. PCR conditions. For some primers, we have empirically found that better and more reliable amplification occurs when the annealing temperature is decreased to 65°C. In these circumstances, rather than two temperature cycling, three temperature cycling with an extension temperature of 72°C should be used (i.e., denaturation [94°C, 1 min], annealing [65°C, 1 min], extension [72°C, 1 min]).
5. Avoiding false positives and negative controls. To test the system, when we purposely introduced large amounts of DNA contaminants (i.e., >10^7 copies) into the RS-PCR reaction mixture, a faint signal was sometimes detected. This was because reverse transcriptase can act as a DNA polymerase during reverse transcription when enough DNA template is present. We therefore recommend that meticulous laboratory technique be maintained to further safeguard against false positives. We recommend

the use of positive displacement pipets or plugged pipet tips, frequent changing of gloves, adding the RNA sample last, aliquoting all reagents, and physical separation of the RS-PCR reagents from the PCR products to minimize chances of carry-over contamination. Some have touted UV irradiation to be useful in decreasing the frequency of false positives *(6,11)*. We routinely use multiple negative controls which include (1) a "no template" control in which water is added rather than an RNA template, and (2) treatment of an aliquot of the RNA sample with RNAse A (0.1 µg, 37°C, 30 min) prior to reverse transcription.

6. PCR carry-over contamination. If false positives recur after a period of contamination-free experiments, carry-over contamination of RS-PCR products that contain the unique tag may be the cause. RS-PCR with new primers that contain a different unique tag (e.g., $d_{17}t'_{30}$ and t'_{30}) will correct this problem. We have found that when meticulous laboratory technique is used, carry-over contamination can be avoided effectively. For example, in one set of experiments that consisted of over 50 different amplifications, it was not necessary to change the sequence of the unique tag.

7. Quantitative RS-PCR. RS-PCR may be adapted for quantitation by using synthetic RNA as an internal standard *(9)*.

Acknowledgments

We wish to thank Drs. Charles T. Roberts, Jr. and Laurie A. Scott for many helpful discussions and comments. This research project was supported in part from a grant from the Juvenile Diabetes Foundation, the Diabetes Research and Education Foundation, and the Mallinckrodt Foundation.

References

1. Kawasaki, E. S., Clark, S. S., Coyne, M. Y., Smith, S. D., Champlin, R., Witte, O. N., and McCormick, F. P. (1988) Diagnosis of chronic myeloid and acute lymphocytic leukemias by detection of leukemia-specific mRNA sequences amplified *in vitro. Proc. Natl. Acad. Sci. USA* **85,** 5698–5702

2. Rappolee, D. A., Mark, D., Banda, M. J., and Werb, Z. (1988) Wound macrophages express TGF-alpha and other growth factors *in vivo*: analysis by mRNA phenotyping. *Science* **241,** 708–712.

3. Rappolee, D. A., Brenner, C. A., Schultz, R., Mark, D., and Werb, Z. (1988) Developmental expression of PDGF, TGF-alpha, and TGF-beta genes in preimplantation mouse embryos. *Science* **241,** 1823–1825.

4. Kwok, S. and Higuchi, R. (1989) Avoiding false positives with PCR. *Nature* **339,** 237,238.

5. Lo, Y.-M, Mehal, W. Z., and Fleming, K. A. (1988) False positive results and the polymerase chain reaction. *Lancet* **2,** 699.

6. Sarkar, G. and Sommer, S. (1990) Shedding light on PCR contamination. *Nature* **343,** 27.

7. Shuldiner, A. R., Nirula, A., and Roth, J. (1990) RNA template-specific polymerase chain reaction (RS-PCR): a novel strategy to reduce dramatically false positives. *Gene* **91,** 139–142.

8. Shuldiner, A. R., Tanner, K., Moore, C. A., and Roth, J. (1991) RNA template-specific PCR (RS-PCR): An improved method that dramatically reduces false positives in RT-PCR. *BioTechniques* **11,** 760–763.

9. Shuldiner, A. R., DePablo, F., and Roth, J. (1991) Two nonallelic insulin genes in *Xenopus laevis* are expressed differentially during neurulation in prepancreatic embryos. *Proc. Natl. Acad. Sci. USA* 88, 7679-7683.

10. Chomczynski, P. and Sacchi, H. (1987) Single-step method of RNA isolation by acid guanidinium thiocyanate-phenol-chloroform extraction. *Anal. Biochem.* **162,** 156–159.

11. Sarkar, G. and Sommer, S. S. (1991) Parameters affecting susceptibility of PCR contamination to UV inactivation. *BioTechniques* **10,** 591–593.

CHAPTER 17

Quantitative Measurement of Relative Gene Expression in Human Tumors

Tetsuro Horikoshi, Kathleen Danenberg, Matthias Volkenandt, Thomas Stadlbauer, and Peter V. Danenberg

1. Introduction

Quantitative measurement of specific mRNA species is of major importance for approaching many fundamental questions in biology. Until now, quantitation of gene expression has usually been done by Northern blotting, but this procedure is relatively insensitive, requiring microgram amounts of RNA. Furthermore, unless linear ranges of RNA concentration are determined, the procedure is semiquantitative at best. Because of the limitations of Northern blotting, various strategies have been developed for quantitation of cDNA by polymerase chain reaction (PCR)-based methods *(1–3)*, most of them utilizing the principle of competitive PCR in which a synthetic segment is coamplified along the target DNA segment. We were interested in comparing the expression of drug target genes in very small samples of human tumor material, such as would be obtained from biopsies or even paraffin blocks. The competitive PCR was not entirely suitable for this purpose because: (1) the "input" RNA or DNA concentration must be known with precision in order to provide a normalization factor, and (2) our results suggested that the competitor and target were not necessarily amplified with the same efficiency even when

From: *Methods in Molecular Biology, Vol. 15: PCR Protocols: Current Methods and Applications*
Edited by: B. A. White Copyright © 1993 Humana Press Inc., Totowa, NJ

the segments to be amplified were the same size *(4)*. To overcome these problems, we developed an alternate method of PCR quantitation with the following key features:

1. The relative rather than absolute levels of gene expression are measured by determining a ratio between PCR products generated by amplifications of the desired target DNA and an endogeneous internal standard gene in separate reactions and then compared with the same ratio in another sample.
2. Linear amplification regions of the target and internal standard genes are determined by serial dilution of the cDNA sample, thereby not requiring the absolute amount of input DNA to be known.
3. A T7 RNA polymerase promoter is attached to one of the PCR primers (usually the 5' primer) so that the amplified cDNA can be converted to RNA, which greatly increases the amount of amplified product. This PCR quantitation method can discriminate less than a twofold difference in gene expression and, although still labor intensive, is more easily used than competitive PCR methods for large numbers of samples. We have used it to determine relative expressions of thymidylate synthase (TS), dihydro-folate reductase, and DT-diaphorase in clinical tumor tissue samples *(5)*.

The PCR quantitation method described here takes advantage of the fact that, in theory, the amount of the PCR product is linearly proportional to the amount of the starting DNA according to the equation $N = N_0(1 + \text{eff})^n$, where N_0 is the number of copies of the starting DNA, N is the amount of DNA after n cycles of amplification, and eff is the amplification efficiency *(3)*. An experimental curve for ß-actin amplification is shown in Fig. 1. As predicted by the aforementioned equation, there is a linear region of amplification. The curve eventually reaches a plateau, indicating that the amplification efficiency declines after a certain extent of reaction presumably owing to reaction product build-up and consumption of primers *(2)*. Figure 2 shows linear regions for the amplification of ß-actin cDNA (the internal standard gene) and TS cDNA (the target gene) from the same solution of human tumor cDNA optimized as to the cycle number used in the PCR. Note the different volumes required to give the same amount of PCR product, indicating a substantial difference in the amount of expression of these two genes. To relate the expression of the gene of interest to that of the internal standard gene, a ratio is determined between the amount of radiolabeled PCR product within each respective linear region:

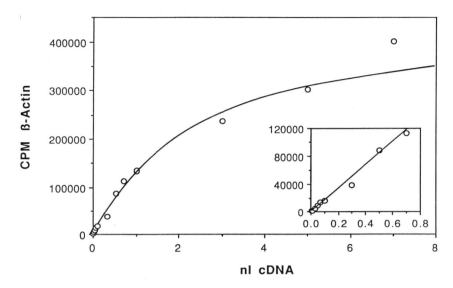

Fig. 1. PCR product formed as a function of input ß-actin cDNA. Aliquots of a cDNA solution from HT-29 cells were PCR-amplified using 5' primer TAATACGACTCACTATAGGGAGA-GCGGGAAATCGTGCGTGACATT (corresponding to the T7 promoter plus the GGGAGA transcription initiation sequence plus bases 2104–2127 of the ß-actin genomic sequence located in exon 3). The 3' primer was GATGGAGTTGAAGGTAGTTTCGTG (corresponding to bases 2409–2432 of the genomic sequence located in exon 4). The PCR product was transcribed with T7 RNA polymerase. After PAGE of the RNA, the bands were visualized by autoradiography, excised, and quantitated by liquid scintillation counting. The inset shows the expanded linear region of the curve.

(CPM of amplified target gene/CPM of amplified internal standard)
× (vol of internal standard/vol of target gene)

The resulting ratio (actually the ratio of the slopes of the linear portions of the concentration vs product curve) compares the amount of radioactivity in the PCR fragments of the target gene and the standard gene generated by an equal volume of the cDNA solution. If the expression of the internal standard gene is constant among different samples, dividing this empirical ratio between the target gene and the internal standard gene from one sample with that from another will give the relative expression of the target genes of interest in the two samples. From the data in Fig. 2, the empirical ratio of TS/ß-actin

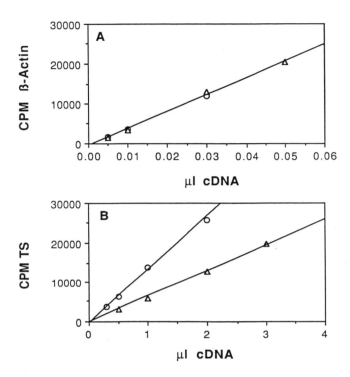

Fig. 2. Comparison of TS gene expression in two tumor specimens. The linear regions of PCR amplification were determined for ß-actin cDNA (**A**) and TS cDNA (**B**) in the cDNA libraries from two colorectal tumor specimens. The 5' primer for TS amplification was TAATACGACTCACTATAGGGAGA-TCCAACACATC-CTCCGCT (T7 promoter plus the GGGAGA transcription initiation sequence plus bases 110–127 of the coding sequence) and the 3' primer was CCAGAACACAC-GTTTGGTTGTCAG (bases 220–243 of the coding sequence).

amplification is 6.4×10^{-4} in one sample and 3.2×10^{-4} in the other sample. Thus, the relative TS gene expression in the two samples is the quotient of these ratios, 2.0.

The overall scheme for the quantitation assay is shown in Fig. 3. The individual steps are:

1. Pulverization of the tumor;
2. Extraction of total RNA;
3. Conversion of the RNA to cDNA with random hexamers;
4. PCR amplification of the target and the internal standard in separate tubes with 5' primers that have a T7 polymerase promoter sequence on the 5' end;

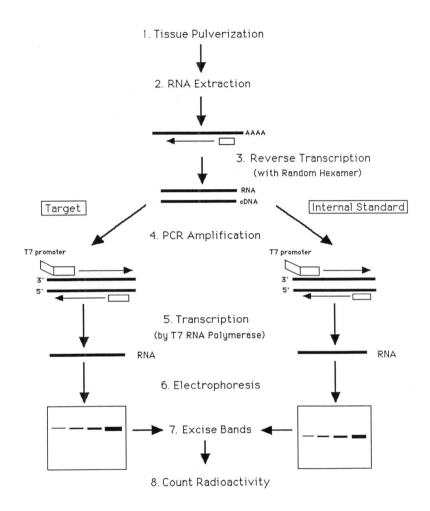

Fig. 3. The overall scheme for PCR quantitation of gene expression.

5. Conversion of the amplified cDNA to RNA by T7 RNA polymerase; and
6. Gel electrophoresis of the reaction mixture and quantitation of the bands.

2. Materials

2.1. RNA Isolation

1. Denaturing solution: $4M$ guanidinium thiocyanate, 25 mM sodium citrate, pH 7, 0.5% sarcosyl, and $0.1M$ β-mercaptoethanol (*see* Note 1).
2. $2M$ Sodium acetate, pH 4.

3. Water-saturated phenol.
4. Chloroform-isoamyl alcohol, 49:1.
5. Isopropanol.
6. 75% Ethanol.

2.2. Reverse Transcription of RNA

1. RNAguard RNase inhibitor (Pharmacia, Piscataway, NJ).
2. 100 mM Dithiothreitol (DTT).
3. Reverse transcription buffer: 250 mM Tris-HCl, pH 8.3, 375 mM KCl, and 15 mM MgCl$_2$.
4. dNTPs: 10 mM in water.
5. Bovine serum albumin (BSA) solution: 3 mg/mL in water.
6. Random hexamers: 50 O.D. U in 0.55 mL of 10 mM Tris-HCl, pH 7.5 and 1 mM EDTA.
7. MMLV reverse transcriptase: 200 IU/μL.

2.3. Polymerase Chain Reaction

1. 10X *Taq* buffer: 500 mM KCl, 100 mM Tris-HCl, pH 8.3.
2. MgCl$_2$ solution: 12.5 mM in H$_2$O.
3. Primers: 12.5 pmol/μL *(see* Note 2).
4. *Taq* polymerase.
5. Mineral oil.

2.4. T7 Polymerase Transcription

1. 10X Transcription buffer: 400 mM Tris-HCl, pH 7.5, 120 mM MgCl$_2$.
2. 0.5M Spermidine.
3. rNTPs: 10 mM in water.
4. RNA guard (Pharmacia).
5. [α-^{32}P]CTP: specific activity of 3000 C$_i$/mmol.
6. 0.5M EDTA.

2.5. Gel Electrophoresis

1. Components for urea-acrylamide gel electrophoresis *(see* Chapters 1 and 4).
2. Gel loading solution: 10M urea, 0.02% of bromophenol blue and 0.02% of xylene cyanol.

3. Methods

The gene quantitation method described here consists of the following separate procedures as depicted in Fig. 3:

1. Isolation of total RNA from human tumor specimens based on Chomczynski and Sacchi *(6);*

2. Reverse transcription of the RNA;
3. PCR amplification of the desired gene segments;
4. T7 RNA polymerase transcription;
5. Quantitation of the transcription product.

3.1. Isolation of Total RNA from Human Tumor Specimens

This RNA isolation procedure has been used successfully with less than 10 mg of tumor tissue.

1. Place frozen tumor specimen under liquid nitrogen for several minutes (*see* Note 3).
2. Remove tumor from the liquid nitrogen and immediately pulverize it with a steel mortar and pestle that has been precooled in dry ice for about 1 h.
3. Transfer the pulverized tumor fragments rapidly before they thaw to a 50-mL disposable tube placed in dry ice.
4. Add 6 mL of denaturing solution to this tube and suspend the pulverized tissue by suctioning the solution up and down with a transfer pipet (e.g., Stockwell Scientific 251-1S, Walnut, CA) until it appears homogeneous. Add 0.6 mL of sodium acetate solution and mix the solution by inversion of the tube.
5. Add 6 mL of water-saturated phenol and 1.2 mL of chloroform-isoamyl alcohol, shake the mixture thoroughly for 10 s, and place on ice for 15 min.
6. Centrifuge the tube for 20 min at 4°C at 2800 rpm ($2000g$). At this point, the RNA is present in the upper aqueous phase whereas DNA and proteins are present in the interphase and the lower chloroform-phenol phase.
7. Transfer the aqueous phase to a fresh 50-mL tube and mix with 6 mL of isopropanol.
8. Place the solution at –20°C for 1 h to precipitate the RNA and then centrifuge for 20 min at $2000g$.
9. Dissolve the RNA pellet in 1.8 mL of the denaturing solution. Transfer the solution to a 15-mL tube and precipitate again with 2 mL of isopropanol for 1 h at –20°C and centrifuge again for 10 min at $2000g$.
10. Resuspend and wash the resulting pellet of RNA in 7 mL of 75% ethanol by suctioning it up and down with a transfer pipet at least 10 times. Centrifuge the resuspended RNA for 10 min at $2000g$ and vacuum-dry the pellet for at least 15 min.
11. Dissolve the RNA in 140 µL of water and immediately proceed to reverse transcription (*see* Note 4).

3.2. Reverse Transcription of the RNA

1. Place a 100-µL aliquot of the aforementioned RNA solution into a sterile 0.5-mL tube along with 5 µL of RNAguard RNAse inhibitor, 20 µL of 0.1M DTT, 40 µL of reverse transcription buffer, 10 µL of the dNTP solution, 5 µL of BSA solution, 1 µL of random hexamers *(see* Note 5*)*, and 10 µL of MMLV reverse transcriptase.
2. Incubate the mixture for 10 min at room temperature, 45 min at 42°C. Denature at 90°C for 3 min.
3. After cooling to 42°C, add an additional 5 µL (1000 U) of MMLV reverse transcriptase and incubate again for 45 min at 42°C and then 10 min at 75°C to destroy reverse transcriptase activity. Store the solution at –80°C.

3.3. PCR Amplification of cDNA Segments

1. To a sterile 0.5-mL tube, add 2.4 µL of 10X *Taq* buffer, 0.5 µL of the dNTP solution, 3.75 µL of MgCl$_2$, 1 µL each of the 3' and 5' primers (12.5 pmol/µL), 0–10 µL of the cDNA solution from the previous procedure *(see* Note 6*)*, and water to make a total vol of 23 µL. Place 40 µL of mineral oil on top of the aqueous layer.
2. Heat the tube for 5 min at 95°C in a thermocycler.
3. Add 0.63 U of *Taq* polymerase in 2 µL of 1X *Taq* buffer. Amplify by PCR using the following cycle profiles:

   30 main cycles   93.5°C, 1 min (denaturation)

                   55°C, 1 min (annealing)

                   72°C, 1 min (extension)

   Final extension step 72°C, 7 min

   Store reaction at room temperature or 4°C.

3.4. T7 Polymerase Transcription
(see *Note 7)*

1. Prepare a T7 transcription master solution by adding the following: 132 µL of water, 25 µL of 10X transcription buffer, 25 µL of the rNTP solution, 2.5 µL of RNAguard, 25 µL of DTT solution, 1 µL of spermidine, 2.5 µL of [α-^{32}P]CTP (3000 C$_i$/mmol). This is sufficient for 10 reactions.
2. To a 0.5-mL tube, add 20.4 µL of the T7 transcription master solution and 3 µL of the PCR reaction mixture obtained in the previous procedure. The PCR solution can easily be removed from beneath the oil layer using long pipet tips.
3. Initiate the reaction by adding 0.68 µL of T7 RNA polymerase solution (69 U/µL) and incubate the mixture for exactly 1 h at 37°C. Add 0.75 µL of the EDTA solution to stop the reaction.

3.5. Electrophoresis and Quantitation of the Transcription Product (see Notes 8 and 9)

1. Add 25 µL of gel-loading solution to the transcription reaction mixture from the aforementioned procedure.
2. Place the reaction mixture in one well of a 6 or 8% polyacrylamide gel containing $8M$ urea. Each $140 \times 140 \times 1.5$ mm gel (Hoefer 600 SE, San Francisco, CA) can conveniently hold up to 10 reaction mixtures.
3. Perform electrophoresis *(see* Chapter 1).
4. Dry the gel after electrophoresis is complete.
5. Prior to autoradiography, apply marker ink mixed with some ^{32}P isotope (the the radiolabeled T7 transcription master solution mixed with loading solution containing dye is suitable). This aids in locating the bands corresponding to the highly labeled RNA product so that they can be excised after autoradiography.
6. Obtain a 15–30 min exposure of film to the gel. Using the autoradiogram as a guide, mark the bands on the gel. Cut the bands out with scissors and count them in a liquid scintillation counter.

4. Notes

1. All solutions are made with DEPC-treated water.
2. PCR primers were designed to have a GC content as close to 50% as possible. We usually designed the primers to span a region of approx 100–200 bases, if possible corresponding to regions of adjacent exons so as to minimize contamination by genomic DNA amplification. Each 5' primer had the sequence TAATACGACTCACTATA-GGGAGA attached to its 5' end. The first 17 bases are the T7 promoter sequence and the last 6 are transcription initiation sequence.
3. Tumor samples should be frozen in liquid nitrogen as soon as possible after surgery or biopsy.
4. Total RNA isolated from the tumors should be converted to cDNA immediately. We found that if the RNA is allowed to stand in solution for any appreciable period of time, even for a few hours, substantial degradation of the RNA could occur.
5. We found that the yield of PCR products for any particular cDNA was better if random hexamers were used in the reverse transcription reaction rather than gene-specific primers or oligo-dT. Also, adding a second cycle of reverse transcription often increased the yield. This can be easily done in the thermocycler.
6. When analyzing a new sample for the first time, it is necessary to determine the volumes of the cDNA falling within the linear range of amplifi-

cations for the genes of interest. We use a calibrated solution of cDNA from HT-29 cells as a linearity standard. That is, we establish the precise volumes of this solution that are at or near the end of the linear amplification regions of the genes of interest. These volumes of the HT-29 solution are then amplified simultaneously with four dilutions of the unknown solution over a broad range (usually 1000-fold). Those dilutions resulting in an amount of PCR product less than or equal to that of the HT-29 solution are assumed to be within the limits of the linear range of amplification for that gene. Once the quantity of the cDNA solution corresponding to the linear range has been established, the determination can be refined by assaying several more aliquots within the linear range. The use of an external linearity standard provides an automatic correction for isotope decay.

7. The T7 polymerase promoter sequence was included on the 5' end of the 5' PCR primer so that the PCR product could be converted to RNA. This modification confers several advantages, the main one being that the transcription step gives another 500–1000-fold amplification thereby gaining an additional dimension of sensitivity for amplification of low abundance mRNA and potentially making the method sensitive enough to analyze very small samples such as tumor cell clusters obtained from touch preparations. The large amount of nucleotide material thus generated permits loading enough RNA onto gels so that the bands can be seen by UV shadowing. The latter technique is done by placing a commercial silica gel thin-layer plate containing a fluorescent indicator underneath the wet polyacrylamide gel. The RNA bands appear purple when exposed to a hand-held 254-nm UV light. This allows a practically instantaneous assessment of the success or failure of the PCR amplification and is especially useful if a new set of primers is being tested. Another advantage of conversion to RNA is that the radiolabeling of the product is not done during the PCR step. The addition of a radiolabeled ribonucleoside triphosphate during the transcription step gives a uniformly labeled RNA with sufficient radioactivity to obtain a good autoradiogram within 10–15 min. If desired, ethidium bromide staining can also be used instead of radioactivity to visualize the bands. Quantitation of the bands is then done by densitometry of the polaroid negative.

8. When quantitation is performed, normalization of the data is required so that the expressions among samples can be compared to a common denominator. This denominator could be the cell number, total RNA, DNA, or protein. However, it is unlikely that these quantities can be determined with precision in preparations from very small quantities of

tissue material. Thus, the expression of an endogenous gene was used as a denominator *(7)*. The advantage of this is that the reference mRNA and the target mRNA share the same circumstances from the beginning of the procedure, thus providing an internal control for the amount of intact RNA successfully isolated and converted to cDNA. If relative expression of a gene in different samples is to be meaningfully compared, the endogenous standard gene should be one whose expression does not vary appreciably. In one study on PCR quantitation, ß-2-microglobulin was used as an internal standard *(7)*, whereas ß-actin expression has often been used as a denominator for the amount of RNA loaded in Northern blots. Our work has shown that use of β-actin expression as the denominator gives an excellent correlation between relative gene expression and enzyme levels *(5,8)*

9. An important feature of this method is that, in contrast to previously published methods *(2)*, the target gene and the endogenous standard are amplified individually in separate tubes. Since the conditions for obtaining a linear increase in amplification products are unlikely to be the same for different segments and primers used in amplification, separate amplification of the target and standard allows the use of individually optimized amplification conditions for each gene (temperatures, the amount of template, the number of cycles). One internal standard gene may then be used as a reference standard for any number of genes from the same cDNA preparation.

References

1. Becker-Andre, M. and Hahlbrock, K. (1989) Absolute quantitation using the polymerase chain reaction (PCR). A novel approach by a PCR aided transcript titration assay (PATTY). *Nucleic Acids Res.* **17,** 9437–9446.
2. Gilliland, G., Perrin, S., Blanchard, K., and Bunn, H. F. (1990) Analysis of cytokine mRNA and DNA: Detection and quantitation by competitive polymerase chain reaction. *Proc. Natl. Acad. Sci. USA* **87,** 2725–2729.
3. Wang, A. M., Doyle, M. V., and Mark, D. F. (1989) Quantitation of mRNA by the polymerase chain reaction. *Proc. Natl. Acad. Sci. USA* **86,** 9717–9721.
4. Volkenandt, M., Dicker, A. P., Banerjee, D., Fanin, R., Schweitzer, B., Horikoshi, T., Danenberg, K., Danenberg, P., and Bertino, J. R. (1992) Quantitation of gene copy number and mRNA using the polymerase chain reaction. *Proc. Soc. Exp. Biol. Med.* **200,** 1–6.
5. Horikoshi, T., Danenberg, K. D., Stadlbauer, T. H. W., Volkenandt, M., Shea, L. C. C., Frosing, R., Ray, M., Gibson, N. W., Spears, C. P., and Danenberg, P. V. (1992) Quantitation of thymidylate synthase, dihydrofolate reductase, and DT-diaphorase gene expression in human tumors using the polymerase chain reaction. *Cancer Res.* **52,** 108–116.

6. Chomczynski, P. and Sacchi, N. (1987) Single-step method of RNA isolation by acid guanidinium thiocyanate-phenol-chloroform extraction. *Anal. Biochem.* **162,** 156–159.

7. Noonan, K. E., Beck, C., Holzmayer, T. A., Chin, J. E., Wunder, J. S., Andrulis, I. L., Gazdar, A. F., William, C. L., Griffith, B., Von Hoff, D. D., and Roninson, I. B. (1990) Quantitation analysis of MDR1 (multidrug resistance) gene expression in human tumors by polymerase chain reaction. *Proc. Natl. Acad. Sci. USA* **87,** 7160–7164.

8. Traver, R. D., Horikoshi, T., Danenberg, K. D., Stadlbauer, T. H. W., Danenberg, P. V., Ross, D., and Gibson, N. W. (1992) NAD(P)H: Quinone oxidoreductase gene expression in human colon carcinoma cells: Characterization of a mutation which modulates DT-diaphorase activity and mitomycin sensitivity. *Cancer Res.* **52,** 797–802.

CHAPTER 18

Identification of Alternatively Spliced mRNAs and Localization of 5' Ends by Polymerase Chain Reaction Amplification

Cheng-Ming Chiang, Louise T. Chow, and Thomas R. Broker

1. Introduction

Messenger RNAs of higher eucaryotes are usually modified posttranscriptionally to contain 5' caps and nucleotide methylation, 3' polyadenylation, and one or more internal splices to remove introns and join exon segments into the mature protein-coding sequences. With the abilities of retroviral reverse transcriptases to create complementary DNA (cDNA) copies of RNA and of thermostable DNA polymerases to amplify specific DNA segments by repeated cycles of denaturation of duplex templates, annealing of complementary oligonucleotide primers, and strand elongation, it is becoming increasingly popular to use this highly sensitive polymerase chain reaction (PCR) to isolate partial cDNA sequences for the purpose of identifying RNA splice sites and inferring the coding capacity. The splice junction information from the partial cDNAs, together with additional biophysical or biochemical information, can then be employed to map the 5' ends of the mRNAs and assign the AUG protein initiation codon and open reading frame to the message.

One of the most remarkable features of eucaryotic genes and particularly those of eucaryotic viruses is the use of alternative splicing to generate families of related mRNAs with shared as well as unique

From: *Methods in Molecular Biology, Vol. 15: PCR Protocols: Current Methods and Applications*
Edited by: B. A. White Copyright © 1993 Humana Press Inc., Totowa, NJ

exons that encode proteins with common and specialized domains. Such genetic versatility is often compounded by the utilization of alternative promoters and alternative 3' polyadenylation sites to create nested sets of overlapping transcripts, which may differ tremendously in relative abundance or as a function of organ development, tissue differentiation, or stage of virus infection. In this chapter, we describe some practical considerations and detailed procedures of PCR with which we have recovered families of cDNAs of spliced, overlapping mRNAs and localized their respective 5' ends (*1–3* and our unpublished results).

1.1. General Strategies for Determining mRNA Splice Sites by PCR Amplification of cDNAs

It is generally essential to know the sequence of the genomic DNA or the cDNA copy of one member of an mRNA family. Degenerate primers based on protein sequences of the same or related systems have also been used successfully in a number of laboratories *(4–6)*. The locations of RNA splice sites can be approximated by electron microscopic analysis of RNA:genomic DNA heteroduplexes *(7,8)*, S1 nuclease protection *(9,10)*, or retrovirus-mediated gene transfer *(3,11)*. In the absence of reasonably accurate splice-site mapping, a variety of pair-wise combinations of 5' and 3' oligonucleotide primers can be tested blindly.

The PCR is performed directly on the first-strand cDNA templates generated by reverse transcription of total or selected poly(A)$^+$ RNA. Primers for reverse transcription can be oligo-dT, random hexamers, or an antisense oligonucleotide complementary to mRNA sequences downstream of the putative splice acceptor site.

For PCR, a 5' sense-strand primer upstream of the potential splice donor site and a 3' antisense-strand primer downstream of the potential splice acceptor site are designed according to standard principles. The primers should be approx 20 nucleotides long to provide specificity, consist of approx 50% G + C so that different primers can be mixed and matched without having to change reaction conditions or prepare new primers, and should not contain self-complementary or cross-complementary sequences, particularly at their 3' ends (*see* Chapter 2).

Candidate PCR products smaller than the genomic DNA or unspliced RNA transcript are further characterized by cloning and DNA sequencing. Since DNA recombination can occur in vitro during PCR amplification *(12)* and aberrant mRNA can be generated in vivo from occasional rearranged genomic templates, the demonstration of the splice consensus sequence /GU——AG\ at the exon/intron\exon junctions *(13)* is essential to confirm the authenticity of splice sites. If determination of the cDNA sequence directly from the PCR products rather than from a cloned PCR product is preferred, then more specific dideoxynucleotide sequencing reactions and clearer sequencing gels can be obtained using sequencing primers internal to the PCR primers.

1.2. Use of PCR to Localize the 5' Ends of mRNAs Among a Family of Related Species

Alternatively spliced mRNAs generated from a DNA sequence that contains overlapping open reading frames may have different 5' ends and AUG protein initiation codons, which dictate the coding potential of the mRNAs. Accordingly, it is essential to correlate the splice sites with the 5' ends of the mRNAs. Localization of the 5' ends of extremely rare messages is hindered by the competing amplification of the more abundant, alternatively spliced transcripts. However, we have been successful in approximating the 5' ends of rare viral mRNAs among a family of related messages using a modified PCR protocol *(3* and our unpublished results).

Following the synthesis of the first-strand cDNAs, one of several sense-strand primers placed immediately downstream of suspected promoters is used in conjunction with an antisense primer spanning the splice junction. The junction primer is complementary to the contiguous (3') 6 and (5') 14 bases (or 10 and 10 bases) in the adjoining exons. The antisense primers will also be partially complementary to other transcripts that share either the splice donor site or the splice acceptor site, but will have several mismatched 5' or 3' nucleotides with all but the targeted spliced message (Fig. 1).

2. Materials

2.1. Enzymes

1. Moloney murine leukemia virus (Mo-MLV) reverse transcriptase (200 U/µL) (BRL/Life Technologies, Gaithersburg, MD).
2. *Taq* DNA polymerase (5 U/µL) (Perkin Elmer/Cetus, Norwalk, CT).
3. Vent DNA polymerase (1 U/µL) (New England Biolabs, Beverly, MA).

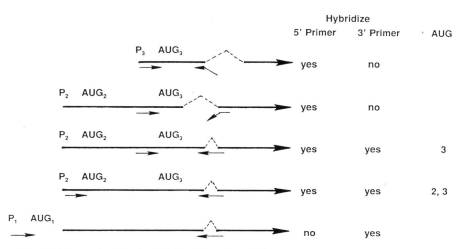

Fig. 1. Strategies for localizing 5' ends of alternatively spliced mRNAs. Large arrows denote mRNAs, with dashes representing introns that are spliced out. Small arrows represent PCR primers with mismatches in the bent-out regions. Under standard set-up and annealing conditions, all primers anneal and produce correct or incorrect products, except for the last combination. However, under the modified high-stringency PCR conditions, only perfectly matched primers-templates generate a product, from which the 5' end can be approximated.

2.2. Reagents

It is important to take the appropriate steps to insure that all reagents remain sterile and RNase-free (*see* Note 1).

1. GITC solution: 4.23M guanidine isothiocyanate, 0.5% sodium N-lauryl sarcosine, 25 mM sodium citrate (pH 7.0), 0.1M β-mercaptoethanol. Sterilize by filtration through a 0.22-μm pore membrane and store at 4°C. Use within 2 wk.
2. CsCl cushion: 5.7M CsCl, 0.1M sodium EDTA (pH 8.0) (density = 1.7 g/mL). Treat with 0.1% (v/v) diethylpyrocarbonate (DEPC) at 37°C overnight and store at 4°C after autoclaving.
3. Phosphate-buffered saline solution A (PBSA): 10 g of NaCl, 0.25 g of KCl, 1.44 g of Na_2HPO_4, and 0.25 g of KH_2PO_4 made to 1 L. Store at 4°C after autoclaving.
4. 3M Sodium acetate, pH 5.2. Treat with DEPC as in step 2.
5. DEPC-treated double-distilled water.
6. 80% Ethanol/0.1M sodium acetate, pH 5.2.
7. 10X RT-PCR buffer: 200 mM Tris-HCl, pH 8.4, 500 mM KCl, 25 mM $MgCl_2$, 1 mg/mL of nuclease-free bovine serum albumin.
8. Random hexamers (100 pmol/μL) (LKB/Pharmacia, Uppsala, Sweden), specific PCR primers or oligo(dT)$_{12-18}$.

9. RNasin (40 U/μL) (Promega, Madison, WI).
10. Mo-MLV reverse transcriptase.
11. 2.5 m*M* dNTPs.
12. *Taq* DNA polymerase.

3. Methods

3.1. Preparation of Total RNA from Cultured Eucaryotic Cells (see Note 2)

1. Wash the cells with PBSA and drain well to remove residual liquid.
2. Add 1 mL of GITC solution to each 100-mm plate and let stand at room temperature for 5 min.
3. Collect all lysates in a 50-mL centrifuge tube with a rubber policeman.
4. Add a few drops of antifoam and shear the DNA by passage through an 18-g needle five times.
5. Add 1/3 to 1/4 tube vol of a CsCl cushion to a 17-mL ultracentrifuge tube (for a Beckman, Palo Alto, CA, SW28 rotor or equivalent), then carefully overlay with the lysates.
6. Balance the tubes with the GITC solution and ultracentrifuge at 26,000 rpm (122,000*g* at r max) at 20°C for 26 h in a Beckman SW28 rotor or the equivalent.
7. Remove about 80% of the liquid from the centrifuge tubes by suction with a Pasteur pipet placed at the center of the meniscus, taking care not to disturb protein debris stuck to the tube wall.
8. Quickly invert the tube to drain the remaining liquid, then cut off and save the bottom of the tube with a razor blade.
9. Place the cut-off base of the tube upside down on absorbent paper to drain and dry for 5 min.
10. Use a sterile 1-mL Eppendorf pipet tip to break the RNA pellet into pieces and transfer them to an autoclaved Eppendorf tube containing 0.5 mL of 80% ethanol/0.1*M* sodium acetate.
11. Rinse the bottom of the ultracentrifuge tube with an aliquot of 80% ethanol/0.1*M* sodium acetate and transfer the solution back to the Eppendorf tube.
12. Centrifuge in a microcentrifuge for 5 min at 4°C to collect the RNA pellet.
13. Pour out the ethanol solution and dry the RNA pellet under vacuum.
14. Add 400 μL of DEPC-treated H$_2$O to dissolve the RNA at 4°C.
15. Precipitate the RNA by adding 40 μL of 3*M* sodium acetate, pH 5.2 and 1 mL of 100% ethanol. Store at –20°C overnight or –70°C for 30 min.
16. Collect the RNA precipitate by centrifugation in a microcentrifuge at maximum speed for 5 min at 4°C.
17. Pour out the supernatant and wash the pellet with 500 μL of 80% ethanol/0.1*M* sodium acetate, pH 5.2.

18. Centrifuge again and dry the pellet under vacuum.
19. Add 400 µL of DEPC-treated H_2O to dissolve the RNA pellet and dilute 2 µL to about 600 µL for an A_{260} measurement.

3.2. Generation of cDNAs by Coupled Reverse Transcription-PCR

1. Set up the reaction mixture for reverse transcription by adding the following components sequentially into a 0.5-mL Eppendorf tube:

10X RT-PCR buffer	2.0 µL
2.5 m*M* dNTPs	8.0 µL
Random hexamer (100 pmol/µL; *see* Note 3) or 0.5 µg of oligo-dT and 0.7 µg for downstream primer (*16*; *see* Note 4)	1.0 µL
RNasin (40 U/µL)	0.5 µL
Heat denatured-RNA (*see* Note 5) and H_2O	7.5 µL
Mo-MLV reverse transcriptase (200 U/µL)	1.0 µL

2. Incubate the reaction mixture at room temperature for 10 min, then at 42°C for 30–60 min.
3. Heat-inactivate the reverse transcriptase and denature the RNA:DNA duplexes at 95°C for 5–10 min, then quickly chill the sample on ice. Collect all liquid by centrifugation at 4°C.
4. Add the following components into the above cDNA preparation:

H_2O	70.5 µL
10X RT-PCR buffer	8.0 µL
Upstream primer (100 pmol/µL)	0.5 µL
Downstream primer (100 pmol/µL)	0.5 µL
Taq DNA polymerase (5 U/µl)	0.5 µL

5. Overlay 100 µL of mineral oil to prevent evaporation during thermal cycling.
6. Amplify by PCR using the following cycle profile:

30 main cycles	94°C, 1 min (denaturation)
	55°C, 90 s (annealing; *see* Note 6)
	72°C, 2 min (extension)
Final extension	72°C, 7 min

   Cool the reaction to 4°C.
7. Remove the mineral oil by extraction once with 100 µL of chloroform. Analyze 5 µL of the upper aqueous phase by agarose gel electrophoresis (*see* Chapter 1).
8. Use 5 µL of the PCR products for a second round of amplification, if necessary, and analyze similarly. Alternatively, DNA recovered from the expected region in the gel can be used as templates for the second round of amplification.

3.3. Modified PCR for the Localization of the 5' Ends of mRNAs

1. Prepare first-strand cDNAs according to steps 1–3 in Section 3.2.
2. Add PCR primers (one of which is the splice junction primer), 10X RT-PCR buffer and H_2O as described in step 4 in Section 3.2., but withhold the addition of DNA polymerase.
3. Overlay the cocktail with 100 µL of mineral oil.
4. Heat the reaction mixture at 90°C for 90 s to denature the primers and templates that have paired specifically or nonspecifically during set-up.
5. Add 2.5 U of *Taq* or Vent DNA polymerase (*see* Note 7).
6. Perform 30 cycles of PCR amplification and analyze the products as described in Section 3.2. except set the annealing temperature at 60–70°C for 20-mer primers with 50% G + C (*see* Note 7). Adjust the temperature depending on the length and G + C composition of the primers.

4. Notes

1. All glassware should be baked at 150°C for 6 h and plasticware should be steam-autoclaved. Water and chemicals should be DEPC-treated and autoclaved whenever possible. Eppendorf tubes for PCR reactions should be silanized.
2. The procedures for preparation of total RNA and RT-PCR were modified from those of MacDonald et al. *(17)* and Kawasaki *(18,19)*. Preparation of tissue RNA and the running conditions for different ultracentrifugation rotors have been described *(17)*. The RNA preparation can be contaminated by genomic DNA. One way to distinguish an RNA template from a DNA template is to employ a pair of primers located in different exons, as in the case of mapping RNA splice sites. However, identification of gene expression from intronless genes is hindered by trace contaminants from genomic sequences. Pretreatment of RNA samples with RNase-free DNase has been used successfully to avoid the pseudo-positive results derived from intronless genes *(21)*.
3. If the target mRNA is of low abundance, the yield of first-strand cDNA is higher when random hexamers are used as primers.
4. The specificity of PCR can be increased if the first-strand cDNA primer is an antisense oligonucleotide situated downstream of the PCR primer.
5. The amount of RNA used for RT-PCR depends on the abundance of the transcript. We normally use 1 µg of total RNA for RT-PCR. However, PCR products of abundant transcripts in as little as 5 pg of total RNA can be detected by autoradiography when ^{32}P-labeled nucleoside triphosphates were used in the PCR reaction *(20)*.
6. The annealing temperature is set for 20-mers with 50% G + C content. Adjust the temperature according to the primer length and G + C con-

A. 55 °C :   B. 70 °C :

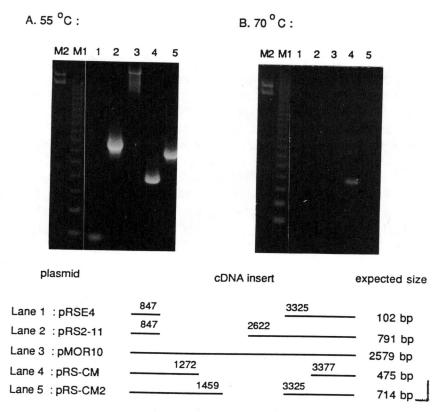

plasmid   cDNA insert   expected size

Lane 1 : pRSE4	847	3325	102 bp
Lane 2 : pRS2-11	847	2622	791 bp
Lane 3 : pMOR10			2579 bp
Lane 4 : pRS-CM	1272	3377	475 bp
Lane 5 : pRS-CM2	1459	3325	714 bp

Fig. 2. Selective amplification of target sequence under stringent conditions. PCR amplification was performed on plasmids containing different HPV-11 cDNAs or genomic DNA (bottom portion) with an antisense junction-spanning primer (3' 1267-1272^3377-3390 5', 14 of which are G + C) and a sense-strand primer (5' 812-831 3', 10 of which are G + C) for 30 cycles using the *Taq* DNA polymerase as described in Section 3. The annealing was conducted at either 55°C (panel **A**) or 70°C (panel **B**). The products as predicted for a perfectly matched (lane 4) or 3'-mis-matched priming events (lanes 1, 2, 3, and 5) were displayed in 1.5% agarose gels. Gaps represent splices, with numbers denoting nucleotide positions of splice donor and acceptor sites. M1, size markers are a $(123)_n$ bp DNA ladder; M2, λ-*Hind* III size markers (2.32 kb, 2.03 kb, and 0.56 kb).

tent based on the approximate equation T_m = 4°C (number of G + C) + 2°C (number of A + T) for oligonucleotides approx 18–24 residues long.

7. Elongation from 3'-mismatched primers has been observed (*14,15* and Fig. 2, panel A). Selective PCR amplification of the targeted template is achieved when the reaction conditions are made more stringent to eliminate nonspecific priming from mismatched primer:template involv-

ing alternatively spliced relatives (Fig. 2, panel B). We also find that, for certain mRNAs, the use of Vent DNA polymerase (New England BioLabs), which has 3' → 5' proofreading exonuclease activity, is also necessary to achieve specific amplification *(3)*. Vent DNA polymerase eliminates the occasional priming from short mismatched 3' branches of imperfectly matched primer:template.

Acknowledgments

This research was supported by NIH Grant CA 36200.

References

1. Rotenberg, M. O., Chow, L. T., and Broker, T. R. (1989) Characterization of rare human papillomavirus type 11 mRNAs coding for regulatory and structural proteins, using the polymerase chain reaction. *Virology* **172,** 489–497.
2. Palermo-Dilts, D. A., Broker, T. R., and Chow, L. T. (1990) Human papillomavirus type 1 produces redundant as well as polycistronic mRNAs in plantar warts. *J. Virol.* **64,** 3144–3149.
3. Chiang, C.-M., Broker, T. R., and Chow, L. T. (1991) An E1M^E2C fusion protein encoded by human papillomavirus type 11 is a sequence-specific transcription repressor. *J. Virol.* **65,** 3317–3329.
4. He, X., Treacy, M. N., Simmons, D. M., Ingraham, H. A., Swanson, L. W., and Rosenfeld, M. G. (1989) Expression of a large family of POU-domain regulatory genes in mammalian brain development. *Nature* **340,** 35–42.
5. Chang, T.-H., Arenas, J., and Abelson, J. (1990) Identification of five putative yeast RNA helicase genes. *Proc. Natl. Acad. Sci. USA* **87,** 1571–1575.
6. Hoey, T., Dynlacht, B. D., Peterson, M. G., Pugh, B. F., and Tjian, R. (1990) Isolation and characterization of the drosophila gene encoding the TATA box binding protein, TFIID. *Cell* **61,** 1179–1186.
7. Chow, L. T., Nasseri, M., Wolinsky, S. M., and Broker, T. R. (1987) Human papillomavirus types 6 and 11 mRNAs from genital condylomata acuminata. *J. Virol.* **61,** 2581–2588.
8. Chow, L. T., and Broker, T. R. (1989) Mapping the genetic organization of RNA by electron microscopy. *Methods Enzymol.* **180,** 239–261.
9. Berk, A. J. (1989) Characterization of RNA molecules by S1 nuclease analysis. *Methods Enzymol.* **180,** 334–347.
10. Muranyi, W., and Flugel, R. M. (1991) Analysis of splicing patterns of human spumaretrovirus by polymerase chain reaction reveals complex RNA structures. *J. Virol.* **65,** 727–735.
11. Rotenberg, M. O., Chiang, C.-M., Ho, M. L., Broker, T. R., and Chow, L. T. (1989) Characterization of cDNAs of spliced HPV-11 E2 mRNA and other HPV mRNAs recovered via retrovirus-mediated gene transfer. *Virology* **172,** 468–477.
12. Meyerhans, A., Vartanian, J.-P., and Wain-Hobson, S. (1990) DNA recombination during PCR. *Nucleic Acids Res.* **18,** 1687–1691.
13. Mount, S. M. (1982) A catalogue of splice junction sequences. *Nucleic Acids Res.* **10,** 459–472.

14. Kwok, S., Kellogg, D. E., McKinney, N., Spasic, D., Goda, L., Levenson, C., and Sninsky, J. J. (1990) Effects of primer-template mismatches on the polymerase chain reaction: human immunodeficiency virus type 1 model studies. *Nucleic Acids Res.* **18,** 999–1005.

15. Nassal, M. and Rieger, A. (1990) PCR-based site-directed mutagenesis using primers with mismatched 3'-ends. *Nucleic Acids Res.* **18,** 3077–3078.

16. Rappolee, D. A. (1990) Optimizing the sensitivity of RT-PCR. *Amplifications* **4,** 5–7.

17. MacDonald, R. J., Swift, G. H., Przybyla, A. E., and Chirgwin, J. M. (1987) Isolation of RNA using guanidinium salts. *Methods Enzymol.* **152,** 219–227.

18. Kawasaki, E. (1989) Amplification of RNA sequences via complementary DNA (cDNA). *Amplifications* **3,** 4–6.

19. Kawasaki, E. (1990) Amplification of RNA, in *PCR Protocols* (Innis, M. A., Gelfand, D. H., Sninsky, J. J., and White, T. J., eds.), Academic, San Diego, CA, pp. 21–27.

20. Mocharla, H., Mocharla, R., and Hodes, M. E. (1990) Coupled reverse transcription-polymerase chain reaction (RT-PCR) as a sensitive and rapid method for isozyme genotyping. *Gene* **93,** 271–275.

21. Grillo, M., and Margolis, F. L. (1990) Use of reverse transcriptase polymerase chain reaction to monitor expression of intronless genes. *BioTechniques* **9,** 262–268.

CHAPTER 19

Utilization of Polymerase Chain Reaction for Clonal Analysis of Gene Expression

Alice L. Witsell and Lawrence B. Schook

1. Introduction

The reverse transcription-polymerase chain reaction (RT-PCR) is a powerful tool when studying gene expression in a limited number of cells. RT-PCR was first described by Veres et al. *(1)* and numerous accounts have since followed. Classic studies of gene expression have utilized Northern blot analysis to monitor expression of transcripts in response to developmental and activational signals. Problems associated with using the Northern blot technique include the necessity for an abundant number of cells, the limited numbers of genes that can be analyzed on a single blot, and that, usually, only one gene can be monitored at a time. Finally, little information on which cells within the cell population being analyzed are expressing the gene(s) is provided by Northern blot analyses. For example, on a Northern blot it is virtually impossible to determine whether all cells contain a transcript for a specific gene(s) or whether all cells exclusively express the particular gene of interest. *In situ* hybridization has been used in conjunction with Northern blot analyses to determine which cells in a population are expressing a particular gene. However, this technique is limited by low sensitivity and by the fact that usually only one gene can be monitored at a time. RT-PCR alleviates these problems because specific transcripts can routinely be detected in RNA quickly isolated *(2)* from

From: *Methods in Molecular Biology, Vol. 15: PCR Protocols: Current Methods and Applications*
Edited by: B. A. White Copyright © 1993 Humana Press Inc., Totowa, NJ

5–10 cells *(3–5)*. The principle of clonal analysis by RT-PCR is to use 50–100 cells that are clonally derived from a single progenitor cell. RNA is then isolated from individual clones, reverse-transcribed into cDNA, and amplified with transcript-specific primers. Two steps are involved in RT-PCR:

1. Isolation and reverse-transcription of the specific transcripts or total mRNA into cDNA.
2. Amplification of specific sequences of transcripts by PCR.

During the PCR step, primers used for amplification of specific sequences are designed in order to amplify sequences of predetermined size. In addition, it is also useful to design primers that span across introns to assure that contaminating DNA is not being amplified. There can be numerous technical difficulties associated with amplifying cDNA sequences from limited transcripts obtained from low cell numbers. We have used an internal nested primer that results in increased specific amplification with decreased background amplification to overcome this problem *(4)*. In demonstrating the RT-PCR technique, we describe in this chapter the approach we have developed to molecularly phenotype colonies of bone marrow-derived macrophage obtained using hematopoietic growth factors *(5)*.

2. Materials

1. Guanidinium isothiocyanate ($4.0M$) containing 25 mM sodium citrate, pH 7.0, 0.5% N-laurylsarkosyl, and $0.1M$ β-mercaptoethanol.
2. Diethylethylpyrocarbonate (DEPC)-treated H_2O. Prepare by adding 1 mL of DEPC to 1 L of H_2O. Mix the solution thoroughly and allow to set 30 min to overnight followed by autoclaving.
3. 10 mg/mL Yeast tRNA.
4. 1.7- and 0.65-mL snap-cap Eppendorf tubes.
5. $2M$ Sodium acetate, pH 4.0.
6. Phenol saturated with DEPC-treated H_2O.
7. Chloroform:isoamyl alcohol (49:1).
8. Phenol:chloroform:isoamyl alcohol (24:24:1) buffered with $T_{10}E_1$ (10 mM Tris-HCl, 1 mM EDTA, pH 8.0).
9. Isopropanol.
10. $T_{10}E_1$ containing $0.2M$ NaCl.
11. 95 or 100% Ethanol.
12. 5X RT-buffer: 250 mM Tris-HCl, pH 8.3, 300 mM KCl, 15 mM $MgCl_2$.
13. Random hexamers (pd[N]$_6$ sodium salt, Pharmacia, Uppsala, Sweden).

14. dNTP. Prepare 1-mL stocks of 40 mM dATP, TTP, dCTP, and dGTP dissolved in DEPC-treated H_2O and brought to pH 7.5 with 25 mM Tris, pH 7.5. Mix all four to yield a 10 mM final concentration. Prepared dNTP solutions can also be purchased commercially.
15. Moloney murine leukemia virus reverse transcriptase (M-MLV RT), 200 U/μL (BRL, Bethesda, MD).
16. 10X PCR-buffer: 100 mM Tris-HCl, pH 8.3, 500 mM KCl, 20–70 mM $MgCl_2$. *See* Note 1 for specific remarks regarding $MgCl_2$ concentration.
17. Sterile H_2O, *not DEPC-treated (see* Note 2).
18. 50 μM of distilled stock solution of each primer (15–24 bases in length).
19. *Taq* polymerase (Perkin-Elmer, Cetus) (5 U/μL).
20. Agarose (NuSieve and SeaPlaque (FMC Bioproducts, Rockland, ME) agarose are optional).
21. Stock (50X) of Tris Acetate EDTA buffer: 242 g of Trizma base, 57.1 mL of glacial acetic acid, and 100 mL of 500 mM EDTA, pH 8.0 made up to 1 L using double-distilled H_2O. The final pH should be 7.6.
22. Loading dye: 0.25% bromophenol blue (w/v), 15% (w/v) Ficoll 400 in double-distilled H_2O.
23. Ethidium bromide (10 mg/mL stock) in double-distilled H_2O.

3. Methods

3.1. RNA Isolation

1. Isolate colonies or aliquots containing 50–50,000 cells directly into 1.7-mL Eppendorf tubes containing 50 μL of 4M guanidinium thiocyanate with 25 mM sodium citrate, pH 7.0, 0.5% N-laurylsarkosyl, 0.1M β-mercaptoethanol, and 5–10 μg of yeast tRNA *(see* Note 3).
2. Sequentially, add 5 μL of 2M sodium acetate pH 4.0, 50 μL of water saturated phenol, and 20 μL of chloroform:isoamyl alcohol (49:1). Vortex the mixture and chill on ice for 15 min followed by centrifugation at 10,000g for 10 min at 4°C.
3. Transfer the aqueous supernatant (50 μL) into a 1.7-mL Eppendorf tube and reextract with phenol:chloroform:isoamyl alcohol (24:24:1) saturated with $T_{10}E_1$ and centrifuge for 5 min in a microfuge at room temperature.
4. Transfer the aqueous supernatant to a separate 1.7-mL tube and precipitate the RNA by addition of 200 μL of isopropyl alcohol for at least 4 h at –20°C. Centrifuge the samples in a microfuge for 10 min at 25°C and carefully discard the supernatant using a 200-μL pipetman so as not to dislodge the white RNA pellet *(see* Note 4).
5. Resuspend the pellet in 50 μL of 0.2M NaCl dissolved in $T_{10}E_1$ and precipitate with 200 μL of 100% ethanol for 3–4 h at –20°C. Centrifuge

the samples for 10 min in a microfuge at room temperature and discard the supernatant without disturbing the pellet. Invert the tubes and allow them to air dry.

3.2. cDNA Synthesis

1. Resuspend the RNA pellet in 12 µL of DEPC-treated water, 4 µL 5X RT buffer, 2 µL of 10 mM dNTPs, and 1 µL of 8 nM hexamers *(see Note 5)*.
2. Denature the samples in a 65°C water bath for 10 min and immediately cool on ice. Pellet the condensate, and then add 1 µL of 200 U/µL M-MLV RT *(see Note 6)*. Incubate at 42°C for 45–60 min, followed by heating to 95°C for 10 min to stop the reaction.

3.3. Polymerase Chain Reaction

1. Prepare the PCR reactions by addition of 2 µL of the 10 mM dNTP stock, 5 µL of 10X buffer, 0.5 µL of 50 µM of each primer, 42 µL sterile-distilled H$_2$O and 1 U of *Taq* polymerase (5 U/µL) to a 0.65-mL Eppendorf tube. The cDNA sample (2 µL of the 20 µL total) is added as the final component of the reaction and overlayed with 2–3 drops of mineral oil. If several samples are to be amplified with the same primer set, the reaction mixture should be made in batch and aliquoted to the appropriate number of tubes followed by addition of the cDNA to individual reaction mixtures *(see Notes 7 and 8)*.
2. Amplify by PCR using the following cycle profile:

    35 main cycles   94°C, 1 min (denaturation)

                         T_m of primers, 45 s (annealing) *(see Note 9)*

                         72°C, 30 s (extension)
3. When <1000 cells are used for an isolation, it is beneficial to use a nested internal primer *(see Note 10)*. After 20 PCR cycles, 1 µL of the internal primer (50 µM) is added directly to the reaction for an additional 20 cycles.
4. Five microliters of the reaction sample is mixed with 1 µL loading dye and products are electrophoresed on a 2.5% agarose gel at 80 V for 90 min followed by visualization with ethidium bromide staining *(see Note 11)*.

4. Notes

1. It is important to titrate the magnesium concentration used in the PCR reactions for each primer set and each RNA isolation. Depending on the quantity of RNA isolated and the concentration of dNTPs, the magnesium concentration may vary owing to chelation of magnesium ions. For example, if isolating RNA from 50 cells with excess carrier tRNA, the magnesium concentration may be very different if RNA is isolated

from 10,000 cells without the added carrier tRNA. Therefore, it is essential to titrate the magnesium concentration between 2.0 and 7.0 mM final concentration with each primer set and following each isolation procedure.

2. DEPC is an inhibitor of *Taq* polymerase thus DEPC-treated H_2O and solutions should be avoided in the PCR steps to ensure that residual DEPC does not inhibit the PCR reaction.

3. A total of 50–100 cells will provide enough RNA for 10 reactions, when 10^5 or more cells are used we have had better success with diluting the cDNA 1–2 orders of magnitude before adding to the PCR reaction. We typically isolate colonies of 50–100 cells directly into 50 µL of guanidinium. This volume should be increased to a maximum of 500 µL for 5×10^6 cells with appropriate changes in other volumes described being made to accommodate the increased cell number.

4. Pellets should be observed after both precipitations during the RNA isolation. The supernate should be aspirated carefully so as to not dislodge the pellet. If RNA pellets are not visible during the precipitations either not enough carrier tRNA was initially added or the solutions contain RNase. Try adding up to 20 µg of tRNA to each sample or precipitate for longer times at –20°C to ensure visible RNA pellets. If pellets are still not visible, then remake solutions being careful to be RNase-free.

5. Random hexamers have been more successful in our phenotyping studies than oligo-dT and a 3' sequence specific primer. Hexamers allow the entire transcript to be copied whereas oligo-dT sometimes falls short of a complete, full length cDNA leaving only the 3' ends copied. Sequence specific primers will only allow specific genes to be copied yet limit this approach to analyzing the presence of a single transcript following PCR gene amplification.

6. The use of M-MLV RT is important with reactions containing large amounts of carrier RNA as avian myeloblastosis virus reverse transriptase (AMV-RT) has been shown to be inhibited by excess carrier RNA whereas M-MLV RT is not inhibited by high amounts of carrier RNA *(6)*, thus M-MLV RT should be used in the type of procedure shown here.

7. Multicolored tubes are commercially available and when a number of distinct genes are to be amplified simultaneously it is beneficial to amplify the individual genes in the same colored tubes as the mineral oil can cause labels to wipe off easily.

8. Both positive and negative controls should be run. The positive control involves amplification of a plasmid carrying the sequence of interest. The negative control is very important in controlling for contamination of the reaction components by plasmid DNA. The reaction should be set up as normal but the cDNA left out of the components. If a positive

PCR reaction product is found, a component is contaminated and the reagents should be remade. When amplifying cDNA, it is optimal to design primers that span exons. Thus, any contaminating DNA will be amplified at a size distinct from the cDNA.

9. The T_m for primers is calculated at 2°C for every A:T pair and 4°C for every G:C pair. Since such a huge excess of primer is used in the PCR reactions, the annealing temperature can generally be increased 5°C above the T_m to decrease background amplification with no adverse effects on the sequence of interest.

10. When <1000 cells are used for an isolation, the PCR reaction can have high background amplification. It is beneficial to add a nested internal primer to the reaction for the last 20 cycles. The nested internal primer (1 μM final concentration) is added directly to the reaction after 20 cycles for an additional 20 cycles. This step greatly increases the sensitivity and clarity of the results. Furthermore, it adds an additional degree of assurance that the product being amplified is in fact not a homologous sequence since three instead of two primers must anneal and provide appropriately sized products.

11. RT-PCR products are run on a 2.5% agarose gel at 80 V for 90 min and visualized by ethidium bromide staining. Replacement of 25% of the agarose with Sea Plaque agarose makes the agarose dissolve much more easily and yields higher resolution of the RT-PCR products.

References

1. Veres, G., Gibbs, R. A., Schrer, S. E., and Caskey, C. T. (1987) The molecular basis of the sparse fur mouse mutation. *Science* **237**, 415–417.
2. Chomczynski, P. and Sacchi, N. (1987) Single step method of RNA isolation by acid guanidinium thiocyanate-phenol-chloroform extraction. *Anal. Biochem.* **162**, 156–159.
3. Rappolee, D. A., Wand, A., Mark, D., and Werb, Z. (1989) Novel method for studying mRNA phenotypes in single or small numbers of cells. *J. Cell. Biochem.* **39**, 1–11.
4. Witsell, A. L. and Schook, L. B. (1990) Clonal analysis of gene expression by PCR. *BioTechniques* **9**, 318–322.
5. Witsell, A. L. and Schook, L. B. (1991) Macrophage heterogeneity occurs through a developmental mechanism. *Proc. Natl. Acad. Sci. USA* **88**, 1963–1967.
6. Gerard, G. F. (1987) Making effective use of cloned M-MLV reverse transcriptase. *Focus (BRL)* **9**, 5,6.

CHAPTER 20

Sequencing DNA Amplified Directly from a Bacterial Colony

Martin A. Hofmann and David A. Brian

1. Introduction

A few hundred bacterial cells obtained by touching a bacterial colony with a sterile toothpick can be used directly in a polymerase chain reaction (PCR) amplification procedure to identify and orient a plasmid insert *(1,2)*. By combining this procedure with one in which asymmetrically amplified DNA is used for sequencing (ref. *3* and Fig. 1), we have demonstrated that DNA amplified from a bacterial colony can be sequenced directly by the dideoxy chain-termination method to yield results as good as those obtained when purified template DNA is used for amplification (ref. *4* and Fig. 2). By end-labeling the primer that is used in limiting amounts during the amplification step and using it for sequencing, an entire insert of 300 nucleotides or less can be sequenced in one step. Inserts of larger size can be sequenced by using labeled primers that bind within the amplified single-stranded DNA sequence. The procedure is rapid and enables one to obtain sequences from as many as 20 clones in a single day.

2. Materials

1. Colonies of transformed bacteria containing plasmid (to date we have used only *E. coli*).
2. Oligonucleotide primers with a minimum length of 20 nucleotides and a minimum G + C content of 50% (*see* Note 1). We routinely use oligonucleotides purified only by size exclusion chromatography

From: *Methods in Molecular Biology, Vol. 15: PCR Protocols: Current Methods and Applications*
Edited by: B. A. White Copyright © 1993 Humana Press Inc., Totowa, NJ

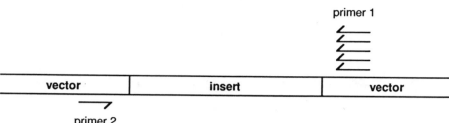

Fig. 1. The asymmetric PCR reaction. In the procedure described in this chapter, primer 2 is the limiting primer and will be 5' end-labeled and used in the sequencing reaction for sequencing the single-stranded DNA product.

in water. Approximately 1 mg of oligonucleotide is purified on a NAP5 Sephadex G-25 column (Pharmacia, Piscataway, NJ), which has a bed volume of 3 mL. Oligonucleotides are quantitated by spectrophotometry (1 A_{260} U = 33 µg/mL) and diluted to a final concentration of 10 µM or 0.1 µM (depending on the use; *see* below) in water. Store at –20°C.

3. 10X PCR buffer: 500 mM KCl, 100 mM Tris-HCl, pH 9.0, 15 mM MgCl$_2$, 0.1% gelatin, 1.0% Triton X-10. Store at –20°C.
4. dNTP mix, each at 1.25 mM in water. Store at –20°C.
5. *Taq* DNA polymerase.
6. 10X kinase buffer: 500 mM Tris-HCl, pH 7.5, 100 mM MgCl$_2$, 50 mM DTT, 1 mM spermidine, 1 mM EDTA.
7. T4 polynucleotide kinase.
8. [γ-^{32}P]ATP (3000 Ci/mmol).
9. BioSpin 6 or BioSpin 30 columns (Bio-Rad, Rockville Centre, NY).
10. Dideoxynucleotide chain termination mixes as described in the following. (Store all at –20°C.)

   G: 250 µM ddGTP. Prepare from 2.5 µL of 10 mM ddGTP, 10 µL of 10X PCR buffer, 1.6 µL of dNTP mix, and 85.9 µL of H$_2$O.
   A: 1.28 mM ddATP. Prepare from 12.8 µL of 10 mM ddATP, 10 µL of 10X PCR buffer, 1.6 µL of dNTP mix, and 75.6 µL of H$_2$O.
   T: 1.92 mM ddTTP. Prepare from 19.2 µL of 10 mM ddTTP, 10 µL of 10X PCR buffer, 1.6 µL of dNTP mix, and 69.2 µL of H$_2$O.
   C: 640 µM ddCTP. Prepare from 3.2 µL of 10 mM ddCTP, 10 µL of 10X PCR buffer, 1.6 µL of dNTP mix, and 82 µL of H$_2$O.

11. Formamide stop solution: 95% deionized formamide, 0.1% bromophenol blue, 0.1% xylene cyanol, and 10 mM EDTA, pH 7.0.

G A T C   G A T C

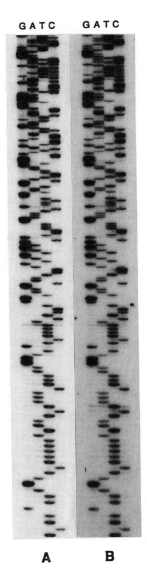

A   **B**

Fig. 2. DNA sequencing gel for DNA amplified directly from bacterial cells (**A**) or from purified DNA (**B**). A cloned 1.7-kb fragment containing the bovine corona-virus nucleocapsid protein gene *(5)* was subcloned into the pGEM-3Z vector (Promega, Madison, WI) and grown in *E. coli* strain JM 109. The reverse and forward (universal) primers for the pGEM system were used as primers 1 and 2, respectively, for asymmetric amplification and DNA sequencing. For (**A**) the toothpick method was used, and for (**B**) 5 ng of purified DNA was used as template in the asymmetric amplification reaction. Reprinted by permission from ref. *4*.

3. Methods
3.1. Asymmetric Amplification of Insert DNA

1. Prepare a 40-µL PCR reaction mix by adding together 33.2 µL of water, 4 µL of 10X PCR buffer, 0.8 µL of dNTP mix, 1 µL of 10 µM primer 1, and 1 µL of 0.1 µM primer 2. (10 pmol of primer 1 and 0.1 pmol of primer 2, the limiting primer, are used in this reaction. End-labeled primer 2 will be used in the sequencing reaction described in Section 3.3.)
2. With a sterile toothpick, add a few cells from an isolated bacterial colony into the reaction mix.
3. Heat for 15 min at 95°C, then chill on ice.
4. Add 10 µL (1.25 U) of *Taq* DNA polymerase in 1X PCR buffer.
5. Cover with two drops mineral oil.
6. Amplify by PCR using the following cycle profile:
   40 main cycles: 90°C, 30 s (denaturation)
                        50°C, 1 min (annealing)
                        72°C, 3 min (extension)

3.2. 5'-End Labeling
of the Sequencing Primer (Primer 2)

This is essentially the forward reaction as described by Sambrook et al. *(6)*.

1. Prepare a 20-µL end-labeling reaction mix by adding together 1 µL of 10 µM primer 2 (10 pmol), 11 µL of water, 2 µL of 10X kinase buffer, 5 µL of [γ-^{32}P]ATP, 1 µL (10 U) of T4 polynucleotide kinase. This makes enough for 10 sequencing reactions.
2. Incubate at 37°C for 30 min, then add 30 µL of H$_2$O to make a total volume of 50 µL.
3. Purify end-labeled primer by passing it through a BioSpin column that has been equilibrated with water according to manufacturer's instructions.
4. Estimate the volume of eluate and use 0.1 vol (x µL used in Section 3.3., step 1 below) (1 pmol of radiolabeled primer) in the sequencing reaction below. One microliter of the eluate can be counted to determine the specific activity of the radiolabeled primer. Approximately 10^6 cpm Cerenkov counts/pmol primer is needed.

3.3. Sequencing Reaction
with the End-Labeled Primer

This is essentially as described by Innis et al. *(3)*.

1. Prepare a 12-µL primer mix by adding together 1.2 µL of 10X PCR buffer, x µL (1 pmol, 10^6 cpm Cerenkov counts) of end-labeled primer 2, 10.6 µL-x µL of water, and 0.2 µL (1.25 U) of *Taq* DNA polymerase.

2. Add 20 μL of PCR mix (from Section 3.1., step 6) containing single-stranded DNA, mix by pipetting 10 times.
3. Immediately distribute 7.5 μL into each of the G, A, T, and C reaction tubes, which contain 2.5 μL each of the respective termination mixes and mix by pipetting five times.
4. Heat the reactions at 72°C for 5 min.
5. Terminate the reactions by adding 4 μL of formamide stop solution. Store at –20°C.
6. For sequencing, the termination mix (10 μL) is heated at 100°C for 3 min and 3 μL/lane is loaded onto a DNA sequencing gel (*see* Note 2).

4. Notes

1. In the experiment shown in Fig. 2, the primers used for amplification were the reverse primer [primer 1; 5'-CACAGGAAACAGCTATGACC-3'] and the forward [universal] primer [primer 2; 5'-GTTGTAAAA-CGACGGCCAGT-3'] for the pGEM [Promega, Madison, WI] system. However, we have used many other primers successfully (ref. 7 and data not shown).
2. The concentration of dNTPs in the asymmetric PCR was kept low (20 μ*M* each) so that residual amounts would not interfere with dideoxynucleotide chain termination reactions (Innis et al., ref. 3). We have learned that when residual amounts do cause a problem (i.e., when short termination products cannot be seen on the sequencing gel), generally as a result of short (<200 nucleotides) inserts in the clone, two approaches can be used to solve this problem: (1) Additional cycles of the PCR (50–60 total) can be run to deplete the dNTPs and (2) The product from the asymmetric reaction (Section 3.1., step 6) can be passed through a BioSpin 6 column.

References

1. Gussow, D. and Clackson, T. (1989) Direct clone characterization from plaques and colonies by the polymerase chain reaction. *Nucleic Acids Res.* **17**, 4000.
2. Sandhu, G. S., Precup, J. W., and Kline, B. C. (1989) Rapid one-step characterization of recombinant vectors by direct analysis of transformed Escherichia coli colonies. *BioTechniques* **7**, 689–690.
3. Innis, M. A., Mayambo, K. B., Gelfand, D. H., and Brow, M. A. D. (1988) DNA sequencing of polymerase chain reaction-amplified DNA. *Proc. Natl. Acad. Sci. USA* **85**, 9436–9440.
4. Hofmann, M. A. and Brian, D. A. (1991) Sequencing PCR DNA amplified directly from a bacterial colony. *BioTechniques* **11**, 30,31.
5. Lapps, W., Hogue, B. G., and Brian, D. A. (1987) Sequence analysis of the bovine coronavirus nucleocapsid and matrix protein genes. *Virology* **157**, 47–57.

6. Sambrook, J. E., Fritsch, E. F., and Maniatis, T. (1989) *Molecular Cloning: A Laboratory Manual* (2nd ed.), Cold Spring Harbor Laboratory, Cold Spring Harbor, NY.

7. Hofmann, M. A. and Brian, D. A. (1991) The 5-prime end of coronavirus minus-strand RNAs contain a short poly(U) tract. *J. Virol.* **65,** 6331–6333.

CHAPTER 21

Use of Polymerase Chain Reaction to Screen Phage Libraries

Lei Yu and Laura J. Bloem

1. Introduction

Isolating a clone from a cDNA or genomic library often involves screening the library by several rounds of plating and filter hybridization. This is not only laborious and time-consuming, but also is prone to artifacts such as false positives commonly encountered in filter hybridization. These problems can be alleviated by using polymerase chain reaction (PCR) in the early rounds of screening prior to conventional filter hybridization. The advantages of PCR screening are threefold: (1) Positive clones are identified by DNA bands of correct sizes in gel, thus avoiding the confusion from false positive spots in filter hybridization; (2) it saves time, especially in initial rounds of screening; and (3) screening of multiple genes can be performed in the same PCR by using appropriate primers for these genes. After the complexity of the phage pool is reduced and the existence of true positives in the pool confirmed, individual clones can be isolated by conventional methods.

The basic method consists of three steps, as shown in Fig. 1:

1. Lifting membrane filters from library plates;
2. Rinsing off phage particles from the filters; and
3. Using the phage eluate for PCR.

After a positive signal is identified from a particular plate, another round of PCR screening may be carried out by repeating these steps

From: *Methods in Molecular Biology, Vol. 15: PCR Protocols: Current Methods and Applications*
Edited by: B. A. White Copyright © 1993 Humana Press Inc., Totowa, NJ

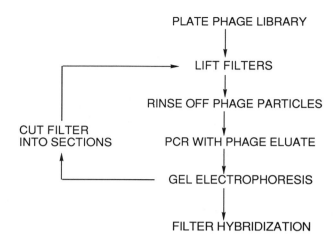

Fig. 1. Basic method of using PCR to screen phage libraries.

with smaller sectors of that plate to reduce further the size of the phage pool. Alternatively, another filter can be lifted from the positive plate and used in conventional filter hybridization.

2. Materials

1. Phage dilution buffer: 100 mM NaCl, 8 mM MgSO$_4$, 50 mM Tris-HCl, pH 7.5, and 0.1% gelatin, sterilize by autoclaving and store at room temperature.
2. Filter membranes (*see* Note 1).

3. Methods

See Notes 2–4.

1. Begin with plated phage cDNA or genomic library with appropriate host bacterial strain on 150-mm agar plates with top agarose using standard protocols (*1*). After the phage plaques have grown to the desired size at 37°C, chill the plates at 4°C for at least 1 h.
2. Place a nitrocellulose filter on the plate, making sure there is no air bubble trapped between the filter and the agar. The filter will be wet in a few seconds if the plate is fresh. For plates stored for a period of time and for later rounds in multiple lifting, longer time may be required to wet the filter completely.
3. Carefully lift the filter with a pair of flat-ended forceps, making sure not to rip the top agar layer. Place the filter, with the phage side down,

onto a sterile 150-mm Petri dish (either the bottom part or the lid) containing 3 mL of phage dilution buffer. Rinse off the phage particles by lifting the filter up and down a couple of times. Finally, lift the filter and let the solution drip for a while (usually 10–20 s is sufficient). Then discard the filter.

4. Transfer 20 µL of phage eluate to a PCR tube, close the cap tightly, place in a boiling water bath for 5 min, chill on ice, and use as template in a 100 µL PCR (*see* Chapter 1 for conditions of the standard PCR and Notes 5 and 6). Save some phage eluate in an Eppendorf tube if you consider doing more PCR later on (*see* Note 7).

5. Analyze the PCR products by agarose gel electrophoresis (*see* Note 8). If a plate has one or more positives, its corresponding lane on the gel will have a band of expected size. This plate is then considered a positive plate.

6. At this point, one can take the positive plates and proceed to screen them for positive phage plaques by conventional filter hybridization. Alternatively, one can carry out another round of PCR screening as described herein to reduce further the complexity of the phage pool on a positive plate.

7. Put a new nitrocellulose filter on a positive plate (*see* Note 9). With a sterile scalpel, cut the filter and the agar underneath it into several sections. Sections can be any number or shape, although we found that the radial-shaped (pie-shaped) sections between four and eight per plate can be handled easily.

8. Lift each filter sector and rinse in phage dilution buffer as in step 2. Owing to the smaller size of the filter sector, rinsing can be done either on smaller Petri dishes such as the 100-mm ones, or on a piece of plastic film such as Saran Wrap™. An aliquot of 0.2–0.5 mL of phage dilution buffer is used for rinsing depending on the size of the filter sector.

9. Do PCR with the section eluate as in step 4.

10. After a sector containing the positive phage signal is identified by PCR, another membrane filter can be lifted from this sector and used in conventional filter hybridization. Alternatively, the top agar of this sector can be scraped off for replating as described in the following.

11. To replate the phage from a positive section, use a rubber policeman to scrape off the top agar into a 50–mL conical tube containing 20 mL of phage dilution buffer. The rubber policeman should be rinsed extensively with sterile water between each scraping to prevent cross-contamination among sections.

12. The tube is vortexed vigorously and the top agar is soaked for 2 h to overnight. The tube is then centrifuged at 8000*g* for 15 min to pack the

agar. Save a portion of the supernatant in an Eppendorf tube and discard the conical tube containing the agar. Determine the phage titer and plate at desired density for another round of PCR screening or conventional filter hybridization.

4. Notes

1. Different brands of noncharged membrane all work well for this procedure, be they nitrocellulose or nylon. However, we have experienced difficulty with charged membranes, presumably because the charges on the membrane interfere with rinsing off of the phage particles.
2. The main advantage of using PCR to screen a phage library is to reduce complexity before starting conventional filter hybridization. Multiple plates of a library can be processed and screened by the method described here. To screen one million phage plaques, for example, plating at 50,000 PFU/plate will give 20 plates. Although considerable time and effort would be needed to screen these plates by conventional filter hybridization, they can be easily screened in a few hours by PCR screening. Furthermore, by cutting a positive plate into eight sections and performing another round of PCR screening, positive phage plaques can be located to a pool of approx 6250 PFU. This pool can be readily handled by conventional filter hybridization.
3. Before screening a library, it is highly recommended to do a PCR with an aliquot of the whole library. This serves two purposes: to ensure that the desired clone is present in the library and to ascertain that the conditions used for PCR can amplify the expected signal. It can be done by heat denaturing 5 µL of the library and using it in a 100-µL PCR. It is also prudent to sequence the amplified fragment to confirm its identity and the specificity of PCR conditions. This can be done either before or in parallel with the PCR screening of the library.
4. Another useful thing to do before screening the library is to estimate the abundance of the desired clone in the library. This can be combined with the whole library amplification by PCR (*see* Note 3) by amplifying various library dilutions in addition to the undiluted library. If the desired clone is present in a pool of 20,000 clones, for example, it would not be necessary to screen one million clones.
5. From time to time, PCR may fail with no apparent reason. To prevent such an occasional "system failure" to be interpreted as being negative, it is desirable to include controls in PCR. Two types of controls may be used: internal and parallel. For internal control, primers can be included in the same PCR tube that will amplify vector sequence, a stretch of the λ arm if the library is in a λ vector. Alternatively, the library can be

spiked with a known clone and the primers for this clone can be used in the same PCR. If an internal control is not desired, a PCR to amplify a known clone in an adjacent tube can be used as a parallel control.

6. In simultaneous screening of multiple genes, several pairs of primers can be used in the same PCR, provided that the expected PCR products are of different lengths and can be distinguished from one another on a gel. The major limitation of this technique is the need for sequence information for making primers before library screening can be undertaken. Nondegenerate primers based on definitive sequence information are best for PCR screening. It may be possible to use degenerate primers to search for related genes, although we have not tried this approach.

7. Because the phage eluate can be stored, the same set of plates can be used for several screenings, either for the same gene or for different genes.

8. If screening for a small fragment, it may be necessary to amplify more than the standard 30 rounds. For a 200-bp fragment, we amplified 35 rounds to achieve a high enough DNA concentration for band visualization on a 3% agarose gel.

9. Because of the sensitivity of PCR, the plated library can be screened several times without losing the positive signal. Also, plates can be stored at 4°C for several weeks before subsequent lifts. We have made five lifts from the same plates, both fresh and after 3-wk storage, and the phage eluate amplified well. The same goes with the phage eluate, which can be stored for several weeks or longer before amplification.

Reference

1. Sambrook, J., Fritsch, E. F., and Maniatis,T. (1989) *Molecular Cloning: A Laboratory Manual* (2nd ed.), Cold Spring Harbor Laboratory, Cold Spring Harbor, NY.

CHAPTER 22

Molecular Cloning of Polymerase Chain Reaction Fragments with Cohesive Ends

Beverly C. Delidow

1. Introduction

The transfection of expressible DNA into cell lines has allowed studies of gene expression and interaction that are often very difficult or impossible using the endogenous genes. The importance of this method is highlighted by the observation that, in a recent issue of *Cell*, 10 out of 14 articles reported data obtained through DNA transfer techniques *(1)*. The production of DNA constructs for transfection requires the availability of cloned DNA sequences including promoters, enhancers, transcribed sequences, and 3' elements. Many investigations also use constructs in which promoters are recombined with heterologous reporter genes, such as luciferase or chloramphenicol acetyl transferase.

Before the development of polymerase chain reaction (PCR) techniques, the easiest method for recombination of DNA fragments was to take advantage of convenient restriction sites in the component DNAs. However, in cases where convenient restriction sites were lacking, recombination required generation of larger fragments, which were then subjected to exonuclease deletion, followed by either blunt ligation of two fragments or of linkers to allow cohesive ligation into a vector. Although useful for producing series of deletion mutants, this technique is both time-consuming and difficult.

From: *Methods in Molecular Biology, Vol. 15: PCR Protocols: Current Methods and Applications*
Edited by: B. A. White Copyright © 1993 Humana Press Inc., Totowa, NJ

In contrast to this extensive manipulation, the ability to generate specific DNA fragments by PCR has allowed this technique to be used to great advantage in the cloning, manipulation, and recombination of specific DNAs. PCR techniques now exist for the generation of DNA libraries, and for generating hybrid DNAs and site-directed mutagenesis (*see* Chapters 27–29).

PCR also provides a means of generating a desired DNA fragment with cohesive ends for easy cloning into a recipient construct. This is accomplished by extending the 5' and 3' oligonucleotide primers to include restriction sites at their 5' ends. We have used this technique to insert the adenovirus major late promoter (MLP) in front of a promoter-less prolactin (PRL) minigene, for use in RNA expression studies. The general scheme of the method is outlined in Fig. 1. The insert and vector may be prepared simultaneously, requiring 2 d. The ligation is carried out during the second night. Bacteria are transformed on the third day and allowed to grow overnight. On the fourth day (and fifth, if required), colonies are screened for the desired transformants by a miniprep procedure. Thus, the entire procedure from DNA preparation to colony screening can be completed in 4–5 d.

2. Materials

2.1. Preparation of PCR-Generated Cohesive-End Insert

1. Source of insert sequence to be amplified. In this case, it is the plasmid pML-SK+ containing the adenovirus MLP sequences.
2. Oligonucleotide primers (*see* Note 1). The primers used for amplifying the MLP are 29-mers containing *Kpn* I sites at their 5' ends for insertion into a convenient *Kpn* site in the vector. The primer sequences are: *Kpn*-MLP-UP, 5'-AA*GGTAC*↓C T TCCGC GGTCC TTCGT ATAGA-3'; and MLP-*Kpn*-DN, 5'-AA*GGTAC*↓C A ACAGC TGGCC CTCGC AGACA-3' (*see* Note 2).
3. Reagents for PCR (*see* Chapter 1): Store at –20°C.
4. DNA grade agarose for electrophoresis (low-melting point [LMP] agarose is optional).
5. E buffer, for running agarose gels (40X stock): $1.6M$ Tris base, $0.8M$ anhydrous sodium acetate, 40 mM EDTA. Adjust pH to 7.9 with glacial acetic acid and 0.2-µm filter. To make 1X buffer, dilute 25 mL of stock to 1 L in distilled water. Store at room temperature.
6. 6X agarose gel loading dye: 0.25% bromophenol blue, 0.25% xylene cyanol, 30% glycerol. Prepare in sterile water and store at room temperature.

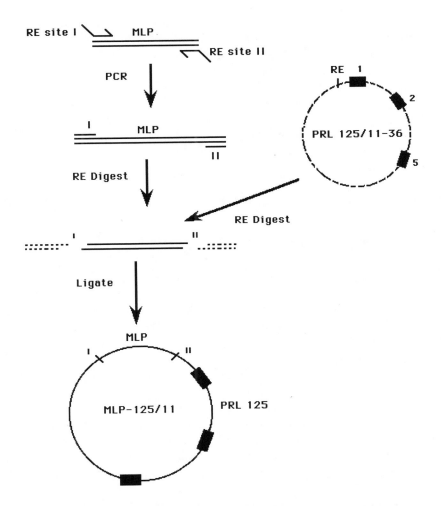

Fig. 1. General scheme for inserting the adenovirus MLP (solid lines) into the plasmid bearing the minigene PRL 125/11-36 (dashed lines). The MLP sequences are amplified using primers containing recognition sites for the restriction enzyme (RE) *Kpn* I (RE sites I and II). The amplified insert DNA and the plasmid are then cut with *Kpn* I (RE digest) and purified. The cohesive ends of the plasmid and insert are joined in a ligation reaction to produce the recombinant construct, MLP-125/11.

7. DNA markers. Several are available. We routinely use a *BstE* II digest of lambda DNA (New England Biolabs, Beverly, MA). This preparation contains 14 DNA fragments, ranging from 8454–117 bp. Store at –20°C.
8. Ethidium bromide (10 mg/mL) in sterile water. Store at 4°C in a dark container. **Ethidium bromide is a potent mutagen. Use a mask and**

gloves when weighing powder. Clean up spills immediately. Wear gloves when handling solutions. Dispose of wastes properly.

9. Restriction enzymes *Hind* III and *Kpn* I, including the manufacturer-recommended buffers (Bethesda Research Laboratories [BRL, Gaithersburg, MD]). Store at –20°C.
10. Buffer-saturated phenol (*see* Chapter 1).
11. CHCl$_3$.
12. 7.5M Ammonium acetate.
13. 3M Sodium acetate, pH 5.2.
14. 95% Ethanol.
15. 70% Ethanol.
16. Sterile water. 0.2-µm filter deionized, distilled water. Store at room temperature.

2.2. Preparation of Recipient Vector

1. Appropriate vector for cloning. The promoter-less PRL minigene PRL-125/11-36 contains 36 bp of 5' flank, 3.5 kb of transcribed sequences, including exons 1, 2, and 5, introns A, B, and part of D, and approx 3 kb of 3' flank. The parent plasmid is pBR322, which confers ampicillin resistance. The *Kpn* I site at position –36 is the site for insertion of the MLP. This construct was generously provided by Dr. Z. D. Sharp (University of Texas Health Science Center).
2. Restriction enzyme *Kpn* I, with the recommended buffer (BRL). Store at –20°C.
3. Bacterial alkaline phosphatase (100 U/µL). Store at –20°C.
4. 1M Tris-HCl, pH 8.0. 0.2-µm filter and store at room temperature.
5. 0.1M ZnCl$_2$. 0.2-µm filter and store at room temperature.
6. 10% SDS. 0.2-µm filter and store at room temperature.
7. Proteinase K. Resuspend in sterile water at a concentration of 20 mg/mL. Store at –20°C.
8. Agarose gel electrophoresis reagents (*see* items 4–8 in Section 2.1.).
9. GeneClean DNA purification kit, Bio 101 (La Jolla, CA) (optional).
10. Reagents for extractions and precipitations (*see* items 10–14 in Section 2.1.).

2.3. Ligation and Bacterial Transformation

1. Prepared vector and insert, each at a concentration of 50–200 ng/µL.
2. T$_4$ DNA ligase, 1 U/µL, with the supplied buffer (BRL). Store at –20°C.
3. 10 mM ATP, in sterile water. Store at –20°C.
4. Competent bacteria: *E. coli*, DH5α (BRL). Store at –70°C (*see* Note 3).
5. TY broth: Prepare by dissolving 16 g of Bacto-tryptone, 10 g of yeast extract, and 5 g of NaCl/L of H$_2$O in an autoclavable flask. Autoclave for 20 min on liquid cycle. Cool and store refrigerated.
6. Ampicillin: Prepare a stock solution of 50 mg/mL in sterile water, store at 4°C.

7. LB agar plates containing ampicillin. Prepare by dissolving 10 g of Bacto-tryptone, 5 g of yeast extract, 10 g of NaCl, and 15 g of agar in 1 L of H_2O in a 2.8-L Fernbach flask. Heat to near boiling, while stirring to dissolve the agar. Autoclave for 20 min on liquid cycle. Allow the solution to cool to 50–55°C before adding ampicillin to 50 µg/mL (1 mL of a 50 mg/mL stock solution). Pour the still molten agar into sterile Petri dishes, using 20–25 mL/plate. If the plates are poured in a sterile hood, they may be allowed to cool uncovered until the agar has set. Cover, invert, and store refrigerated. Antibiotic-containing agar has a shelf-life of about 1 mo. If longer storage is required, prepare agar plates without antibiotic and spread with ampicillin (25 µL of a 50 mg/mL stock) just before use.

2.4. Plasmid Minipreps
for Screening Transformant Colonies

1. TY broth (item 5 in Section 2.3.) containing 50 µg/mL ampicillin.
2. 80% glycerol (v/v, in water), sterile-filtered. Store at room temperature.
3. GTE buffer: 25 mM Tris-HCl, pH 8.0, 50 mM glucose, 10 mM EDTA. Sterilize by 0.2-µm filter and store at 4°C.
4. Lysis buffer: 0.2M NaOH, 1% SDS. Prepare fresh.
5. KOAc buffer: 3M potassium acetate, 2M acetic acid. Sterilize by 0.2-µm filter and store at room temperature.
6. TE, pH 8.0: 10 mM Tris-HCl, pH 8.0, 1 mM EDTA. Sterilize by 0.2-µm filter and store at room temperature.
7. Restriction enzymes *Sst* II and *Xba* I, with the recommended buffer (BRL). Store at –20°C.

3. Methods
3.1. Preparation of PCR-Generated
Cohesive-End Insert

1. Linearize the plasmid pML-SK$^+$ with *Hind* III (*see* Note 4). To 5 µg of plasmid, add 5 µL of 10X React 2 buffer and 1 µL (5–10 U) of *Hind* III in a total volume of 50 µL. Incubate for 1 h at 37°C.
2. Prepare five standard 100-µL PCR reactions each containing 200 ng (2 µL) of linearized plasmid, 100 pmol of each restriction site-tagged primer, 200 µM of each deoxynucleotide and 2.5 U of *Taq* polymerase in 1X PCR reaction buffer (*see* Chapter 1). Cover the reaction mix with a drop of light mineral oil to prevent evaporation.
3. Amplify by PCR using the following cycle profile (*see* Note 5):

17–25 main cycles	94°C, 1 min
	55°C, 2 min
	72°C, 1 min
Final extension	72°C, 5 min

4. Add 200 µL of $CHCl_3$ to the amplified DNA. This causes the mineral oil to sink to the bottom in the organic phase. Spin briefly in a microfuge at top speed to ensure phase separation and transfer the upper phase to a clean 1.5-mL microfuge tube.

5. Precipitate the DNA by addition of 0.5 vol (50 µL) of $7.5M$ ammonium acetate plus 2.5 vol (375 µL) of 95% ethanol. Incubate for 10 min at room temperature, or for 30 min on ice (*see* Note 6). Collect the DNA by centrifugation in a microfuge at top speed (12,000 rpm) for 15 min.

6. Decant the supernatants. Add 50 µL of 70% ethanol gently to the side of each tube and allow to flow over the pellets. Spin again for 1–2 min at top speed in a microfuge, then decant the supernatant and place the inverted tubes on a clean lab wipe to drain dry (5 or 10 min should be sufficient). Resuspend each pellet in 22 µL of sterile water (*see* Note 6).

7. To confirm the size of the amplified DNA (approx 300 bp), take 2 µL of each sample to run on a 1% agarose gel. To each 2-µL aliquot, add 8 µL of sterile water and 2 µL of 6X loading dye. Mix well and spin briefly in a microfuge to collect all the liquid. Prepare an aliquot of DNA markers (2 µg) in the same total volume. Perform electrophoresis as described in Chapter 1.

8. Digest each of the remaining 20 µL of DNA samples with *Kpn* I (20 U) in a total volume of 40 µL of the 1X supplied buffer at 37°C for 1 h.

9. Pool the cut samples (200 µL total volume) in a 1.5-mL microfuge tube, add an equal volume of phenol/chloroform (1/1) and mix well to extract. Spin the tube in a microfuge at top speed for 2 min.

10. Remove the upper aqueous phase to a clean tube and add an equal volume of $CHCl_3$. Again mix well and centrifuge to separate the phases. Remove the aqueous phase to a fresh tube once more and repeat the $CHCl_3$ extraction.

11. Transfer the final, clean aqueous phase to a new tube and precipitate with ammonium acetate (100 µL) and ethanol (750 µL) (*see* Notes 7 and 8).

12. Pellet the DNA by centrifugation in a microfuge at top speed for 15 min. Resuspend the pellet in 500 µL of sterile water and measure the A_{260} of this solution directly. Our reactions yielded a DNA content of 25 µg, or 50 ng/µL.

13. Run 5 µL of this DNA on a 1% agarose gel to confirm the estimated concentration and that the insert is intact. If gel analysis shows that there are other contaminating sequences present, the desired insert may be isolated by purification out of agarose gels (*see* Note 8).

14. Store the prepared insert at −20° until it is to be used.

3.2. Preparation of the Recipient Vector

1. Linearize 10 μg of the plasmid PRL-125/11-36 with *Kpn* I (10 U) at 37°C for 1 h, as for the insert (step 7 in Section 3.1.).
2. Inactivate the *Kpn* I by heating at 65°C for 10 min.
3. Remove the enzyme by phenol/CHCl$_3$ extraction, as described earlier (steps 8 and 9 in Section 3.1.).
4. Ethanol precipitate the DNA by addition of sodium acetate to 0.3M and 2.5 vol of 95% ethanol and incubate at –70°C for 30 min, or at –20°C overnight. Collect the DNA by centrifugation at top speed in a microfuge for 30 min at 4°C. Resuspend the pellet in 46 μL of sterile water (*see* Note 9). If phosphatase treatment is not necessary, the vector may be resuspended at a concentration of 100 ng/μL.
5. Because the restriction cut generates compatible ends, the vector must be treated with bacterial alkaline phosphatase (BAP), to remove 5' phosphates and reduce recircularization of the plasmid during the ligation reaction. To 46 μL of plasmid, add 2.5 μL of 1M Tris-Cl, pH 8.0, 0.5 μL of 0.1M ZnCl$_2$ and 1 μL of BAP. Incubate at 60°C for 30 min.
6. To remove the BAP, add 1 μL of 10% SDS and 2.5 μL of proteinase K (20 mg/mL), and incubate at 37°C for 30 min.
7. Add 35 μL of sterile water and 10 μL of 3M sodium acetate (100 μL total volume), then extract once with phenol/CHCl$_3$ and twice with CHCl$_3$, as described earlier (steps 8 and 9 in Section 3.1.).
8. Precipitate the DNA by adding 2.5 vol (250 μL) of 95% ethanol and incubating for 30 min at –70°C or overnight at –20°C (*see* Note 6). Recover the DNA by centrifugation, and resuspend the pellet at a concentration of approx 100 ng/μL.
9. Run a sample of the prepared vector on an agarose minigel to confirm that cutting was complete. If the cut was incomplete or if a piece of DNA larger than approx 100 bp was removed, the linearized vector may be gel-purified out of agarose using the GeneClean kit (BIO 101) (*see* Note 10).
10. Store the prepared vector at –20°C until it is to be used.

3.3. Ligation and Bacterial Transformation

1. Combine the insert and vector in several ratios (I:V): 1:1, 3:1, 5:1, and 0:2 (control) in 1X ligation buffer plus 1 μL ATP in a volume of 29 μL. The total amount of DNA added should be approx 0.5–1 μg. Add 1 μL (1 U) of T$_4$ DNA ligase (30 μL final volume) and incubate overnight at 16°C (*see* Note 11).
2. The entire ligation reaction may be used to transform competent *E. coli* DH5α. Thaw frozen competent bacteria on ice. To the ligation prod-

ucts (30 µL) in a 1.5-mL microfuge tube, add 100-µl of competent cells. As controls for the efficacy of the antibiotic and the efficiency of transformation, include samples of bacteria transformed with 30-µL of TE buffer and with 200 ng of intact plasmid, respectively. Tap the tubes gently to mix and incubate on ice for 30–45 min.

3. Heat-shock the bacteria by incubating the tubes in a 42°C water bath for 2 min, then place them on ice for several minutes. *See* Note 12.
4. Transfer the bacteria to a large tube (at least 15-mL capacity) containing 2 mL of 2X TY broth, *without* selection antibiotic, and grow for 1 h at 37°C, with vigorous shaking (225 rpm).
5. Spread several aliquots (50–200 µL) of each sample of the transformed bacteria onto separate LB agar plates containing 50 µg/mL ampicillin. Cover and invert the plates and incubate them overnight at 37°C.
6. Reserve any remaining cells in TY broth at 4°C overnight. Should no growth occur on the plates, these cells may be pelleted gently (2 min at half speed in a microfuge) and resuspended in 100–200 µL of TY for plating onto fresh plates.

3.4. Plasmid Minipreps
for Screening Transformant Colonies (5)

1. The plates containing bacteria transformed with TE should be free of colonies; those containing bacteria transformed with the intact plasmid should have many colonies. This assures you that both the transformation and the antibiotic are working.
2. The plates containing cells transformed with the linearized, dephosphorylated vector only (0:2 control) should have very few or no colonies. If these plates have many colonies, it indicates that the vector was either inefficiently linearized, or not completely dephosphorylated. A new aliquot of vector should be prepared to try again.
3. The plates containing cells transformed with the ligated constructs should contain many more colonies than those of cells transformed with linearized, dephosphorylated vector. Some of these colonies will bear recombinant plasmids and some will probably bear the recircularized vector. These colonies may be screened by direct PCR sequencing (*see* Chapter 20) or by preparation of plasmid DNA by the following miniprep procedure.
4. Touch a sterile loop or toothpick to each colony to be screened and inoculate each into a 2-mL aliquot of 2X TY broth containing 50 µg/mL ampicillin. It is convenient to do minipreps of sample numbers corresponding to multiples of the number of spaces available in the lab microfuge.
5. Incubate at 37°C, with shaking (225 rpm), for 4–16 h, until the broth becomes moderately turbid from bacterial growth.

6. Transfer 0.4 mL of each sample into a 1.5-mL microfuge tube containing 0.1 mL of 80% glycerol. Mix and freeze on dry ice. Store these aliquots at –70°C as the future source of the desired transformants.

7. Transfer the remainder of each sample to another 1.5-mL tube, and centrifuge at 6000 rpm for 2 min in a table-top microfuge. Aspirate the supernatants. Leave the caps open for the next two steps.

8. Resuspend each pellet in 50 µL of GTE buffer by vortexing. Incubate at room temperature for 5 min.

9. Add 100 µL of Lysis buffer to each tube while vortexing. Incubate on ice for 5 min.

10. Add 75 µL of KOAc buffer to each tube, close the caps, vortex to mix well, and incubate on ice for 5 min.

11. Centrifuge at top speed (12,000 rpm) for 10 min in a table-top microfuge.

12. Transfer the supernatants to fresh 1.5-mL tubes containing 0.5 mL of 95% ethanol. Mix well and centrifuge again at top speed for 10 min.

13. Decant the supernatants and air-dry the DNA pellets for 20–30 min. Resuspend the DNA in 50 µL of TE, pH 8.0.

14. To determine which plasmids contain the desired insert, the DNAs may be digested with the restriction enzymes used for cloning, or with combinations of enzymes to aid in determining the orientation of the insert. The orientation of the MLP in the *Kpn* I site of pPRL-125/11-36 is determined by restricting miniprep DNA with *Sst* II, which cuts at the 5' end of the MLP, and *Xba* I, which cuts at position +490 within the PRL minigene. Because *Sst* II and *Xba* I use the same buffer (React 2, BRL), these digestions may be carried out simultaneously. For each sample to be analyzed, combine 3–5 µL of miniprep DNA with 2 µL of 10X reaction buffer, 1 µL of *Sst* II and 1 µL of *Xba* I in a total volume of 20 µL. Incubate at 37°C for 30–60 min.

15. Analyze the digested DNA samples by agarose gel electrophoresis (*see* Chapter 1). The restriction cuts produce an 850-bp fragment if the promoter is in the forward orientation, or a 550-bp fragment if it is reversed. Plasmids without the MLP will be linearized at the *Xba* I site and will appear as bands of about 9 kb (*see* Fig. 2).

4. Notes

1. The spacer of several bases is required by many restriction enzymes to allow efficient cutting at the end of a piece of DNA (2,3; *see* Chapter 2).

2. It is possible to use different restriction sites in the two primers, thereby generating a PCR product that may be cloned into a compatibly cut vector in only one orientation. With directional cloning it is not necessary to treat the vector with phosphatase before the ligation because it cannot recircularize.

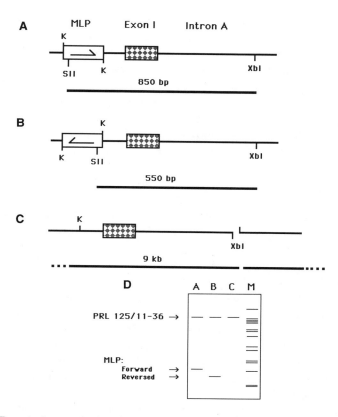

Fig. 2. Restriction analysis of recombinant plasmids produced by ligation of *Kpn* I-tagged MLP into PRL 125/11-36. In **A, B,** and **C** the three possible forms of the plasmid DNAs are diagrammed, with the fragments generated by restriction digestion with *Sst* II (SII) and *Xba* I (XbI). The *Kpn* I insertion site is also indicated (K). The *Sst* II site is 5 bp from the 5' end of the MLP (13 bp from the 5' end of the *Kpn*-tagged MLP). The *Xba* I site is at position +490, within the first intron of the PRL minigene. The MLP and its orientation are indicated by the box containing an arrow. The first exon of the PRL minigene is represented by the checkered box. **A.** Insertion of the MLP in the forward (desired) direction. Restriction digestion of this construct produces two DNA fragments, one containing the 300-bp MLP plus approx 550 bp of the PRL minigene (approx 850 bp), and one containing the majority of the plasmid (approx 8.5 kb). **B.** Insertion of the MLP in the reverse orientation produces a plasmid fragment and one of approx 550 bp. **C.** If the insertion of the MLP fails and PRL 125/11-36 recircularizes, *Sst* II will not digest the DNA and the plasmid will be linearized by *Xba* I, producing a single band of about 9 kb. **D.** Schematic drawing of 1% agarose gel electrophoresis of the DNA fragments generated in **A–C**. The plasmid bands (8.5–9 kb) in lanes **A–C** appear to comigrate because these large fragments are not resolved on 1% agarose under the gel conditions required to see the much shorter insert fragments. The lane marked M refers to DNA markers, a *BstE* II-digest of 1 DNA.

3. Competent *E. coli* DH5α bacteria may be purchased in small volumes from BRL. Alternatively, bacteria may be grown up in quantity and made competent for transfection by several methods *(4,5)*. We use a CaCl$_2$ procedure *(4)*.

4. Note that the enzyme used to linearize the plasmid must not cut within the region to be amplified. If no sites outside this region are available, it is possible to use a plasmid template for PCR without linearizing it. If an intact plasmid is used, the first cycle of PCR should begin with a slightly longer denaturation step (2–5 min) to ensure access of the primers to the template. The remaining cycles are run with a 1-min denaturation time.

5. Since *Taq* polymerase has no 3' to 5' exonuclease editing function *(6)*, it is important not to over-amplify, which may introduce mutations. There are several other thermostable polymerases now available with higher fidelity than *Taq* polymerase *(6,7)*. However some polymerases with editing ability (e.g., Vent™ DNA polymerase, New England Biolabs) also remove the 3' ends of the primers (manufacturer's information). In restriction site-tagged primers, this activity may remove too large a part of the sequence complementary to the target DNA to allow efficient amplification. We used 25 cycles in this particular experiment, but in the interest of minimizing mutations, one should start with 17–20 cycles.

6. If necessary, both insert and vector DNA may be held at –20°C overnight at any step at which they are in ethanol, or in aqueous solution in the absence of enzymes.

7. Ammonium acetate/ethanol precipitation is useful because it does not precipitate very small DNAs, such as primers or end-linkers from restriction digestion. However, some ligase preparations have reduced activity in the presence of ammonium ions *(5)*. Therefore, it may be necessary to wash the DNA pellet with 70% ethanol, then resuspend it in water and precipitate again, using sodium acetate (to 0.3*M*) or NaCl (to 0.25*M*) as the salt. Alternatively, the insert DNA may be purified away from the cut ends out of solution *(see* Note 8).

8. If the insert DNA is greater than approx 200 bp in length it may also be purified out of agarose gel after electrophoresis, or out of solution, using the GeneClean kit by Bio 101. Because GeneClean does not recover very small pieces of DNA (100 bp or less) it also serves to remove the end-linkers generated by restriction cutting of insert or plasmid DNA. In our hands the recovery of DNA is very good by both these methods (at least 80%) and the DNA is clean enough for direct use in the cloning reactions. However, if the DNA is purified out of an ethidium-stained agarose gel, care must be taken to minimize the exposure of the DNA

to UV light (e.g., during excision of the band on a UV lightbox), which can nick nucleic acids.

9. As an alternative to phenol/chloroform extraction and precipitation, the cut vector may be purified away from both the enzyme and small DNA fragments by using the GeneClean kit as described earlier.

10. If both the vector and insert must be gel purified, the ligation step may be simplified by electrophoresing both on a gel of LMP agarose, combining small portions of the appropriate bands and carrying out the ligation in the LMP agarose *(4)*. To use LMP agarose, pour and load the gel as for regular agarose. To prevent melting during the electrophoresis, LMP gels may be run at low voltage or in a cold room.

11. Incomplete ligation reactions may be forced to completion after the overnight incubation by bringing the reactions to room temperature and adding 1 µL each of 10 mM ATP and ligase. Continue the incubation at room temperature for another 4 h and then use the DNA to transform competent bacteria.

12. Timing and temperature at this step are important. Heat shocking the cells for too long or at too high a temperature may reduce transformation efficiency or kill the cells.

Acknowledgments

Supported by NIH grant DK43064.

References

1. *Cell* (1992) **68,** no. 1, 1–76.
2. Kaufman, D. L. and Evans, G. A. (1990) Restriction endonuclease cleavage at the termini of PCR products. *BioTechniques* **9,** 304–306.
3. Crouse, J. and Amorese, D. (1986) Double digestions of the multiple cloning site. *Focus* **8,** 9.
4. Ausubel, F. M., Brent, R., Kingston, R. E., Moore, D. D., Smith, J. A., Seidman, J. G., and Struhl, K. (1987) *Current Protocols in Molecular Biology.* Wiley Interscience, New York.
5. Sambrook, J., Fritsch, E. F., and Maniatis, T. (1989) *Molecular Cloning. A Laboratory Manual.* Cold Spring Harbor Laboratory, Cold Spring Harbor, NY.
6. Eckert, K. A. and Kunkel, T. A. (1990) High fidelity DNA synthesis by the Thermus aquaticus DNA polymerase. *Nucleic Acids Res.* **18,** 3739–3744.
7. Mattila, P., Korpela, J., Tenkanen, T., and Pitkänen, K. (1991) Fidelity of DNA synthesis by the Thermococcus litoralis DNA polymerase—an extremely heat stable enzyme with proofreading activity. *Nucleic Acids Res.* **19,** 4967–4973.

CHAPTER 23

Rapid (Ligase-Free) Subcloning of Polymerase Chain Reaction Products

Alan R. Shuldiner and Keith Tanner

1. Introduction

The polymerase chain reaction (PCR) is a versatile, widely used method for the production of a very large number of copies of a specific DNA molecule *(1,2)*. For some applications, it is advantageous to subclone the PCR product into a plasmid vector for subsequent replication in bacteria *(3–6)*. Subcloning the PCR product into a plasmid vector has several advantages: (1) the amplified fragment can be sequenced with greater reliability, (2) only one allele is sequenced per clone, and (3) the vector containing the PCR product may be used for other molecular biological experiments, e.g., in vitro transcription, radiolabeling, and further amplification in bacteria.

Although conventional strategies such as blunt-end or sticky-end ligation can be very successful for subcloning most DNA fragments, the subcloning of DNA generated by PCR is often very difficult *(7–9)*. We describe a rapid and versatile method to subclone PCR products directionally into a specific site of virtually any plasmid vector *(8,9)*. Ligase-free subcloning of PCR products has several advantages over conventional strategies: (1) it does not require DNA ligase, (2) it requires only four primers, two of which are complementary to the plasmid vector, and therefore may be used repeatedly, and (3) it may be accomplished in a single day. Typically, hundreds to thousands of colonies per transformation are obtained (approximate efficiency 5×10^3 to 5×10^4 colonies/microgram of PCR insert). With this method, PCR products of up to 1.7 kb in length have been subcloned successfully.

From: *Methods in Molecular Biology, Vol. 15: PCR Protocols: Current Methods and Applications*
Edited by: B. A. White Copyright © 1993 Humana Press Inc., Totowa, NJ

1.1. Theory of the Method

1.1.1. Plasmid Linearization and Primer Design

The first step in ligase-free subcloning is to linearize the plasmid vector at the desired site with the appropriate restriction endonuclease(s). The second step is to perform PCR on genomic DNA or the cDNA of interest. Like conventional PCR, the primers (primers a and b in Fig. 1) must contain sequences at their 3' ends (approx 20–25 nucleotides) that are complementary to opposite strands of the target sequence at a predetermined distance from each other. For ligase-free subcloning, primers a and b must also contain sequences at their 5' ends, approx 24 nucleotides in length (designated the 5' addition sequences), that are identical to each of the 3' ends of the linearized plasmid (opened and closed boxes in Fig. 1; also, *see* example in Fig. 2). Since the two 3' ends of the linearized plasmid are different from each other, the PCR fragment may be subcloned directionally simply by choosing the appropriate 5' addition sequence for each primer.

1.1.2. PCR Amplification and Ligation by Overlap Extension

PCR amplification is accomplished using primers a and b, which results in large amounts of the PCR product containing the 5' addition sequences at each end. Next, the PCR product is freed from excess primers by ultrafiltration, and divided into two tubes each containing the linearized plasmid vector. A second PCR reaction is performed. Tube 1 also contains primer a, the same primer that was used during the initial amplification of the PCR product, and primer c, a primer that is complementary to the strand of the plasmid vector that contains the primer a addition sequence (the [+] strand in Fig. 1), in a region internal to the primer a addition sequence (Figs. 1 and 2). Similarly, tube 2 contains primer b, the same primer that was used for the initial amplification of the PCR product, and primer d, a primer that is complementary to the strand of the plasmid vector that contains the primer b addition sequence (the [–] strand in Fig. 1), in a region internal to the primer b addition sequence (Figs. 1 and 2).

During the first cycle of the second PCR, denaturation and annealing result in hybridization of the 3' ends of the PCR product (the complements of the 5' addition sequences) to the complementary 3' ends of the linearized plasmid (Fig. 1). During the extension step of the

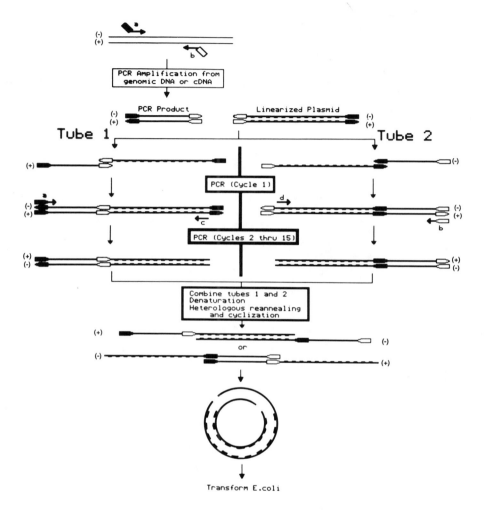

Fig. 1. Schematic of ligase-free subcloning. The 5' addition sequences of primers a and b are designated by closed and open boxers respectively. DNA sequences corresponding to the PCR-amplified product are shown as straight lines, whereas DNA sequences corresponding to the plasmid vector are shown as hatched lines.

first cycle, these overlapping 3' ends act as templates for each other. Extension results in the "ligation" of the PCR fragment to the linearized plasmid (ligation by overlap extension) *(10,11)*.

Amplification during subsequent PCR cycles (cycles 2–20) with the respective primers in each tube results in large amounts of linear double-stranded DNA that contain the plasmid vector ligated to the

Nucleotide sequence of pGEM4Z linearized with SmaI:

primer d

5'-GGGGATCCTCTAGAGTCGACCTGCaggcatgcaagcttgtctccc**tatagtgagtcgtattagagc**...(3 Kb)...ttggatttaggtgacactatag**AATACGAATTCGAGCTCGGTACCC**-3'

primer a

3'-**CCCCTAGGAGATCTCAGCTGGACG**tccgtacgttcgaacagaggatatcacctcagcataatctcg...(3 Kb)...**aacgtaaatcgactgatatc**TTATGCTTAAGCTCGAGCCATGGG-5'

primer b   primer c

5' addition sequences should be identical to the 3' ends (upper case bold and underlined nucleotides):

Primer a: 5'-AATACGAATTCGAGCTCGGTACCC.....3'

Primer b: 5'-GCAGGTCGACTCTAGAGGATCCCC.....3'

Primers c and d should be complementary to opposite strands of the plasmid internal to the 5' addition sequences
(lower case bold and underlined nucleotides):

Primer c: 5'-tatagtgtcacctaaatccaa-3'

Primer d: 5'-tatagtgagtgctattagagc-3'

(Note that primers c and d do not need to be immediately adjacent to primers a and b.)

Fig. 2. Example of primer design for ligase-free subcloning into the *Sma* I site of pGEM4Z (Promega Biotec, Madison, WI).

PCR insert at one end (tube 1 in Fig. 1), or to the other end (tube 2 in Fig. 1). Although not shown in the figure, hybridization may also occur between the 5' ends of the PCR product and the complementary 5' ends of the linearized plasmid. However, these hybrid products cannot act as templates for *Taq* DNA polymerase.

1.1.3. Denaturation, Heterologous Annealing, and Cyclization

After the second PCR reaction, the reaction mixtures in tubes 1 and 2 are combined, and denaturation of double-stranded DNA into single-stranded DNA is accomplished by treating with alkali. After neutralization and dilution of the denatured DNA, the single-stranded DNA may anneal into several different products. Complementary DNA strands may reanneal to form the same double-stranded products that were present in tubes 1 and 2 before the denaturation step (not shown in Fig. 1). Alternatively, single-stranded DNA products from tube 1 may anneal with single-stranded DNA products from tube 2 resulting in DNAs that are partly double-stranded, and contain long single-stranded 5' or 3' overhangs (heterologous annealing in Fig. 1). Since these long 5' or 3' overhangs are complementary to each other, annealing at low DNA concentrations results in cyclization.

Although the cyclized product contains two nicks (noncovalently linked ends), it may be used directly to transform competent *E. coli*. Once inside the bacterial cell, the two nicks are covalently joined, and the recombinant plasmid is replicated.

2. Materials

1. Plasmid vector.
2. Appropriate restriction endonuclease(s) and buffer(s) for linearization of the plasmid vector.
3. Oligonucleotide primers *(see* Note 1). Prepare the four oligonucleotide primers required for ligase-free subcloning by synthesizing chemically with the last dimethoxytrityl (DMT) group off. Cleave the oligonucleotide from the resin by treatment with concentrated ammonia at room temperature, and deprotect by heating to 55°C in ammonia overnight *(see* Chapter 1). The oligonucleotides may be used for PCR after desalting through a G-25 spin column (Boehringer Mannheim Biochemicals, Indianapolis, IN). Alternatively, the last DMT group may be left on, and the oligonucleotides may be purified with NENsorb Prep columns (New England Nuclear, Boston, MA) according to the manufacturer's directions.

4. 10X GeneAmp PCR buffer: 500 m*M* KCL, 100 m*M* Tris-HCl, pH 8.3 at 25°C, 15 m*M* MgCl₂, and 0.1 mg/mL of gelatin. GeneAmp is a trademark of Perkin-Elmer/Cetus (Norwalk, CT).
5. dNTPs: 10 m*M* each.
6. *Taq* polymerase.
7. Paraffin oil.
8. Autoclaved, distilled, deionized water.
9. Phenol equilibrated in TE buffer (10 m*M* Tris-HCl, pH 8.0 at 25°C, 1 m*M* EDTA).
10. Chloroform:isoamyl alcohol: 24:1, v/v.
11. tRNA: 10 mg/mL in sterile water.
12. Sterile, 5*M* NaCl.
13. Autoclaved sterile Eppendorf tubes.
14. Vacuum dessicator.
15. Ultrafree-MC-100 and Ultrafree-MC-30 ultrafiltration devices (Millipore, Bedford, MA) (*see* Note 2).
16. 50 m*M* Tris-HCl buffer, pH 6.6 at 25°C.
17. DNA size marker: *Hae*III digest of PhiX174.
18. 2*M* NaOH.
19. Competent DH5-alpha *E. coli* (Bethesda Research Laboratories, Gaithersburg, MD; theoretical transformation efficiency >1 × 10⁹ transformants/ μg PUC19).
21. LB agar plates with the appropriate antibiotic.
22. 10,000X ethidium bromide solution: 5 mg/mL in water. Keep in dark container at 4°C for up to 12 mo. Wear gloves when handling ethidium bromide.
23. High-melting-point agarose.
24. NuSeive GTG agarose (FMC Bioproducts, Rockland, ME).
25. 95% Ethanol.
26. 70% Ethanol.

3. Methods

3.1. Plasmid Linearization

1. Linearize 1–2 μg of the plasmid vector by digestion with the appropriate restriction endonuclease(s). Incubate the plasmid vector, the appropriate buffe,r and restriction enzyme(s) in a final volume of 20 μL at the appropriate temperature for 2–3 h in an autoclaved Eppendorf tube.
2. Add 30 μL of sterile water and 1 μL of tRNA (10 mg/mL) to the reaction mixture.
3. Extract by adding 25 μL of phenol that has been preequilibrated with TE buffer. Vortex. Add 50 μL of chloroform:isoamyl alcohol. Vortex.

Centrifuge in a microfuge at room temperature for 30 s to separate the phases. Remove the upper (aqueous) phase (approx 50 µL) and place in a clean autoclaved Eppendorf tube.

4. Repeat step 3 once.
5. Precipitate the linearized plasmid DNA by adding 12.5 µL of 5M NaCl and 125 µL of cold 95% ethanol. Incubate at –20°C for at least 3 h.
6. Recover the linearized plasmid DNA by centrifugation in a microfuge (12,000g) at 4°C for 30 min. Pour off the supernatant and wash the pellet with 250 µL of cold 70% ethanol. Carefully decant the supernatant, remove excess liquid from the walls of the Eppendorf tube with a cotton swab, and vacuum-dry the pellet in a vacuum dessicator or Speed-Vac for 10 min.
7. Resuspend the pellet in 20 µL of sterile water. Confirm that the plasmid was linearized and determine the approximate DNA concentration by running 10% (2 µL) of the linearized plasmid DNA on a 1% agarose gel with 100 ng of uncut plasmid and a known amount of a *Hae*III-digest of PhiX174 DNA size marker (*see* Chapter 1). DNA should be visualized by staining in ethidium bromide solution for 10–15 min at room temperature and inspection (or photography) during UV transillumination using appropriate precautions (*see* Note 3). The intensity of staining of the sample DNA and the size marker DNA may be compared to each other so that the amount of sample DNA can be roughly quantified.

3.2. PCR Amplification and Ligation by Overlap Extension

1. Assemble the PCR reaction in a final volume of 100 µL containing 1X GeneAmp PCR buffer, 200 µM of each dNTP, 100 nM each of primers a and b, 1.5 U/tube of *Taq* polymerase, and DNA template (*see* Note 4). Overlay with 50 µL of paraffin oil.
2. Amplify by PCR using the following cycle profile (*see* Note 5):

Initial denaturation	94°C, 5 min
25–35 main cycles of PCR	55°C, 1 min (annealing)
	72°C, 1 min (extension)
	94°C, 1 min (denaturation)
Final extension	72°C, 10 min

3. To assess the purity of the PCR product, remove 18 µL of the PCR reaction mixture, add 2 µL of 10X loading buffer, and perform electrophoresis on a composite gel consisting of 1% agarose and 2% NuSeive GTG agarose (*see* Chapter 1). Estimate the amount of DNA as described in step 7 in Section 3.1.
4. Add 320 µL of water to the remaining 82 µL of the PCR reaction mix-

ture and filter through an Ultrafree-MC-100 device by centrifugation at 2000*g* at room temperature until the retentate volume is about 20–25 μL (approx 3–5 min), being careful not to overfilter *(see* Note 6).

5. After the first ultrafiltration, add 400 μL of water to the retentate, and repeat ultrafiltration in the same device to a final retentate volume of approx 20–25 μL, again being careful not to overfilter.
6. Prepare a 100-μL PCR reaction mixture in two separate 0.5-mL Eppendorf tubes as follows:

	Tube 1	Tube 2
10X GeneAmp PCR buffer	10 μL	10 μL
dATP (10 mM)	2 μL	2 μL
dTTP (10 mM)	2 μL	2 μL
dCTP (10 mM)	2 μL	2 μL
dGTP (10 mM)	2 μL	2 μL
Primer a (5 μM)	2 μL	—
Primer c (5 μM)	2 μL	—
Primer b (5 μM)	—	2 μL
Primer d (5 μM)	—	2 μL
Linearized plasmid (50 ng/μL)	2 μL	2 μL
PCR product (50 ng/μL)	2 μL	2 μL
Water	73.6 μL	73.6 μL
Taq Polymerase (5 U/μL)	0.4 μL	0.4 μL
Total volume	100 μL	100 μL

Overlay with approx 50 μL of paraffin oil.
7. Amplify by PCR (20 cycles) as described in step 2 of Section 3.2., except increase the extension time of the main cycles to 1.5 min.

3.3. Denaturation, Heterologous Annealing, and Cyclization

1. After thermocycling is completed, start water boiling in preparation for steps 3 and 4 *(see* below).
2. Combine 15 μL of the PCR reaction mixture from tube 1 with 15 μL of the PCR reaction mixture from tube 2. Denature the DNA by adding 3.4 μL of 2M NaOH. Vortex and incubate for 5 min at room temperature.
3. During the incubation, transfer 400 μL of Tris-HCl buffer (50 mM, pH 6.6 at 25°C) to a 1.5-mL Eppendorf tube, and place in the boiling water bath.
4. Add the denatured PCR products from step 2 (this section) to the heated Tris-HCl buffer. Vortex briefly and place into the boiling water bath for 5 min.
5. Cool to 60°C over approx 15 min *(see* Note 7), and incubate at 60°C for 3–24 h in order to accomplish heterologous annealing and cyclization.

6. Following heterologous annealing and cyclization, allow the reaction mixture from step 5 to cool to room temperature, and place the entire contents in an Ultrafree-MC-30 device. Concentrate to a final volume of approx 10–20 µL by centrifugation at 2000*g* at room temperature (approx 3–5 min). Check the retentate volume frequently to avoid over-filtration and irreversible loss of DNA *(see* Note 6).

7. Transfer the concentrated cyclized product into an autoclaved 0.5-mL Eppendorf tube. The concentrated cyclized product may be stored indefinitely at –20°C for transformation at a future time, or may be used immediately. Transform 100 µL of competent DH5-alpha *E. coli* according to the manufacturer's directions with 1–5 µL of the cyclized product.

8. After transformation, plate the *E. coli* onto LB agar plates containing the appropriate antibiotic, and incubate overnight at 37°C. If the plasmid encodes beta-galactosidase (LacZ), colonies containing the appropriate recombinant plasmid may be chosen by blue-white selection on agar plates that also contain X-gal (50 µg/plate). Since this approach may be misleading *(see* Note 8), we recommend that plasmid minipreparations be prepared from several white colonies *(12; see* Chapter 22), and the presence and size of the DNA insert be confirmed by restriction endonuclease cleavage and gel electrophoresis. Alternatively, colony hybridization with the appropriate radiolabeled probe *(13)*, or PCR of plasmid DNA with primers that flank the DNA insert *(14)* may be used to select colonies with plasmids containing the desired DNA insert.

4. Notes

1. Once the plasmid vector has been linearized with the appropriate restriction endonuclease(s), the oligonucleotide primers required for ligase-free subcloning may be designed. Primers a and b should be approx 39–49 bases in length with 20–25 bases at their 3' ends that are complementary to opposite strands of the target DNA at a predetermined distance from each other (Fig. 1). In addition, primers a and b must also contain 19–24 nucleotides at their 5' ends (the 5' addition sequences) that are identical to each of the 3' ends of the linearized plasmid vector (Figs. 1 and 2). In most of our experiments, 5' addition sequences were arbitrarily chosen to be 24 nucleotides in length. However, 5' addition sequences 19 nucleotides in length may be used if the temperature of the second-stage PCR is decreased to 37°C. 5' addition sequences 14 nucleotides in length or less did not on a work reliably. Primers c and d should be approx 21 nucleotides in length, and be complementary to opposite strands of the plasmid vector internal to each of the 5' addition sequences (Figs. 1 and 2).

2. Centricon-100 and Centricon-30 ultrafiltration devices (Amicon, Danvers, MA) may be substituted for Ultrafree-MC-100 and Ultrafree-MC-30 devices respectively. When Centricon devices are used, sodium azide, which is used as a preservative, must first be removed from the membrane by treatment with sodium hydroxide according to the manufacturer's directions.
3. On a 1% agarose gel, the linearized plasmid will run slower than the uncut (supercoiled) plasmid. If digestion is complete, it is generally not necessary to gel-purify the linearized plasmid.
4. When genomic DNA is used as starting template, 0.1–0.5 µg/tube will suffice. If plasmid DNA is being used as starting template, 0.1–1 ng/tube is sufficient. If RNA is being used as starting template, reverse transcription must first be performed followed by PCR as outlined.
5. PCR conditions should be optimized for the specific application.
6. During centrifugation, frequently check the volume of the retentate since overfiltration can result in irreversible binding of DNA to the filter, and loss of the PCR product.
7. This is most conveniently done by using a 500-mL beaker filled approximately half-way with water as the boiling water bath, and transferring the entire beaker directly into a conventional water bath that has been preset at 60°C. This will result in cooling at the appropriate rate, and automatic equilibration at 60°C.
8. In our hands, the percentage of antibiotic-resistant colonies that contain the desired DNA insert varies from 50–100%. If primer-dimers form during the initial PCR, and are not removed prior to ligation by overlap extension, they contain the proper 5' addition sequences, and are subcloned into the vector with very high efficiency resulting in a higher background (i.e., colonies with recombinant plasmids that do not contain the desired DNA insert). If large quantities of primer-dimers are present, we recommend that the conditions of the PCR be modified to minimize these primer artifacts (i.e., decrease the primer concentration and/or decrease the amount of *Taq* polymerase). Alternatively, the PCR product may be gel-purified prior to ligation by overlap extension.

Acknowledgments

We thank Jesse Roth, Charles T. Roberts, Jr., Steven Lasky, and Domenico Accili for their helpful comments on the manuscript. This research project was supported in part by grants from the Diabetes Research and Education Foundation, the Juvenile Diabetes Foundation, and the Mallinckrodt Foundation.

References

1. Saiki, R. K., Scharf, S., Faloona, F., Mullis, K. B., Horn, G. T., Erlich, H. A., and Arnheim, N. (1985) Enzymatic amplification of beta-globin genomic sequence and restriction site analysis for diagnosis of sickle cell anemia. *Science* **230,** 1350–1354.
2. Saiki, R. K., Gelfand, D. H., Stoffel, S., Scharf, S. J., Higuchi, R., Horn, G. T., Mullis, K. B., and Erlich, H. A. (1988) Primer-directed enzymatic amplification of DNA with a thermostable DNA polymerase. *Science* **239,** 487–491.
3. Scharf, S. J., Horn, G. T., and Erlich, H. A. (1986) Direct cloning and sequence analyses of enzymatically amplified genomic sequences. *Science* **233,** 1076–1078.
4. Lee, C. C., Wu, X., Gibbs, R. A., Cook, R. G., Muzny, D. M., and Caskey, C. T. (1988) Generation of cDNA probes directed by amino acid sequence: cloning of urate oxidase. *Science* **239,** 1288–1290.
5. Higuchi, R. (1989) Using PCR to engineer DNA, in *PCR Technology* (Erlich, H. A., ed.), Stockton, New York, pp. 61–70.
6. Scharf, S. J. (1990) Cloning with PCR, in *PCR Protocols* (Innis, M. A., Gelfand, D. H., Sninsky, J. J., and White, T. J., eds.), Academic, San Diego, pp. 84–91.
7. Kaufman, D. L. and Evans, G. A. (1990) Restriction endonuclease cleavage at the termini of PCR products. *BioTechniques* **9,** 304–306.
8. Shuldiner, A. R., Scott, L. A., and Roth, J. (1990) PCR induced subcloning polymerase chain reaction (PCR) products. *Nucleic Acid Res.* **18,** 1920.
9. Shuldiner, A. R., Tanner, K., Scott, L. A., and Roth, J. (1991) Ligase-free subcloning: a versatile method of subcloning polymerase chain reaction (PCR) products in a single day. *Anal. Biochem.* **194,** 9–15.
10. Higuchi, R., Krummel, B., and Saiki, R. K. (1988) A general method of *in vitro* preparation and specific mutagenesis of DNA fragments: study of protein and DNA interactions. *Nucleic Acid Res.* **16,** 7351–7367.
11. Horton, R. M., Hunt, H. D., Ho, S. N., Pullen, J. K., and Pease, L. R. (1989) Engineering hybrid genes without the use of restriction enzymes: gene splicing by overlap extension. *Gene* **77,** 61–68.
12. Del Sal, G., Manfioletti, G., and Schneider, C. (1989) A one-tube DNA minipreparation suitable for sequencing. *Nucleic Acids Res.* **16,** 9878.
13. Maas, R. (1983) An improved cloning hybridization method with significantly increased sensitivity for detection of single genes. *Plasmid* **10,** 296–298.
14. Gussow, D. and Clackson, T. (1989) Direct clone characterization from plaques and colonies by the polymerase chain reaction. *Nucleic Acid Res.* **17,** 4000.

CHAPTER 24

Use of Polymerase Chain Reaction for Making Recombinant Constructs

Douglas H. Jones and Stanley C. Winistorfer

1. Introduction

The capacity to recombine and modify DNA are underpinnings of the recombinant DNA revolution. The polymerase chain reaction (PCR) *(1,2)* provides a rapid means for the site-directed mutagenesis of DNA and for the recombination of DNA *(1–9)*. Recently, two methods have been introduced that permit site-directed mutagenesis and DNA recombination without any enzymatic reaction in vitro apart from DNA amplification *(5–9)*. The first method is accomplished by using separate PCR amplifications to generate products, such that when these products are combined, denatured, and reannealed, they form double-stranded DNA with single-stranded ends that are designed to anneal to each other to yield circles, an application termed *recombinant circle PCR* (RCPCR; *see* Chapter 27).

More recently, a PCR-based method was introduced that uses in vivo recombination to generate site-specific mutants and recombinant constructs *(7–9)*. This method is termed *recombination PCR* (RPCR), and selected applications of this method is the topic of this chapter. In the RPCR method, the PCR is used to add homologous ends to DNA, and in this respect, RPCR is similar to RCPCR. These homologous ends mediate recombination in vivo following transfection of *E. coli* with linear PCR products, resulting in the formation of DNA joints in vivo. If these recombinant circles contain plasmid sequences that permit replication, and a selectable marker such as an antibiotic resistance

From: *Methods in Molecular Biology, Vol. 15: PCR Protocols: Current Methods and Applications*
Edited by: B. A. White Copyright © 1993 Humana Press Inc., Totowa, NJ

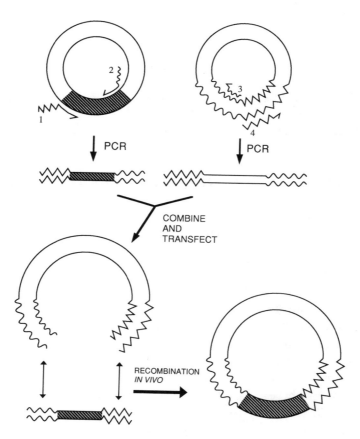

Fig. 1. Diagram showing DNA recombination using recombination PCR (RPCR). The primers are numbered hemiarrows. The insert is the cross-hatched region. Smooth circles represent the DNA strands of the donor plasmid. Circles with wavy and jagged portions represent DNA strands of the recipient plasmid. Reprinted by permission from *BioTechniques* **10**, 62–66.

gene, *E. coli* can be transformed by the recombinant of interest. The mechanisms underlying this recombination between short stretches of homology in bacterial strains used routinely in cloning are not defined, but this does not diminish the power of this method for genetic engineering. RPCR provides a rapid method for the recombination and/or site-directed mutagenesis of DNA with very few steps. The application of this method for DNA recombination is illustrated in Fig. 1, and the steps of this method are briefly outlined in the following:

1. The DNA segment that is to be inserted into the recipient construct is amplified with primers 1 and 2. In Fig. 1, the template DNA from which the insert is amplified is a plasmid. In a separate PCR amplification, the recipient plasmid is amplified with primers 3 and 4. Portions of primers 1 and 2 are complementary to primers 4 and 3, respectively (or 3 and 4, depending on the orientation of the insert desired in the recombinant construct). In this figure, the 5' regions of primers 1 and 2 contain regions that are homologous to the recipient plasmid sequences to which primers 4 and 3 anneal. Therefore, portions of the 5' regions of primers 1 and 2 are complementary to primers 4 and 3. The only requirement for this method is that primers 1 and 2 must have regions of complementarity to primers 4 and 3 (or 3 and 4). If these oligonucleotides are also sufficiently complementary to their respective templates to undergo PCR amplification, the regions of complementarity of one primer to another do not need to anneal to either donor or recipient template sequences. Therefore, this strategy can be used for mutagenesis as well as for recombination (*see* Figs. 2 and 3).

2. Following PCR amplification of each product, the two products are separated from supercoiled plasmid template by agarose gel electrophoresis, and each product is removed from the gel and purified by glass bead extraction. Alternatively, if each template can be linearized outside the region to be amplified by restriction enzyme digestion prior to PCR amplification, agarose gel purification of the PCR products is not necessary.

3. These two products are combined and a portion of this mixture is transfected into MAX efficiency competent *E. coli* (BRL, Gaithersburg, MD). As stated, if each supercoiled plasmid template is linearized by restriction enzyme digestion outside the region to be amplified prior to PCR amplification, the two crude PCR products can be cotransfected directly into *E. coli*, bypassing the gel purification and glass bead extraction steps.

The sum goal of the two amplifications is to yield two PCR products with each end of one product homologous to a distinct end of the other PCR product. Since the amplifying primer sequences are incorporated into the ends of a PCR product, so long as primers 1 and 2 contain regions that are complementary to regions of primers 3 and 4 (or 4 and 3), the PCR products will contain ends that are homologous to each other, and these primer-determined DNA ends do not need to be determined by the original donor or recipient templates. Therefore, it is clear that this RPCR method can be used not only to generate recombinant constructs, but also for the site-directed mutagenesis of two distal sites concurrently (Fig. 2), or for the rapid site-specific mutagenesis of single sites (Fig. 3), as described previously (*8*).

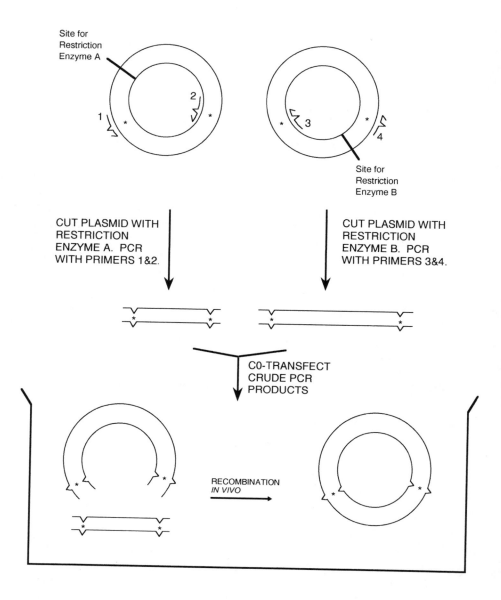

Fig. 2. Diagram showing site-specific mutagenesis of two distal sites using recombination PCR (RPCR). The primers are numbered hemiarrows. Asterisks designate the mutagenesis sites. There is no purification of the PCR products. Notches designate point mismatches in the primers and resulting mutations in the PCR products. Reprinted by permission from *Technique* **2**, 273–278.

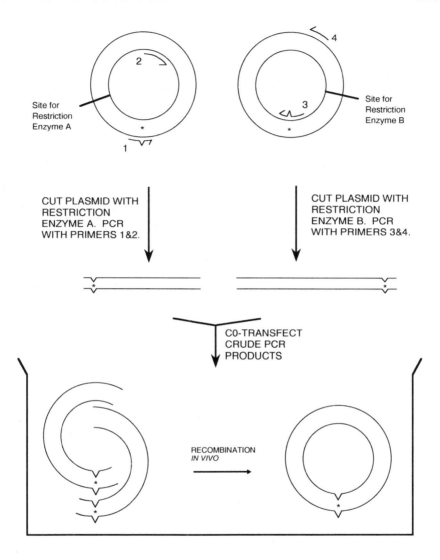

Fig. 3. Diagram showing site-specific mutagenesis of a single site using recombination PCR (RPCR) with four primers. The primers are numbered hemiarrows. The asterisk designates the mutagenesis site. Primer 2 is complementary to primer 4. Restriction enzyme sites A and B bracket the insert. Notches designate point mismatches in the primers and resulting mutations in the PCR products. There is no purification of the PCR products. For each additional single site-specific mutagenesis reaction, only a new primer 1 and 3 need to be synthesized, and the same cut templates can be used. Reprinted by permission from *Technique* **2,** 273–278.

2. Materials

1. 2X PCR stock mixture: Prepare 25 μL aliquots in 0.5-mL microfuge tubes containing 1.25 U of *Taq* polymerase, 400 μ*M* of each dNTP, 100 m*M* KCl, 20 m*M* Tris-HCl, pH 8.3, and 3 m*M* $MgCl_2$. Store at –20°C for up to 8 mo.
2. PCR primers. *See* Note 1.
3. Long thin micropipet tips.
4. Agarose gel components (*see* Chapter 1).
5. Geneclean (Bio 101, La Jolla, CA).
6. TE buffer: 10 m*M* Tris-HCl, pH 8.0, 1 m*M* EDTA.
7. MAX Efficiency HB101 or DH5α Competent *E. coli* (BRL, Life Technologies, Gaithersburg, MD). Once a tube is thawed it should not be reused.
8. SOC media (*see* ref. *10*, p. A2). Prepare by dissolving 20 g of bacto-tryptone, 5 g of bacto-yeast extract and 0.5 g of NaCl in 950 mL of deionized H_2O. Add 10 mL of 250 m*M* KCl. Adjust pH to 7.0 with 5*N* NaOH. Bring final volume to 1 L with H_2O and autoclave for 20 min at 15 lb/sq in on liquid cycle. Allow to cool to 60°C, then add 5 mL of sterile 2*M* $MgCl_2$ and 20 mL of sterile 1*M* glucose.
9. Top agar: 7 g/L of bacto-agar (*see* ref. *10*, p. A4). Autoclave for 20 min at 15 lb/sq in on liquid cycle.
10. LB plates with 100 μg/mL Ampicillin (*see* ref. *10*, p. A4).

3. Methods

1. Add 2 ng of template plasmid and 25 pmol of each primer to the 25-μL aliquot of 2X PCR stock solution and adjust volume to 50 μL with distilled water. Layer 50 μL of mineral oil on top of each reaction mix prior to amplification.
2. Amplify by PCR using the following cycle profile (*see* Note 2):

Initial denaturation	94°C, 1 min
A variable number of main cycles	94°C, 30 s (denaturation)
	50°C, 30 s (annealing)
	72°C, 1 min/kb (extension)
Final extension	72°C, 7 min

Store samples at 4°C if required.
3. Remove each PCR product by inserting a long thin micropipet tip through the mineral oil layer and drawing up the sample.
4. If a supercoiled template is used, dissolve 1% standard high melting point agarose in 1X TAE by boiling, and then place the bottle with the agarose in a 55°C water bath to allow the agarose to cool to 55°C. Pour the gel, mix each entire PCR product with an electrophoresis loading buffer, and resolve through standard high-melting-point 1% agarose in

TAE buffer with 0.5 µg/mL ethidium bromide. Electrophoresis should be carried out until each PCR product has traveled at least 4 cm, in order to separate adequately the PCR products from the supercoiled plasmid template (*see* Note 3).

5. Remove the band of interest with a razor blade, and extract from the agarose using GeneClean according to manufacturer's instructions. Warm the new wash solution (contained in the GeneClean kit) from – 20°C storage to at least 4°C prior to use in order to facilitate dispersion of the glass particles during washing (*see* Note 4).

6. Suspend each PCR product in 30 µL of TE buffer.

7. Transform *E. coli* according to the manufacturer's (BRL) protocol with the following modifications: use 50 µL of *E. coli* for each sample transformed, as this is effective and less expensive than the 100 µL recommended. Use 25 µL of cells for the control plates d and e below, so that only one BRL tube, which contains 200 µL of bacteria, needs to be used per reaction (*see* Note 5).

8. Set up the following transformations:
   a. 2.5 µL of PCR 1 + 2.5 µL of PCR 2
   b. 2.5 µL of PCR 1 + 2.5 µL of TE buffer
   c. 2.5 µL of PCR 2 + 2.5 µL of TE buffer
   d. 0.5 ng of a supercoiled template in 5 µL of TE buffer
   e. 5 µL of TE buffer
   See Note 6.
   Transform with approx 0.3–10 ng of the amplified recipient plasmid with an approx 14–25M excess of the amplified insert when using the GeneCleaned DNA (*see* Note 7).

9. If the PCR templates have been linearized by restriction enzyme digestion outside the region to be amplified prior to 14 cycles of amplification, no purification of PCR products is necessary. Insert the long thin micropipet tip through the mineral oil, withdraw 2.5 µL from each PCR tube (typically 15–50 ng), combine the two samples, and transform *E. coli* with the 5 µL (*see* Note 8).

10. Following incubation at 37°C in a shaker for 1 h, plate the entire sample onto an LB plate containing 100 µg/mL of Ampicillin. In order to keep the sample on the plate, add 2 mL of top agar, prewarmed to 42°C, to each sample immediately prior to pouring it onto the plate.

11. Grow bacterial clones overnight at 37°C (*see* Notes 9 and 10).

4. Notes

1. Primers 1 and 2 in the recombination protocol (Fig. 1) have 3' template annealing ends that are 22-nucleotides long and 5' modifying ends that are 30-nucleotides long. The 30 nucleotide long 5' ends contain the regions

that are complementary to primers 4 and 3, such that PCR amplification with primers 1 and 2 results in a product with 30 bp of homology with the product of primers 3 and 4. Each primer that is not designed to modify a DNA end is 20–22 nucleotides long (primers 3 and 4 in Fig. 1, and primers 2 and 4 in Fig. 3). The primers that introduce point mutations (primers 1, 2, 3, and 4, in Fig. 2; and primers 1 and 3 in Fig. 3) have been designed so that 18–20 nucleotides are 5' to the mismatched nucleotide and 7–11 nucleotides are 3' to the mismatched nucleotide. These parameters are essentially identical to those recommended for mutating primers in the point mutagenesis RCPCR protocol from Chapter 27. In the two-point mutagenesis protocol, these primers have resulted in PCR products where each end of one product contains 38–41 bp of homology to an end of the other PCR product. The lower limits of homology necessary for retrieving a high percentage of recombinants using this protocol have not been determined.

2. If supercoiled template plasmid is used, use 18–25 amplification cycles, and if restriction enzyme linearized plasmid is used as the template (linearized outside the region to be amplified) use 14 amplification cycles.

3. If the PCR templates are not linearized outside the region to be amplified prior to PCR amplification, the products must be separated from the PCR template by agarose gel electrophoresis. If these products do not undergo sufficient electrophoretic separation, a high number of background colonies will result from transformation by contaminating supercoiled PCR template.

4. GeneClean recovery is 25–90%.

5. Since the transformation efficiency is low, highly competent bacteria must be used. Only MAX efficiency competent *E. coli* (transfection efficiency $>5 \times 10^8$/μg of monomer pBR322) have been used. Using restriction enzyme-digested templates, the transformation efficiency has been 1–9 colonies/ng total DNA transfected, with a molar ratio of the short PCR product to the long PCR product of 2:1 to 3:1 (in Figs. 1 and 2 , the short PCR product is produced by primers 1 and 2 and the long PCR product is produced by primers 3 and 4).

6. The yield of colonies from plate a is approx 10X that from plate b or c, confirming a high percentage of recombinants in plate a. Plate d is a transformation control, and should yield a thick lawn of colonies. Plate e is an antibiotic control, and should yield no colonies since the bacteria that have not been transformed are sensitive to Ampicillin.

7. If the PCR template is not linearized prior to PCR amplification, transformation with a 1:1 molar ratio of amplified insert to amplified recipient plasmid instead of a 14:1 molar ratio of amplified insert to amplified recipient plasmid decreases the proportion of clones with the recombi-

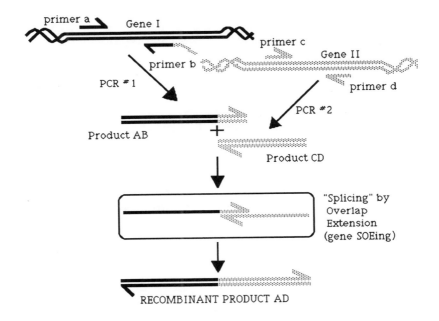

Fig. 2. Gene splicing by overlap extension (SOEing). Here, products AB and CD are derived by two different genes. SOEing primer b has sequences added to its 5' end such that the right end of AB is made complementary to the left end of CD. This allows their sequences to be joined by overlap extension, or "SOEn" together.

the two strands act as primers on one another. Extension of this overlap by DNA polymerase creates the full-length mutant molecule AD, which has the mutation at an arbitrary distance from either end.

In the final overlap extension step, two separate PCR-generated sequences (AB and CD) are joined together. If AB and CD are made from different genes, then product AD is a recombinant molecule *(7,8)*. Figure 2 illustrates the concept of PCR-mediated recombination, or gene SOEing. Here, extra sequence is added to the 5' end of primer b, which results in a short segment of gene II being added to the right-hand end of the PCR product amplified from gene I (product AB). This causes the two intermediate products to have an overlap region of common sequence, so that they can be joined together by overlap extension. The two strands having the overlap at their 3' ends (the "productive" strands, shown in the boxes in Figs. 1 and 2) each act as both a primer and a template to produce a giant "primer dimer," which is the recombinant molecule.

When should you use gene SOEing? For site-directed mutagenesis, overlap extension is simple, and has advantages over other methods in that

1. Recombination and mutagenesis can be performed simultaneously *(7)*;
2. Essentially all of the product molecules are mutated (i.e., 100% mutation efficiency; *6*); and
3. The product is produced in vitro (without having to grow it up in a plasmid or phage) and may be used directly in experiments *(5,9)*.

As a method of DNA recombination, gene SOEing is tremendously useful in situations where no leeway can be given to use nearby restriction sites. Protein engineering projects provide excellent examples *(7,8,10,11)*. Two major drawbacks of recombination by SOEing are the cost of the primers and the potential for introducing random errors with PCR (*see* Notes 3, 5, and 6). Therefore, gene SOEing is most practical in "complicated" constructions where there are no convenient restriction sites.

2. Materials

1. Thermal cycling ("PCR") machine.
2. *Taq* DNA polymerase.
3. 10X PCR buffer: 500 mM KCl, 100 mM Tris-HCl, pH 8.3.
4. 10 mM MgCl$_2$.
5. dNTPs.
6. Primers (*see* Section 3.1. and Note 1).
7. DNA templates (*see* Note 2).
8. Agarose gel electrophoresis supplies and equipment (*see* Chapter 1).
9. GeneClean (Bio101, La Jolla, CA), or your favorite method for purifying DNA from agarose gels (*see* Chapter 1).

3. Methods

3.1. Primer Design (see Note 3)

1. Unlike a regular PCR primer, a SOEing primer (such as b in Fig. 2) has to include two sequence regions capable of hybridizing to a template well enough to act as primers. The first is the priming region at the 3' end of the oligonucleotides which allows it to act as a PCR primer. The second is the overlap region at the 5' end of the oligonucleotide; this allows the strand of the PCR product that is complementary to the oligonucleotide to act as a primer in the overlap extension reaction. As shown in Fig. 2, the priming region of primer b contains a sequence from gene I, whereas the overlap region contains a sequence from gene II which is complementary to primer c.

For examples, some primers from published overlap extension projects *(8)* are shown in the following *(see* Note 4):

```
                    priming
          |—region—| |-overlap region-|
3'-gggtccgggtgtgcttctgctgtaactccg-5' (SOEing primer b)
    5'-gaagacgacattgaggc-3' (primer c)
        |-priming region-|
```

2. In mutagenesis, there is only one template, so the priming and overlap regions may completely coincide *(6)*. The mismatches are placed in the center of the oligo so that both its 3' end and the 3' end of the product made with it can act as primers.

3. My colleagues and I have adopted the convention of designing both the priming and overlap regions of SOEing primers to have an estimated T_m of around 50°C using the following formula *(12)*:

$$T_m = [(G + C) \times 4] + [(A + T) \times 2] \quad (in \, °C)$$

Using SOEing primer b as an example, the T_m for its overlap region is 52°C ([3 + 6] × 4] + [(2 + 6) × 2]) and for its priming region is 46°C ([8 + 2] × 4] + [(0 + 3) × 2]). The mismatches in a mutagenic oligo do not count when estimating T_m. These T_m estimates, though crude, are generally conservative approximations of the annealing temperature needed for an oligo to be used in PCR (e.g., a 50°C annealing step will work with these primers). They lead to priming and overlap regions being somewhere around 13–20 bases long, depending on the GC content. The "~50°C rule" has resulted in quite reliable primers, but it does not represent a systematic effort to determine the absolute minimum lengths that SOEing primers can be *(see* Note 5).

4. The normal considerations in primer design also apply, such as avoiding complementarity between or within primers, and so forth *(see* Chapter 2).

5. The flanking primers a and d should also be capable of priming at 50°C, and may include restriction sites at the ends to facilitate cloning of the product.

6. "Megaprimer" reactions. Modifications of the gene-SOEing protocol make it possible to use only one primer at the recombinant joint *(13,14)*. Rather than using the symmetrical approach shown in Fig. 2, in which the recombination event is formation of a giant primer-dimer, the top strand of AB can be used as a megaprimer in place of primer c directly on template gene II to make the final product AD. For recombination of two sufficiently different genes, only the recombinant product can be amplified using primers a and d, since each of these primers will match one of the template genes but not the other. Therefore, a and d can be

included in the reaction to drive the synthesis of the productive strands, and to amplify the product as soon as it forms *(13,15)*. For megaprimer mutagenesis, however, primer a cannot be included in the final reaction containing primer d because a and d would simply amplify the wild-type sequence from the template. This means that product AB must be added in sufficient quantity to supply all of the megaprimer strand necessary to generate AD *(14)*. For further discussion of megaprimer mutagenesis, *see* Chapter 28.

3.2. PCR Generation of Intermediate Products AB and CD (Fig. 2)

1. Set up two separate PCRs as follows (*see* Note 6):

Ingredients	PCR 1	PCR 2
10X buffer II	10 µL	10 µL
10 mM MgCl$_2$	~15 µL	~15 µL
10X dNTPs	10 µL	10 µL
5' primer (10 µM) a,	10 µL	c, 10 µL
3' primer (10 µM) b,	10 µL	d, 10 µL
template	gene I, ~0.5 µg	gene II, ~0.5 µg
H$_2$O	to 100 µL	to 100 µL
Taq polymerase	~3 U	~3 U
Product	AB	CD

2. Amplify by PCR using the following cycle profile:

    20–25 main cycles   94°C, 1 min (denaturation)
                           50°C, 1 min (annealing)
                           72°C, 1 min (extension)

3.3. Purification of Intermediates (see Note 7)

1. Run the PCR products on an agarose gel to size-purify them (*see* Note 8).
2. The method you use to extract the DNA from the agarose also depends on the size of the piece of DNA with which you are working. Purify fragments larger that ~300 bp using GeneClean. Be aware that GeneClean requires the use of TAE (not TBE) electrophoresis buffer.
3. Smaller fragments are not recovered efficiently by this method, so you should use another procedure, such as electroelution. Cut a well in front of the band, run the band into the well, and recover the DNA in the well with a pipet (this procedure is described in ref. *16*).
4. Recover the DNA from the buffer by ethanol precipitation. Add 0.1 vol of 3M sodium acetate, pH 5.2, 1 µL of yeast tRNA (10 mg/mL) as a carrier, and 2.5 vol of cold 95% ethanol. Incubate on dry ice for 15 min, and then spin at approx 10,000g for 15 min.

3.4. Generation of the Recombinant Product AD

1. Set up the overlap extension (SOE) reaction as follows (*see* Note 9):

Ingredients	
10X buffer II	10 μL
10 mM MgCl$_2$	~15 μL
10X dNTPs	10 μL
5' primer (10 μM) a,	10 μL
3' primer (10 μM) d,	10 μL
Intermediate 1	AB
Intermediate 2	CD
H$_2$O	to 100 μL
polymerase	~3 U

2. Amplify by PCR using the following cycle profile (*see* Note 10):
   20–25 main cycles   94°C, 1 min (denaturation)
                               50°C, 1 min (annealing)
                               72°C, 1 min (extension)

4. Notes

1. Primer purification. The final step in oligonucleotide synthesis by the phosphoramidite method involves treatment in NH$_4$OH. The ammonium hydroxide should be evaporated off using a SpeedVac (Savant, Farmingdale, NY). (Please note that ammonia vapors are extremely hard on vacuum pumps! Either install an appropriate chemical trap in your vacuum line or use a water-powered aspirator instead.) Resuspend the dried primer in water and desalt it over a Sephadex G25 column. Prepacked columns are available for this purpose (NAP-10 columns, Pharmacia LKB, Uppsala, Sweden). No special buffer is needed: Distilled water works fine. Extensive (and expensive!) primer purification schemes, such as acrylamide gel electrophoresis or HPLC, are not necessary. If you plan to purchase primers commercially, note that, as of this writing (6/91), they are available already desalted for ~$4.50/base (BioSynthesis Inc., Denton, TX). There is no need to pay more than this.

2. The starting templates contain the gene sequences that you want to recombine into a tailor-made molecule. Any template suitable for PCR, such as reverse-transcribed RNA, can be used for gene SOEing (e.g., *see* ref. *10*). However, because high template concentrations minimize the probability of the polymerase introducing errors into the sequences (*see* Notes section), your starting templates will probably be cloned genes in plasmids. My coworkers and I have used both cesium chloride and alkyline lysis-purified plasmids with equal success (*see* ref. *16* for plasmid purification protocols).

3. The single simplest and most complete way to mess up a SOEing reaction is to have mistakes in the primer sequences (R.H., personal experience!). As PCR reactions go, amplifying an insert from a plasmid that is present in microgram amounts is like falling off the proverbial log. If this reaction cannot be made to work after the normal titrations of Mg^{2+}, template, and so forth, then something is drastically wrong, and you should recheck the design of your primers. Similarly, if products AB and CD, when mixed together in near-microgram amounts in a SOE reaction, completely fail to produce a recombinant product, the sequences of the primers in the overlap region should be immediately suspect. Writing out the sequence of the desired product and making sure that the SOEing primers each match one strand of the desired product at the recombination point is helpful.

4. The overlap region does not all have to be added to one primer. If, for example, instead of adding a 17-bp overlap region to primer b we had added 9 bases complementary to primer c to the 5' end of b and 8 bases complementary to b to the 5' end of c, the overlap between AB and CD would still have been 17 bp (*see* ref. *7*). This approach avoids using very long primers. However, since making a 40- or 50-mer is now routine, there is usually no need to split the overlap region.

5. The optimal length to design the overlap region has not been settled. Regions as long as 164 bp (*15*) or as short as 12 bp (*17*) have been reported. Although the "50°C rule" reliably produces overlaps that are long enough to work, it does not indicate the minimum workable length. For example, the 12-bp-long overlap region reported in ref. *17* has an estimated T_m of 34°C, although it was used at an annealing temperature of 50°C. On theoretical grounds, however, merely producing the recombinant molecule is not the only consideration to bear in mind. A very short overlap region might lead to an inefficient SOEing reaction, requiring the final product to be amplified through more rounds of PCR. This in turn may increase the error frequency.

6. Errors introduced by polymerase. Studies have demonstrated that *Taq* polymerase is capable of high fidelity DNA synthesis under PCR conditions (*18*). Clones produced by a single round of overlap extension have error frequencies of around 1 in 4000 bases (*6*), whereas more complicated constructs involving random SOEing reactions lead to slightly higher frequencies (~1 in 1800; ref. *7*). Because of the possibility that random mutations will be introduced by polymerase errors, several precautions are in order. First, the highest concentration of template plasmid consistent with amplification should be employed (this will probably be around 500 ng of plasmid in 100 μL). This minimizes the number of rounds of replication required to produce enough product

with which to work, and gives the polymerase fewer opportunities to make errors. Since the reaction plateaus after producing a certain amount of product, it is probably not necessary to minimize the number of heating/cooling cycles to which the samples are subjected. Second, the lowest concentration of magnesium compatible with amplification should be used, as error rates increase with increasing $[Mg^{2+}]$ *(18)*. Finally, for many applications, it is advisable to sequence the final product to ensure that it is free of errors. For this reason, a "cassette" approach, in which PCR manipulations are performed on a small recombinant segment, which is then ligated into a vector containing the remaining portions of the construct, may in some cases be preferable to SOEing directly into a vector (*see* Chapters 24 and 27). Around 300–500 bp is a convenient size for a cassette because it is large enough to handle conveniently yet small enough to sequence quickly. It should be pointed out that other thermostable DNA polymerases (i.e., Vent polymerase, New England Biolabs, Beverly, MA) are reportedly capable of significantly higher fidelity synthesis than *Taq*, and can be used for overlap extension *(19)*.

7. Ho et al. *(6)* found that gel purification of the intermediate products AB and CD led to a cleaner reaction and increased product yield. Purification of intermediates may be most important when the initial template concentration is high. Gel purification removes not only the template plasmids, but also open-ended primer extension products, which, coming from the template, may be longer than the PCR product. These open-ended products may not be obvious on an ethidium-stained gel because they are of indeterminate length and possibly single-stranded, but they have the potential to generate unwanted side products. Nevertheless, other workers have successfully used less extensive purification schemes *(5)* or none at all *(10,13)*.

8. The percent agarose you use depends on the size of the products you are isolating; smaller fragments need a higher percentage of agarose. For up to 1% agarose, "regular" agarose is fine, but for higher percentages, NuSieve (FMC BioProducts, Rockland, ME) gives better resolution. Please note that NuSieve is used to supplement regular agarose: The first 1% agarose is the regular variety, and only the additional percentage is NuSieve (up to a total of 4%).

9. Using large quantities of the intermediates should minimize polymerase errors (*see* Note 6). About 25% of what you recover from the gel should be plenty if the PCRs worked well, and this will leave you some extra in case you have to repeat it. The recombinant may now be cloned, or otherwise used, like a normal PCR product.

10. Related applications. Although the concept is simple, overlap extension is a tremendously powerful technology, and the reader is encour-

aged to spend some time contemplating modifications and applications. A general theoretical review of the subject is given in ref. *15*. Some of the more important and thought-provoking technical developments related to synthetic uses of PCR are given in refs. *14,20–22; see also* Chapters 24, 26, 27, and 28.

Acknowledgments

I am deeply grateful to my colleagues Steffan Ho, Jeff Pullen, Henry Hunt, Zeling Cai, and Larry Pease, for making it possible for me to participate in the development of this technology, and to Bianca Conti-Tronconi for the use of her computer.

References

1. Mullis, K. B. and Faloona, F. A. (1987) Specific synthesis of DNA in vitro via a polymerase-catalysed chain reaction. *Methods Enzymol.* **155,** 335–350.
2. Mullis, K., Faloona, F., Scharf, S., Saiki, R., Horn, G., and Erlich, H. (1986) Specific enzymatic amplification of DNA in vitro: the polymerase chain reaction. *Cold Spring Harbor Symp. Quant. Biol.* **51,** 263–273.
3. Kadowaki, H., Kadowaki, T., Wondisford, F. E., and Taylor, S. I. (1989) Use of polymerase chain reaction catalysed by Taq DNA polymerase for site-specific mutagenesis. *Gene* **76,** 161–166.
4. Vallette, F., Mege, E., Reiss, A., and Milton, A. (1989) Constuction of mutant and chimeric genes using the polymerase chain reaction. *Nucleic Acids Res.* **17,** 723–733.
5. Higuchi, R., Krummel, B., and Saiki, R.K. (1988) A general method of in vitro preparation and specific mutagenesis of DNA fragments: study of protein and DNA interactions. *Nucleic Acids Res.* **16,** 7351–7367.
6. Ho, S. N., Hunt, H. D., Horton, R. M., Pullen, J. K., and Pease, L. R. (1989) Site-directed mutagenesis by overlap extension using the polymerase chain reaction. *Gene* **77,** 51–59.
7. Horton, R. M., Hunt, H. D., Ho, S. N., Pullen, J. K., and Pease, L. R. (1989) Engineering hybrid genes without the use of restriction enzymes: gene splicing by overlap extension. *Gene* **77,** 61–68.
8. Horton, R. M., Cai, Z., Ho, S. N., and Pease, L. R. (1990) Gene splicing by overlap extension: Tailor-made genes using the polymerase chain reaction. *BioTechniques* **8,** 528–535.
9. Kain, K. C., Orlandi, P. A., and Lanar, D. E. (1991) Universal promoter for gene expression without cloning: expression-PCR. *BioTechniques* **10,** 366.
10. Davis, G. T., Bedzyk, W. D., Voss, E. W., and Jacobs, T. W. (1991) Single Chain Antibody (SCA) encoding genes: one-step construction and expression in eukaryotic cells. *Biotechnology* **9,** 165–169.
11. Daughtery, B. L., DeMartino, J. A., Law, M.-F., Kawka, D. W., Singer, I. I., and Mark, G. E. (1991) Polymerase chain reaction facilitates the cloning, CDR-grafting, and rapid expression of a murine monoclonal antibody directed against the CD18 component of leukocyte integrins. *Nucleic Acids Res.* **19,** 2471–2476.

12. Suggs, S.V., Hirose, T., Miyake, T., Kawashima, E. H., Johnson, M. J., Itakura, K., and Wallace, R. B. (1981) Use of synthetic oligo-deoxyribonucleotides for the isolation of cloned DNA sequences, in *Developmental Biology Using Purified Genes* (Brown, D. D. and Fow, C. F., eds.), Academic, New York, pp. 683–693.
13. Yon, J. and Fried, M. (1989) Precise gene fusion by PCR. *Nucleic Acids Res.* **17,** 4895.
14. Sarkar, G. and Sommer, S. S. (1990) The "megaprimer" method of site-directed mutagenesis. *BioTechniques* **8,** 404–407.
15. Horton, R. M. and Pease, L. R. (1991) Recombination and mutagenesis of DNA sequences using PCR, in *Directed Mutagenesis: A Practical Approach.* (McPherson, M. J., ed.) IRL, Oxford, pp. 217–247.
16. Sambrook, J., Fritsch, E. F., and Maniatis, T. (1989) *Molecular Cloning: A Laboratory Manual* (2nd ed.) Cold Spring Harbor Laboratory, Cold Spring Harbor, NY.
17. Yolov, A. A. and Shaborova, Z. A. (1990) Constructing DNA by polymerase recombination. *Nucleic Acids Res.* **18,** 3983–3986.
18. Eckert, K. A. and Kunkel, T. A. (1990) High fidelity DNA synthesis by the Thermus aquaticus DNA polymerase. *Nucleic Acids Res.* **18,** 3739–3744.
19. Hanes, S. D. and Brent, R. (1991) A genetic model for interaction of the homeodomain recognition helix with DNA. *Science* **251,** 426–430.
20. Rudert, R. A. and Trucco, M. (1990) DNA polymers of protein binding sequences generated by PCR. *Nucleic Acids Res.* **18,** 6460.
21. Shuldiner, A. R., Scott, L. A., and Roth, J. (1990) PCR-induced (ligase-free) subcloning: a rapid reliable method to subclone polymerase chain reaction (PCR) products. *Nucleic Acids Res.* **18,** 1920.
22. Jones, D. H. and Howard, B. H. (1990) A rapid method for site-specific mutagenesis and directional subcloning by using the polymerase chain reaction to generate recombinant circles. *BioTechniques* **8,** 178–183.

CHAPTER 26

Use of Polymerase Chain Reaction for the Rapid Construction of Synthetic Genes

Patrick J. Dillon and Craig A. Rosen

1. Introduction

Although the polymerase chain reaction (PCR) *(1,2)* is invaluable for the cloning and manipulation of existing DNA sequences, PCR also makes it possible to create new DNA fragments consisting of a nucleic acid sequence that is specified entirely by the investigator. In this chapter we describe a simple two-step PCR method for the rapid construction of synthetic genes *(3)*. This method is based on early observations by Mullis et al. *(4)* in which multiple overlapping oligonucleotides could be used to generate synthetic DNA through several sequential rounds of Klenow-based PCR amplification. The method described in this chapter utilizes the thermostabile *Taq* polymerase and allows for the generation of synthetic genes in as little as 1 d. This method has proven useful in studies in which synthetic genes were constructed for the HIV-2 Rev protein *(3,5)* and the Wilms' tumor locus zinc finger protein *(6)*. Furthermore, this method has been successfully employed in extensive mutagenesis of the HIV-1 rev response element *(7)*.

The ability to generate synthetic DNA has many applications dependent on the design of the construct. Often eukaryotic genes are poorly expressed in bacterial expression systems. One cause of poor gene expression in bacteria is owing to the codons present in the eukaryotic gene. One can design a synthetic gene that contains codons that are preferentially utilized in the organism to be used for protein expres-

From: *Methods in Molecular Biology, Vol. 15: PCR Protocols: Current Methods and Applications*
Edited by: B. A. White Copyright © 1993 Humana Press Inc., Totowa, NJ

sion. This can be accomplished by reverse-translating the amino acid sequence of the desired protein into a nucleic acid sequence containing codons obtained from codon usage tables that have been generated for the organism of choice. In addition to changes in codon structure, the synthetic gene can be designed to include convenient restriction sites for cloning experiments or the absence of specific restriction sites to facilitate further cloning. Other examples for the use of designing synthetic genes include: (1) the generation of unique chimeric constructs to study structure–function relationships and domain–swapping effects for a variety of related and unrelated proteins; (2) large scale alterations or mutational analysis of motifs present in either proteins or transcriptional elements (e.g., promoters, terminators, and so forth); (3) the creation of unique or novel promoters or proteins; and (4) saturation mutagenesis of genes through the use of random nucleotide incorporation or the use of deoxyinosine in the design of gene sequence.

The principles of this two-step PCR method for the construction of synthetic genes are outlined in Fig. 1. In this method, two sequential PCR reactions are used, the first PCR reaction generates a template DNA corresponding to the synthetic gene, which is then amplified in a second PCR reaction. Before starting this procedure, the investigator must design the construct and determine the nucleic acid sequence of the desired synthetic gene. Once this has been accomplished, oligonucleotides that span the length of the gene must be designed and synthesized. In general, an even number of oligonucleotides should be synthesized and should contain overlaps that are between 15 and 30 nucleotides in length. The orientation of the oligonucleotides should be similar to that in panel A of Fig. 1. It is imperative that the outermost oligonucleotides correspond to opposite strands and be positioned so that they will extend inward toward each other over the gene. The number and length of individual oligonucleotides will vary according to the size of the synthetic DNA to be generated. Typically, oligonucleotides should be between 60 and 125 nucleotides in length. For example, four oligonucleotides can be used to synthesize a 325-bp DNA, whereas eight oligonucleotides can be used to generate a 765-bp construct. For this method, crude oligonucleotide preparations are used and it is not necessary for any additional purifications of the oligonucleotides.

Once the oligonucleotides have been obtained, the first step of the method is to mix the overlapping oligonucleotides in a standard PCR

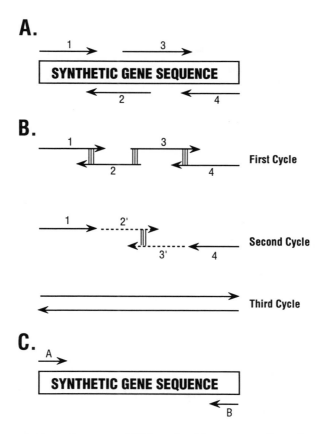

Fig. 1. Description of two-step PCR method for construction of synthetic genes. **(A)** Schematic of design and orientation of overlapping oligonucleotides for first PCR reaction. **(B)** Diagram of oligonucleotide extensions during initial cycle of first PCR. **(C)** Schematic of design and orientation of flanking primers used in the second PCR reaction.

reaction. Panel B shows how four overlapping oligonucleotides would be extended through the first few cycles of PCR. The first PCR should be carried out with enough cycles to generate a double-stranded PCR product that spans the full length of the synthetic gene. The second step in the method is to take a small aliquot of the first PCR reaction and amplify the synthetic gene in a second-strand PCR reaction that contains short flanking primers (A,B) as illustrated in panel C. The sequence of the flanking primers should contain restriction sites to facilitate cloning. This procedure provides ample amounts of DNA for subsequent cloning into appropriate vectors. This method is an extremely powerful tool for the manipulation of nucleic acid and construction of synthetic genes.

2. Materials

1. 10X PCR buffer: 500 mM KCl, 100 mM Tris-HCl, pH 8.0, 15 mM MgCl$_2$.
2. 10X dNTP solution: 2 mM each dATP, dCTP, dGTP, and dTT.
3. *Taq* DNA polymerase.
4. Sterile water.
5. Sterile mineral oil.
6. Overlapping oligonucleotides that span the length of the DNA segment to be synthesized (*see* Note 1). An even number of oligonucleotides should be used and contain overlaps of at least 15 nucleotides (*see* Note 2).
7. Flanking oligonucleotide primers that contain suitable restriction sites for cloning (*see* Chapters 2 and 22).
8. Agarose gel for analysis of PCR products (*see* Chapter 1).

3. Methods

1. Set up first PCR reaction as follows:
   a. 10 μL of 10X PCR buffer;
   b. 10 μL of 10X dNTP solution;
   c. 0.5 μg each of overlapping oligonucleotides;
   d. 2.5 U of *Taq* DNA polymerase;
   e. Sterile water to a final volume of 100 μL;
   f. Overlay sample with 50 μL of sterile mineral oil.
2. Amplify by PCR using the following cycle profile (*see* Notes 3, 4, and 5):

Initial denaturation	94°C, 5 min
10 main cycles	94°C, 1 min (denaturation)
	55°C, 1 min (annealing)
	72°C, 1 min (extension)
Final extension	72°C, 5 min

3. Set up second PCR reaction as follows:
   a. 10 μL of 10X PCR buffer;
   b. 10 μL of 10X dNTP solution;
   c. 1 μL of first PCR reaction as template;
   d. 1 μg of each flanking primer;
   e. 2.5 U *Taq* polymerase;
   f. Sterile water to final volume of 100 μL;
   g. Overlay sample with 50 μL of sterile mineral oil.

4. Run the second PCR using the following cycle profile:

Initial denaturation	94°C, 5 min
25 main cycles	55°C, 1 min
	72°C, 1 min
	94°C, 1 min
Final extension	72°C, 5 min

5. Analyze 10 µL of the first and second PCR reactions by agarose gel electrophoresis (*see* Chapter 1). A faint smear should be present in the first PCR reaction, and a band corresponding to the size of the desired product should be present in the second PCR reaction (*see* Note 6).
6. Digest the product from the second PCR reaction and clone into a suitable vector (*see* Note 7 and Chapter 22).

4. Notes

1. It is not necessary to purify oligonucleotides when using this method. In addition, there is no need to phosphorylate the oligonucleotides since no ligation steps are used in this protocol.
2. Although it is suggested that an even number of overlapping oligonucleotides be used, an odd number may be used as long as the outermost oligonucleotides are on opposite strands and will extend inward toward each other.
3. The number of cycles needed for the first PCR reaction can be varied depending on the number of oligonucleotides used. In theory, only 3 cycles should be necessary for full-length template synthesis using 4 oligonucleotides, whereas 4 cycles would be necessary if 8 oligonucleotides were used.
4. The flanking primers should not be included in the first PCR reaction since their addition results in the generation of many different-sized products that do not amplify well in the second PCR reaction.
5. It should be noted that the nucleic acid sequences of the overlaps may influence the annealing temperatures used during the first PCR reaction.
6. This method has been successful for the generation of synthetic constructs over 750 bp in length.
7. When using this protocol for generating synthetic constructs, it is advisable to sequence the final product to assure that the sequence is correct. The error rate for this method should approximate that observed for other PCR protocols using *Taq* polymerase.

References

1. Saiki, R. K., Gelfand, D. H., Stofel, S.,Scharf, S. J., Higuchi, R., Horn, G. T., Mullis, K. B., and Ehrlich, H. A. (1988) Primer-directed enzymatic amplification of DNA with a thermostable DNA polymerase. *Science* **239,** 487–491.
2. Saiki, R. K., Scharf, S., Faloona, F., Mullis, K. B., Horn, G. T., Ehlrich, H. A.,

and Arnheim, N. (1985) Enzymatic amplification of β-globin genomic sequences and restriction site analysis for diagnosis of sickle cell anemia. *Science* **230,** 1350–1354.

3. Dillon, P. J. and Rosen, C. A. (1990) A rapid method for the construction of synthetic genes using the polymerase chain reaction. *BioTechniques* **9,** 298,299.

4. Mullis, K., Faloona, F., Scharf, S., Saiki, R., Horn, G., and Erlich, H. (1986) Specific enzymatic amplification of DNA in vitro: the polymerase chain reaction. *Cold Spring Harbor Symp. Quant. Biol.* **51,** 263–273.

5. Dillon, P. J., Nelbock, P., Perkins, A., and Rosen, C.A. (1990) Function of the human immunodeficiency virus types 1 and 2 Rev proteins is dependent upon their ability to interact with a structural region present in the env gene mRNA. *J. Virol.* **64,** 4428–4437.

6. Rauscher III, F. J., Morris, J. F., Journay, O. E., Cook, D. M., and Curran, T. (1990) Binding of the Wilms' tumor locus zinc finger protein to the EGR-1 consensus sequence. *Science* **250,** 1259–1262.

7. Olsen, H. S., Beidas, S., Dillon, P. J., Rosen, C. A., and Cochrane, A. W. (1991) Mutational analysis of the HIV-1 Rev protein and its target sequence, the rev response element. *J. Acquired Immun. Defic. Syndrome.* **4,** 558–567.

CHAPTER 27

Recombinant Circle Polymerase Chain Reaction for Site-Directed Mutagenesis

Douglas H. Jones

1. Introduction

Site-directed mutagenesis permits modification of the functional characteristics of specific proteins and the characterization of regulatory DNA elements. Since the amplifying primers used in the polymerase chain reaction (PCR) are incorporated into the product, PCR can be used to modify the ends of DNA segments *(1,2)*. Site-directed mutagenesis is accomplished by introducing a mismatch into one of the oligonucleotides used to prime the PCR amplification. This oligonucleotide, with its mutant sequence, is incorporated into the PCR product. Sequences can also be attached to the ends of a DNA segment by using primers whose 5' ends do not anneal to the original template.

In this chapter, I describe a method in which the PCR is used for site-directed mutagenesis without any enzymatic reaction in vitro apart from DNA amplification. In this method, DNA joints are generated in vitro by using separate PCR amplifications to generate products that when combined, denatured, and reannealed form double-stranded DNA with single-stranded ends. These single-stranded ends are designed to anneal to each other to yield circles, and the application is termed *recombinant circle PCR* (RCPCR) *(3–5)*. The recombinant circles of DNA, containing the mutation of interest, can be transfected into *E. coli* and repaired in vivo. If these recombinant circles contain plasmid sequences that permit replication and a selectable phenotype, such as an antibiotic resistance gene, *E. coli* can be transformed without any

From: *Methods in Molecular Biology, Vol. 15: PCR Protocols: Current Methods and Applications*
Edited by: B. A. White Copyright © 1993 Humana Press Inc., Totowa, NJ

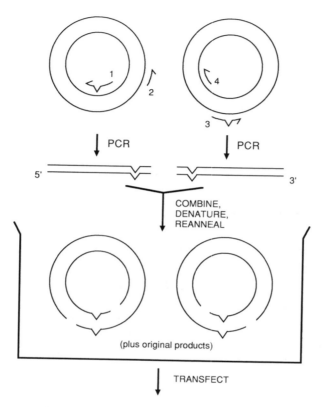

Fig.1. Diagram showing the generation of a point mutation using recombinant circle PCR (RCPCR). The primers are numbered hemiarrows. Notches designate point mismatches in the primers and resulting mutations in the PCR products. Reprinted by permission from *Nature* **344,** 793,794. Copyright © 1990 Macmillan Magazines Limited.

additional cloning steps. Therefore, RCPCR permits the rapid site-directed mutagenesis of DNA. The method is best visualized using the schematic diagrams in Figs. 1 and 2.

Figure 1 illustrates point mutagenesis using RCPCR. The plasmid, containing the insert of interest, is linearized and simultaneously mutated in two separate PCR amplifications. In each of the two amplifications, the identical region is mutated. However, the positioning of the two primers in one PCR amplification is different from the positioning of the two primers in the other PCR amplification. Specifically, the breakpoint of the plasmid in one PCR amplification is different

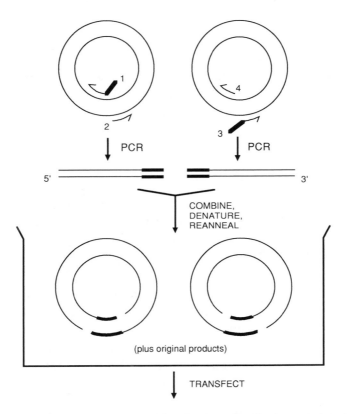

Fig. 2. Diagram showing the generation of an insertional mutagenesis using recombinant circle PCR (RCPCR). The primers are numbered hemiarrows. The heavy lined 5' regions of primers 1 and 3 are the complementary strands of the segment that is inserted into the plasmid. Reprinted by permission from *Nature* **344,** 793,794. Copyright © 1990 Macmillian Magazines Limited.

from the breakpoint in the other PCR amplification. This breakpoint is determined by the positioning of the two 5' ends of the primers on the plasmid, and represents the region of linearization of the plasmid. The breakpoint can also be viewed as the "cut site" of the circular plasmid template resulting from the PCR amplification of a linear product.

Figure 2 illustrates insertional mutagenesis using RCPCR. In this insertional mutagenesis protocol, primers 1 and 3 have 5' ends that are composed of the complementary strands of the segment that is to be inserted. Following PCR amplification, these complementary ends are attached to opposite ends of the plasmid sequence. Therefore, following the combining, denaturing, and reannealing of the two PCR products, double-stranded products will form with single-stranded ends that are complementary to each other, and will anneal to yield recombinant circles. The DNA joint that holds together the recombinant circles is formed by the inserted sequence.

Briefly, site-directed mutagenesis using RCPCR comprises the following steps:

1. Two separate PCR amplification reactions with concurrent mutagenesis;
2. Separation of each entire PCR product from supercoiled plasmid template by agarose gel electrophoresis, followed by glass bead extraction;
3. Combining, denaturing, and reannealing the two PCR products, resulting in the generation of the recombinant circles;
4. Transfection of this reannealed mixture into highly competent *E. coli* without prior purification or extraction of the recombinant circles.

This method was introduced as a method for site-directed mutagenesis and as a means for creating DNA joints without any enzymatic reaction in vitro apart from PCR amplification. Site-directed mutagenesis can be accomplished without use of restriction enzymes or ligation in vitro, and without sequential amplification reactions. RCPCR has been used for the site-directed mutagenesis of genes in plasmids ranging from 2.7–6.1-kb long with >80% of clones containing the mutation of interest *(4)*. The transformation rate of the bacterial is low. Therefore, the PCR products must be adequately separated from the supercoiled template, and highly competent bacteria must be used.

2. Materials

1. 2X PCR stock mixture: Prepare 25-µL aliquots in 0.5-mL microfuge tubes containing 1.25 U of *Taq* polymerase, 400 µ*M* of each dNTP, 100 m*M* KCl, 20 m*M* Tris-HCl, pH, 8.3, 3 m*M* MgCl$_2$. Store at –20°C for up to 8 mo.
2. PCR primers (*see* Note 1).
3. Long thin micropipet tips (Costar #4853, Costar Corporation, Cambridge, MA).
4. Agarose gel components (*see* Chapter 1).
5. Ethidium bromide. Prepare a 5 mg/mL stock solution. Store in dark at 4°C for up to 6 mo. Wear gloves when handling ethidium bromide.
6. GeneClean (Bio 101, La Jolla, CA).
7. TE buffer: 10 m*M* Tris-HCl, pH 8.0, 1 m*M* EDTA.

8. 5*M* NaCl.
9. MAX Efficiency HB 101 or DH5α Competent *E. coli* (BRL, Life Technologies, Gaithersburg, MD). Once a tube is thawed it should not be reused.
10. SOC media (ref. *6*, p. A2). Prepare by dissolving 20 g of bacto-tryptone, 5 g of bacto-yeast extract, and 0.5 g of NaCl in 950 mL of deionized H_2O. Add 10 mL of 250 m*M* KCl. Adjust pH to 7.0 with 5*N* NaOH. Bring final volume to 1 L with H_2O and autoclave for 20 min at 15 lb/sq in. on liquid cycle. Allow to cool to 60°C, then add 5 mL of sterile 2*M* $MgCl_2$ and 20 mL of sterile 1*M* glucose.
11. Top agar: 7 g/L of bacto-agar (ref. *6*, p. A4). Autoclave for 20 min at 15 lb/sq in. on liquid cycle.
12. LB plates with 100 µg/mL Ampicillin (ref. *6*, p. A4).

3. Method

1. Amplify and mutate 2 ng of plasmid in two separate PCR amplifications (*see* Fig. 1 or 2), by adding 25 pmol of each primer to a tube containing the 2X PCR stock solution. Layer 50 µL of mineral oil on top of each reaction mix prior to amplification.
2. Amplify by PCR using the following cycle profile:

Initial denaturation	94°C, 1 min
14–18 main cycles	94°C, 30 s (denaturation)
	50°C, 30 s (annealing)
	72°C, 1 min/kb (extension)
Final extension	72°C , 7 min

   Store samples at 4°C. See Note 2.
3. Remove each PCR product by inserting a long thin micropipet tip through the mineral oil layer and drawing up the sample. Dissolve 1% standard high-melting-point agarose in 1X TAE buffer by boiling, and then place the bottle with the agarose in a 55°C water bath to allow the agarose to cool to 55°C. Pour the gel, mix each entire PCR product with an electrophoresis loading buffer, and electrophorese through standard high-melting-point 1% agarose in TAE buffer with 0.5 µg/mL ethidium bromide. Electrophoresis should be carried out until each PCR product has traveled at least 4 cm, in order to separate adequately the PCR products from the supercoiled plasmid template (*see* Note 3). Each PCR product will typically contain 30–100 ng of DNA, and therefore will be easily visualized by UV light.
4. Remove the band of interest with a a razor blade, and extract each PCR product from the agarose using GeneClean. Warming the NEW wash solution (contained in the GeneClean kit) from –20°C storage to at least

4°C prior to use facilitates dispersion of the glass particles during washing. Suspend each PCR product in 30 μL of TE buffer. GeneClean recovery is 25–90%.

5. Combine 25 μL of each of the two purified PCR products and add 1 μL of 5*M* NaCl, resulting in an annealing buffer with 0.1*M* NaCl. Denature the mixture at 94°C for 3 min, reanneal at 50°C for 2 h, and then place at RT.

6. Transfect MAX efficiency competent *E. coli.* (BRL; *see* Note 4) with 1–5 μL of the annealing mixture that now contains recombinant circles following the manufacturer's protocol with the following modifications: (1) Use 50 μL of *E. coli* for each sample transfected, as this is effective and less expensive than the 100 μL recommended. (2) After incubation at 37°C in a shaker for 1 h, do not dilute the sample prior to plating. Plate the entire sample onto an LB plate containing 100 μg/mL Ampicillin.

7. In order to keep the sample on the plate, add 2 mL of top agar, prewarmed to 42°C, to each sample immediately prior to pouring it onto the plate. Once an aliquot of bacteria is thawed, it is not used subsequently. The proportion of clones containing the mutation is typically >80%. *See* Notes 5 and 6.

4. Notes

1. The primers that introduce point mutations are designed with approx 20 nucleotides 5' to the mismatched nucleotide and 6–10 nucleotides 3' to the mismatched nucleotide. For a given primer length, the farther the point mutagenesis site is toward the 3' end, the longer the stand overlap in the joint that holds together each recombinant circle. These recommendations result in a strand overlap of 41 bp (20 bp + the substituted bp + 20 bp), while maintaining 3' ends of sufficient length for efficient priming *(7)*. Nonmutating primers are approx 22-bp long. Primers that introduce an insertional mutation have 5' ends that are complementary to each other *(see* Fig. 2, primers 1 and 3), and these 5' ends have been 35-nucleotides long resulting in a DNA joint with 35 bp of strand overlap. Transformation of *E. coli* using recombinant circles with short lengths of strand overlap has not been tested. Each primer has about 50% G/C content.

   Primers should be designed with care in order to ensure obtaining a detectable PCR. In general, if the entire PCR sample is loaded onto the gel, carefully designed primers will yield a visible product with the low number of cycles recommended. Use of a low number of cycles diminishes the probability of encountering a PCR-induced base misincorporation *(2)*. The eleven 3' nucleotides of proposed primers are examined by computer analysis for unwanted annealing sites to the construct. Since small gaps do not diminish the transformation efficiency *(3)*, there is considerable flexibility in choosing the annealing region for a given primer.

2. Amplify products that are <5 kb for 14–18 cycles and products that are >5 kb for 18 cycles.
3. If the PCR products do not undergo sufficient electrophoretic separation, there will be a high number of background colonies. This results from transformation by contaminating supercoiled PCR template.
4. Since the transformation efficiency is low, highly competent bacteria must be used. Following the transfection of DNA segments 2.7–3.1 kb into library efficiency HB101 competent *E. coli* from BRL ($>1 \times 10^8$ transformants/µg of monomer pBR322 DNA), an average of 217 colonies (71–400) with the mutation have been obtained per nanogram total DNA transfected, revealing a roughly 1000-fold diminished transfection efficiency per nanogram when compared to supercoiled pBR322 *(5)*. Since the transformation efficiency of larger constructs is expected to be lower, the transfection of large constructs (6.1 kb) was done into MAX efficiency HB101 competent *E. coli* ($>5 \times 10^8$ transformants/µg of monomer pBR322 DNA) resulting in 29–52 colonies with the mutation per nanogram transfected DNA *(5)*. The most common cause of lack of success with the method is use of bacteria that are not extremely competent.
5. Recombinant circles can contain two nicks, a nick and a gap, or two gaps. A recombinant circle with gaps of 116 and 86 bp has been tolerated (Brian Volpe, personal communication), and an upper limit of gap size has not been determined. The ability to use gapped circles increases the flexibility of the method by broadening the choice of primer annealing sites. It lessens the cost of doing several separate mutagenesis reactions on a given insert by allowing the use of conserved outside primers (primers 2 and 4) while using pairs of mutating primers that anneal to different regions of the insert.
6. There is always the possibility of a sequence error in a single clone following PCR amplification. Therefore, one may choose to test more than one clone in a functional assay, or clone a restriction fragment containing the mutated or recombined region of interest into a construct that has not undergone PCR amplification.

Acknowledgments

This work was supported in part by the Roy J. Carver Charitable Trust.

References

1. Mullis, K., Faloona, F., Scharf, S., Saiki, R., Horn, G., and Erlich, H. (1986) Specific enzymatic amplification of DNA in vitro: The polymerase chain reaction. *Cold Spring Harbor Symposia Quant. Biol.* **51,** 263–273.
2. Saiki, R. K., Gelfand, D. H., Stoffel, S., Scharf, S. J., Higuchi, R., Horn, G. T.,

Mullis, K. B., and Erlich, H. A. (1988) Primer-directed enzymatic amplification of DNA with a thermostable DNA polymerase. *Science* **239,** 487–491.

3. Jones, D. H. and Howard, B. H. (1990) A rapid method for site-specific mutagenesis and directional subcloning by using the polymerase chain reaction to generate recombinant circles. *BioTechniques* **8,** 178–183.

4. Jones, D. H., Sakamoto, K., Vorce, R. L., and Howard B. H. (1990) DNA mutagenesis and recombination. *Nature* **344,** 793,794.

5. Jones, D. H. and Winistorfer, S. C. (1991) Site-specific mutagenesis and DNA recombination by using PCR to generate recombinant circles *in vitro* or by recombination of linear PCR products *in vivo. Methods-A Companion to Methods in Enzymology* **2,** 2–10.

6. Sambrook, J., Fritsch, E. F., and Maniatis, T. (1989) *Molecular Cloning: A Laboratory Manual* (2nd ed.), Cold Spring Harbor Laboratory, Cold Spring Harbor, NY.

7. Sommer, R. and Tautz, D. (1989) Minimal homology requirements for PCR primers. *Nucleic Acids Res.* **17,** 6749.

CHAPTER 28

Site-Directed Mutagenesis by Double Polymerase Chain Reaction

Megaprimer Method

Sailen Barik

1. Introduction

The "megaprimer" method *(1)* based on polymerase chain reaction (PCR) is one of the simplest and most versatile procedures of site-specific in vitro mutagenesis available to date. The method utilizes three oligonucleotide primers and two rounds of PCR performed on a DNA template containing the cloned gene that is to be mutated. The rationale of the method is shown schematically in Fig. 1 where A and B represent the "flanking" primers that can map either within the cloned gene or outside the gene (i.e., within the vector sequence) and M represents the internal "mutant" primer containing the desired base change. The first round of PCR is performed using the mutant primer (e.g., M1 in Fig. 1) and one of the flanking primers (e.g., A). The double-stranded product (A-M1) is purified and used as one of the primers (hence the name "megaprimer") in the second round of PCR along with the other flanking primer (B). The wild type cloned gene is used as template in both PCR reactions. The final PCR product (A-M1-B) containing the mutation can be used in a variety of standard applications, such as cloning in expression vectors and sequencing, or in more specialized applications, such as production of the gene message in vitro if primer A (or the template sequence downstream of primer A) also contains a transcriptional promoter (e.g., that of SP6 or T7 phage).

From: *Methods in Molecular Biology, Vol. 15: PCR Protocols: Current Methods and Applications*
Edited by: B. A. White Copyright © 1993 Humana Press Inc., Totowa, NJ

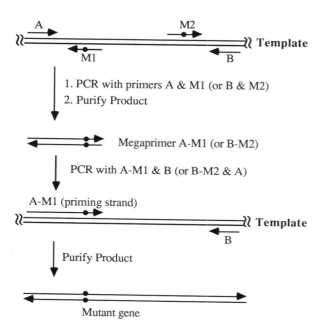

Fig. 1. The megaprimer method of site-specific mutagenesis. The primers A, B, M1, and M2 (as well as the priming strand of the megaprimer, AM1) are indicated by single lines with arrowhead, whereas the double lines represent the template. The dot in M1 and M2 indicates the desired mutations (base changes) to be introduced into the product via the megaprimer. *See* text for details.

Both primers A and B are usually designed to contain convenient restriction sites so that the final, mutant PCR product can be restricted and cloned. When a gene is cloned within two unique sites (a multicloning site, for example) of a vector, primers A and B can be made at these sites. The mutant product can then be restricted and ligated at the same two sites in the same vector, and the final clone will retain the original flanking sequences of the gene. This is especially important when sequences upstream of A (e.g., a Shine-Dalgarno sequence or an upstream activation sequence, UAS) or downstream of B (e.g., a polyadenylation site or an RNase processing site) are essential for gene expression or regulation and must remain unaltered.

More often than not, the gene in question codes for a protein, and studies of structure–function relationship require the generation of a battery of mutant proteins altered at specific amino acid residues. In such cases, primers A and B can be kept constant and a variety of mutant primers (M1, M2, and so on) can be used to produce the various mutants. The reciprocal combination of primers (i.e., M2 and B in the first round of PCR; then A and megaprimer M2-B in the second round) can also be used provided primer M2 is of the opposite sense of B (Fig. 1). Choice of primer location is described in more detail under Methods.

Since the megaprimer can be quite large (it may approach the size of the whole gene) and incorporated internal to the gene, successful and error-free PCR in the second round often requires special considerations. The reader is therefore strongly urged to go through the whole chapter, including the Notes section before proceeding with the actual experiment.

Note that the double-stranded megaprimer is directly used in the second round of PCR; prior separation of the two strands being unnecessary. Melting of the megaprimer is essentially achieved in the denaturation step of the PCR cycle itself. Although both strands of the megaprimer will anneal to the respective, complementary strands of the template, the basic rules of PCR amplification automatically ensures that only the correct strand (one that extends to the other primer, B, in Fig. 1) will be amplified into the double-stranded product. Under some conditions, particularly with large megaprimers (≥1 kb), self-annealing of the megaprimer tends to reduce the yield of the product *(2)*. In order to avoid this, use of higher amounts of template (in microgram range, as opposed to nanogram quantities used in standard PCR) in the second PCR has been recommended in the method; this improves the final product yield in many cases *(2)*, presumably because extra template strands anneal to both strands of the megaprimer, thereby preventing renaturation of the megaprimer.

2. Materials

1. DNA template containing the cloned gene (e.g., in pUC or pGEM vector) to be mutated.
2. Oligonucleotide primers A and B: the "upstream" primer A in the message sense and the "downstream" primer B in the antimessage sense. Include restriction sites, preferably unique, in these primers so that the final product can be restricted and cloned (*see* also Section 3.1. and Note 1).
3. Reagents for PCR (*see* Chapter 1).
4. A system for purifying the PCR products. An electroelutor (IBI, New Haven, CT) is used here. *See* Index for other methods in this volume.

5. Reagents for agarose gel electrophoresis (*see* Chapter 1).
6. 8*M* Ammonium acetate, 0.01% bromophenol blue.
7. TE buffer: 10 m*M* Tris-HCl, pH 7.5, 1 m*M* EDTA.
8. CHCl₃.
9. TE-saturated phenol.
10. 95% Ethanol.
11. 70% Ethanol.

3. Method
3.1. Primer Design

1. For technical reasons described here, avoid making megaprimers that approach the size of the final, full-length product (gene) AB (Fig. 1). If M1 is too close to B, it will make separation of AB and AM1 (leftover megaprimer) difficult after the second round of PCR. Ideally, the megaprimer should be shorter than the full-length gene by more than 200–500 bp, depending on the exact length of the gene. (Example: If the gene AB is 2 kb, megaprimer AM1 can be up to ~1.5-kb long, since 2 kb and 1.5 kb can be separated reasonably well in agarose gels. However, if the gene is 8 kb, the megaprimer should not be bigger than, say, 7 kb, since 8- and 7-kb fragments would migrate so close to each other.)
2. When the mutation is to be created near B, make an M primer of the opposite polarity, e.g., M2 and synthesize BM2 megaprimer (rather than AM2). *See* Fig. 1.
3. When the mutation is at or very near the 5' or 3' end of the gene (say, within 1–50 nucleotides), there is no need to use the megaprimer method! One can simply incorporate the mutation in either A or B primer and do a straightforward PCR. In borderline situations, such as when the mutation is, say, 120 nucleotides away from the 5' end of the gene, incorporation of the mutation in primer A may make the primer too big to synthesize; or else, it will make the megaprimer AM1 too short to handle conveniently. In such a case, simply back up primer A a few hundred bases further into the vector sequence in order to make AM1 megaprimer longer. In general, remember that primers A and B can be located anywhere on either side of the mutant region, and try to utilize that flexibility as an advantage when designing these primers.
4. In addition to the standard rules of primer design described in Chapter 2 (such as, a near 50% GC content, sequence specificity, extra "clamp" sequence for restriction, absence of self-complementarity, and so forth), attention should be paid to the following aspects. As stated before, primers A and B should contain unique restriction sites for ease of cloning. As regards the M primer, two things are important. First, the muta-

tional mismatch should not be too close to the 3' end of the primer. Mismatch at the very 3' nucleotide of the primer will virtually abolish amplification by *Taq* polymerase. For best results, the mismatch should be at least four bases away from the 3' end of the primer. Second, as described in detail in Note 1, the 5' end of the M primer should preferably be located such that there is at least one (two or more is better) T residue in the template strand of the same sense just upstream of this end of the primer. If a T is not available, try to have the "wobble" base of a codon just upstream of the 5' end of the M primer so that substitution of this base with A will code for the same amino acid (*see* Note 1).

3.2. PCR 1: Synthesis of Megaprimer

1. Assemble a standard 100-μL PCR reaction (*see* Chapter 1) containing 5–50 ng of template DNA (containing the cloned gene) and 0.2–0.4 μg each of primers A and M1 (assuming a primer length of 25 nucleotides).
2. Amplify by PCR using the following cycle profiles (*see* Notes 2 and 3):

Initial denaturation	94°C, 6 min
30–35 main cycles	94°C, 2 min (denaturation)
	55°C, 2 min (annealing)
	72°C, appropriate time (*n* min).
Final extension	72°C, 1.5 × *n* min

   To achieve a better yield of megaprimer, do four such PCRs, 100 μL each, in separate tubes, as described.
3. Remove oil overlay as follows. Freeze PCR tubes, then thaw just enough so that the oil overlay melts but the aqueous reaction stays frozen. Remove as much oil as possible. Then use a drawn-out round tip to transfer ~80 μL of the lower, aqueous layer carefully to a clean Eppendorf tube, wiping the outside of the tip to remove any adhering oil. It is important to remove the oil completely, otherwise, the sample will float up when loaded in horizontal agarose gels in step 4 (this section). Pool about 320 μL from four reactions. Reduce volume to ~40 μL in SpeedVac (Savant Instruments, Farmingdale, NY) concentrator (or other means of evaporation or lyophilization).
4. Gel-purify megaprimer in standard agarose gels made in TBE in presence of ethidium bromide (*see* Chapter 1). Load in two or three wells in a midsize horizontal gel. Use an appropriate concentration of agarose (0.7–1.2%) depending on the length of the megaprimer product to be purified (*see* Note 4). Perform electrophoresis until a good separation of the megaprimer and the small primer has been achieved.
5. Locate the megaprimer band by UV light, cut out the gel slices, place all slices of the same megaprimer in one slot of the electroelutor that is

already filled with TE buffer or TBE buffer (*see* instruction manual of the electroelutor). Make sure that the apparatus is leveled, that buffers in the two chambers are connected, and that all buffer flow has stopped. Clear out any air bubble that may be trapped in the V-shaped grooves and put 75 µL of 8*M* ammonium acetate containing 0.01% BPB in each groove. Electrophorese at 150 V for required period (approx 15–30 min), as judged by the disappearance of the DNA band from the gel slice into the V groove (monitored by a hand-held UV light). Do not disturb the apparatus during elution.

6. Discontinue electrophoresis, carefully drain buffer out of both chambers, collect 400 µL of ammonium acetate solution containing DNA from each V groove and transfer into a sterile clean microcentrifuge tube. Add 2–4 µg of carrier tRNA at this point to improve recovery. *See* Notes 5 and 6.

7. Clean the DNA by the usual phenol-chloroform extraction (*see* Chapter 1), precipitate with 2.5 vol (1 mL) of 95% ethanol (do not add extra salt), followed by centrifugation at maximum speed in a microcentrifuge for 15 min.

8. Wash DNA pellet with prechilled 70% ethanol and dissolve in 20 µL of sterile water. Expect 30–60% recovery of DNA irrespective of size.

3.3. PCR 2

1. Assemble a second PCR reaction with 100 µL final vol containing 1–2 µg (note the higher than usual amount, *also see* Note 5) of plasmid template (the same template that was used in Section 3.2., step 2), 0.2–0.4 µg primer B, 20–30 pmol of gel-purified megaprimer (3–5 µg for a 250-bp DNA), and standard concentration of buffer and nucleotides (*see* Chapter 1). Generally, expect to use up all the megaprimer recovered in step 4 in this PCR in order to achieve a good yield of the product! Use the maximum allowable temperature for annealing in the thermal cycle, as dictated by the smaller primer B. Ignore the megaprimer for annealing considerations, since its T_m will be too high for the smaller primer.

2. Amplify by PCR using the cycle profiles described in step 2 of Section 3.2. using 25 main cycles.

3. Purify the mutant PCR product as described in Section 3.2., steps 4–7. It is now ready for restriction, ligation, and so forth by the use of standard procedures. *See* Notes 8 and 9.

4. Notes

1. The problem of nontemplated insertions and its solution will be published in detail elsewhere (Barik, manuscript in preparation). In brief, *Taq* polymerase has a tendency to incorporate nontemplated residues,

particularly A, at the 3' end of the daughter polynucleotide strand at a certain frequency *(3)*. These are then copied and amplified into the double-stranded product. This is generally not a problem in standard PCR where the termini of the product are usually cleaved off by restriction enzymes for cloning purposes. However, in the megaprimer method, the whole megaprimer is directly incorporated into the final product. Therefore, nontemplated A residues in the megaprimer will eventually show up in a certain percentage of the final product and cause a mutation that may be undesirable. The frequency of such "error" is usually low and megaprimers with a mismatch at the 3' end will not prime well; however, the frequency may be appreciable in some cases (Barik, manuscript in preparation). There are two kinds of solutions to this problem; one kind, exemplified by a and b below, does not introduce the nontemplated base or removes it; the other kind, described in c, does not remove the nontemplated base but tolerates the alteration.

a. Use the thermostable Vent DNA polymerase (New England Biolabs, Beverly, MA), which has a 3' exonuclease activity; follow the manufacturer's recipe. A typical 100-µL PCR will contain: 20 mM Tris-HCl, pH 8.8 (at 25°C), 10 mM KCl, 10 mM (NH$_4$)$_2$SO$_4$, 2 mM MgCl$_2$, 0.1% Triton X-100, 100 µg/mL acetylated BSA, 200 µM of each dNTPs, 2 U of Vent polymerase, and standard amounts of template and primers. The temperature and time values of the thermal cycles are identical to those for *Taq* polymerase. Vent polymerase is a relatively recent product and may need some optimization.

b. Treat the purified megaprimer with a DNA polymerase that has a 3'→5' exonuclease activity. Following the first round of PCR reaction, remove oil as described in step 3 of Method. Add DTT to a final concentration of 1 mM (even if the PCR buffer had DTT). Add 2.5 U of T4 DNA polymerase or 7 U of Klenow enzyme *(4)*. Incubate at 37°C for 5 min with the T4 enzyme or at 15°C for 30 min with Klenow. Inactivate the polymerases by heating at 65–70°C for 5–10 min and then proceed to purify the megaprimer after reducing volume, gel-purifying, and so forth as in step 4 of Method. Since all exonuclease digestions are hard to control, this method is *not* highly recommended and should be used as a last resort. Instead try the following method.

c. Tolerate the alteration (If you can't beat 'em, join 'em): *This is the method of choice*. It relies on clever primer design and does not require any extra step (in preparation). There are two ways of achieving this. As an example, suppose the relevant region of the wild-type sequence is (the amino acids are shown at the bottom in single-letter codes):

5'— AAA CTG CCA ACT CCG TCA TAT CTG CAG —3'

3'— TTT GAC GGT TGA GGC AGT ATA GAC GTC —5'
    K   L   P   T   P   S   Y   L   Q

and the *Ser* (TCA) is to be mutated to *Ala* (GCA). A mutant primer in the message sense (like M2 in Fig. 1) may have the sequence $^{5'}$CA ACT CCG GCA TAT CTG CAG $^{3'}$ (the boldface G being the mutant base). However, when this M2 primer and primer B is used in PCR, the nontemplated A incorporated at the M2 end of the product (megaprimer) will result in the sequence:

5'*TCA* ACT CCG GCA TAT CTG CAG —

3'*AGT* TGA GGC CGT ATA GAC GTC —

(the nontemplated A/T is italicized). When incorporated into the final product, this megaprimer will produce the following mutant (the italicized amino acids are altered from the wild-type sequence):

5'— AAA CTG TCA ACT CCG GCA TAT CTG CAG —3'
    K   L   *S*   T   P   *A*   Y   L   Q

resulting in an undesired Ala → Ser change (boldface). To avoid this, make the following M2 primer: $^{5'}$G CCA ACT CCG GCA TAT CTG CAG $^{3'}$ so that there is a T residue upstream of the 5' end of M2 in the template sequence; any extratemplated T in this strand of the megaprimer will therefore match with the T residue in the wild type sequence and will not cause any mutation.

When no T residues are available, use the wobble base of a codon. This is possible when the primary purpose of the clone is to produce a protein product; thus, substitution of a codon with another, synonymous codon is permissible (make sure that the resultant change in the nucleotide sequence is acceptable in terms of introduction or loss of restriction sites, and so forth). Now, make the following M2 primer: $^{5'}$ACT CCG GCA TAT CTG CAG $^{3'}$, so that the codon upstream of it is CCA. The nontemplated T will change this codon to CCT; however, since they both code for proline, the protein will remain unaltered.

2. As a rule, elongation time (at 72°C) in a PCR cycle should be proportional to the length of the product. An approximate guideline is 1 min of elongation/kb, i.e., 100 nucleotides = 10 s; 500 nucleotides = 40 s; 1 kb = 1 min 40 s; 2 kb = 3 min, and so on.

3. Annealing temperature is primarily governed by the base composition of the primers. A golden rule is to calculate the T_m of the primer as

follows: Add 2°C for each A or T, and 4°C for each G or C, then deduct 4°. Example: for a 22-nucleotide primer with 10 G+C and 12 A+T, T_m is $[(10 \times 4 + 12 \times 2)] = 64°C$; therefore, anneal at 60°C. However, the upper limit of the annealing temperature for any primer is 72°C, since it is the elongation temperature of the *(Taq)* polymerase.

4. Use 0.8–1.2% agarose for products ranging from 100–300 bp; 0.7% for 0.5–1 kb; 0.6–0.7% for 1 kb and higher, and so forth. For making megaprimers in the 100–300-bp size range, it is important that the template preparation does not contain too much tRNA, which otherwise tend to mask the megaprimer in agarose gel and make visualization and purification of the megaprimer difficult.

5. If in the second round of PCR, higher concentration of template still does not help (very rare), it is advisable to make a single-stranded megaprimer (the priming strand, Fig. 1) by using the technique of "asymmetric" PCR in the first round. This is done by using a limiting amount of primer M1 (20–40-fold less, i.e., 5 pmol) while using the standard amount of primer A (~0.3 mg) in an otherwise identical PCR. A smaller number of cycles (20 cycles) and a slightly lower annealing temperature (50–55°C) are recommended *(5)*. The major problem in single-stranded DNA (ssDNA) PCR is the difficulty in locating the ssDNA in the gel: it does not bind EtBr well, migrates faster than the its double-stranded counterpart, and produces diffused or multiple bands. However, once standardized with a particular ssDNA product, the procedure can be used routinely for the same DNA. Note that it will be difficult to remove extratemplated nucleotides, if any, from ssDNA megaprimers; thus, plan c in Note 1 should be used in conjunction.

6. The electroelution procedure for purification of PCR products described here has been chosen as the single, most versatile method for such purpose. Several workers may prefer other methods for more specialized and routine applications, a few of which are as follows:

   a. Centricon-30 spin filtration column (Amicon) may be used directly with the PCR reaction to separate the megaprimer product from the small PCR primers provided the megaprimer is sufficiently bigger than the smaller primers. Megaprimer purified by this method will also contain the template; however, this is of no concern since the same template is going to be used in the second round of PCR. Follow the recommended procedure *(6)*.

   b. Freeze-squeeze method *(7)*: The method works best for fragments <500 bp. The agarose gel slice containing the DNA fragment is taken in an Eppendorf tube, frozen in a dry ice-ethanol bath (10 min) or in a –70°C freezer (20 min) and then spun at room temperature in a microcentrifuge for 15 min. The recovered (30–70%) megaprimer DNA can be used directly in PCR.

c. The GeneClean method (Bio 101, La Jolla, CA) utilizes the property of DNA to bind to glass (specially prepared, fine glass powder is supplied in the kit), however, it works best for DNA segments >500 bp. It is a rather elaborate procedure and is outside the scope of this chapter. The reader must follow the detailed instructions that come with the kit. A somewhat modified version has been published *(1)*.

7. In a recent method *(8)*, DNA fragments in low-melting-point agarose slices have been directly used in PCR reaction apparently without any problem. Since a good quantity of the megaprimer is important for the second PCR, it might be worth trying to use it in a similar manner. This will bypass the need to recover it from the gel. Further optimization of the method is advised.

8. As in any cloning procedure, the final mutants obtained by the megaprimer method must be confirmed by DNA sequencing. This can be done either by directly sequencing the PCR product (Chapters 13 and 14) or after cloning the mutant product in plasmid vectors. When using the dideoxy method, PCR primer A or B can be used as sequencing primers as well.

9. Other methods of site-specific mutagenesis exist that are based on PCR. These are discussed in detail in Chapters 24–27.

References

1. Sarkar, G. and Sommer, S. S. (1990) The "megaprimer" method of site-directed mutagenesis. *BioTechniques* **8,** 404–407.
2. Barik, S. and Galinski, M. (1991) "Megaprimer" method of PCR: Increased template concentration improves yield. *BioTechniques* **10,** 489,490.
3. Clark, J. M. (1988) Novel non-templated nucleotide addition reactions catalyzed by procaryotic and eucaryotic DNA polymerases. *Nucleic Acids Res.* **16,** 9677–9686.
4. Maniatis, T., Fritsch, E. F., and Sambrook, J. (1982) *Molecular Cloning: A Laboratory Manual.* Cold Spring Harbor Laboratory, Cold Spring Harbor, NY, pp. 113–119.
5. Finckh, U., Lingenfelter, P. A., and Myerson, D. (1991) Producing single-stranded DNA probes with the Taq DNA polymerase: A high yield protocol. *BioTechniques* **10,** 35–39.
6. Higuchi, R., Krummel, B., and Saiki, R. K. (1988) A general method of in vitro preparation and specific mutagenesis of DNA fragments: Studies of protein and DNA interactions. *Nucleic Acids Res.* **16,** 7351–7367.
7. Stoflet, E. S., Koeberl, D. D., Sarkar, G., and Sommer, S. S. (1988) Genomic amplification with transcript sequencing. *Science* **239,** 491–494.
8. Zintz, C. B. and Beebe, D. C. (1991) Rapid re-amplification of PCR products purified in low melting point agarose gels. *BioTechniques* **11,** 158–162.

CHAPTER 29

Generation of a Polymerase Chain Reaction Renewable Source of Subtractive cDNA

W. Michael Kuehl and James Battey

1. Introduction

Differential (+/-) first-strand cDNA screening methods identify clones corresponding to mRNAs that are expressed at a higher level in one of a pair of phenotypically different cells. This approach is limited by the fact that screening of libraries with labeled first-strand cDNAs synthesized from unfractionated mRNA can detect clones containing sequences representing approx 0.1% or more of the complexity of mRNA (i.e., mRNAs present at greater than about 200 copies per cell since a typical mammalian cell line contains approx 250,000 mRNAs).

To enhance the sensitivity of this approach, a number of laboratories *(1–4)* developed methodology for preparing subtractive cDNA probes: (1) First-strand cDNA from one cell type is hybridized to an excess of mRNA from a closely related cell type; (2) the small fraction of single-stranded cDNA is separated from the bulk of double-stranded mRNA-cDNA hybrid; and (3) small fragments of nonhybridizing cDNA are removed from the subtractive cDNA. Depending on the extent of relatedness of the mRNAs from the parental and subtractive partner cell lines, cDNA sequences unique to the parental cell line might be enriched as much as 50-fold in the subtractive cDNA. As a result of this enrichment, the subtractive cDNA probe could be used to identify clones containing differentially expressed sequences comprising approx 0.002% or more of the complexity of cellular mRNAs (i.e., greater than approx 5 copies of

From: *Methods in Molecular Biology, Vol. 15: PCR Protocols: Current Methods and Applications*
Edited by: B. A. White Copyright © 1993 Humana Press Inc., Totowa, NJ

mRNA per cell). The subtractive cDNA could also be used to prepare subtractive cDNA libraries, so that many fewer clones need to be screened.

There are increasing numbers of important genes that have been identified using subtractive cDNA technology, but the method requires large amounts of mRNA, is technically demanding, and results in very low yields of short cDNA fragments. A number of novel approaches, many of which involve preparation of cDNA libraries from each subtractive partner, appear to avoid some of these problems *(5–11)*. In this chapter, we describe a method that incorporates PCR technology, which partially solves the need for large amounts of mRNA but also permits more flexibility in using the PCR renewable subtracted cDNA as a probe or for cloning *(11)*. A similar method has been used successfully for genomic DNA subtractions *(12)*.

The following is an outline of the PCR/subtractive cDNA method presented:

1. Synthesize first-strand cDNA from parental cell mRNA;
2. Hybridize cDNA to excess of mRNA from subtractive partner;
3. Remove cDNA:mRNA duplex from negatively selected single-stranded cDNA by hydroxyapatite (HAP) chromatography;
4. Hybridize single-stranded cDNA to mRNA from parent or related cell;
5. Remove single-stranded cDNA from positively selected cDNA:mRNA duplex by HAP chromatography;
6. Synthesize second-strand cDNA from mRNA:cDNA template with RNase H and DNA polymerase I;
7. "Polish" and kinase ends of double-stranded cDNA with T4 DNA polymerase and T4 kinase;
8. Add amplification adapters with T4 ligase;
9. Remove excess adapters and small cDNAs by Sepharose 4B chromatography;
10. PCR amplify subtractive cDNA;
11. PCR chase reaction to ensure subtractive cDNA is mostly homoduplexes;
12. Size-fractionate amplified, subtractive cDNA on Sepharose 4B.

2. Materials

2.1. First-Strand cDNA Synthesis

1. 5 µg of parental mRNA, selected twice on oligo(dT) cellulose (*see* Notes 1–3).
2. MMLV reverse transcriptase (RT), 200 U/µL.
3. 5X RT buffer: 0.25M Tris-HCl, pH 8.3, 0.375M KCl, 15 mM MgCl$_2$, 50 mM DTT.

4. Random hexamer at 0.25 µg/µL.
5. Adapter-dT$_{17}$ primer
   (e.g., 5'-GGACTCGAGGTATCGATGCTTTTTTTTTTTTTTTTTTTT-3',
   which has *Xho* I and *Cla* I restriction sites) at 0.1 µg/µL.
6. Mixture of 10 mM of each dNTP (i.e. dGTP, dATP, dCTP, TTP), pH 7.0.
7. [α-^{32}P]dCTP, 3000 C$_i$/mmol.
8. RNAsin, 30 U/µL (Promega, Madison, WI).
9. BSA, nuclease free, 10 mg/mL.
10. 0.5M EDTA, autoclaved.
11. 2M Tris-HCl, pH 7.4, autoclaved.
12. 2N HCl.
13. 10N NaOH.
14. Phenol.
15. CHCl$_3$.
16. TE buffer: 10 mM Tris-HCl, pH 7.5, 0.1 mM EDTA.
17. Sephadex G-50 column or spin column equilibrated with TE.
18. 100% Trichloroacetic acid.
19. 3M Sodium acetate, pH 6.0, autoclaved.
20. Ethanol.
20. Siliconized 1.5-mL microfuge tubes (e.g., PGC Scientifics, Gaithersburg, MD, #505-201).

2.2. Hybridization of mRNA to cDNA

1. 50 µg of subtractive partner mRNA (selected twice on oligo [dT] cellulose).
2. 50 µg of poly(A)- subtractive partner RNA for carrier, and also providing rRNA for subtraction.
3. 5X Hybridization buffer: 3M NaCl, 100 mM Tris-HCl, pH 7.7, 10 mM EDTA (autoclaved).
4. 1% SDS.
5. Mineral oil.
6. Siliconized 0.5-mL microfuge tube (e.g., PGC Scientifics #505-195).

2.3. Separation of cDNA and cDNA:mRNA by Hydroxyapatite Chromatography

1. Hydroxyapatite powder (Bio-Rad HTP, Richmond, CA).
2. 0.5M Na phosphate buffer, pH 6.8 (PB). Prepare by mixing equal volumes of 0.5M monosodium phosphate and 0.5M disodium. phosphate.
3. 5M NaCl.
4. Appropriate dilutions of PB buffer in 150 mM NaCl (*see* Section 3.3.2. and Note 4).
5. Chromatography system that can be used at 60°C (*see* Section 3.3.2. and Note 5).

6. Labeled test DNAs: single-stranded = first-strand cDNA (Section 3.1.); double-stranded = φX174 *Hae* III fragments labeled with Klenow enzyme and [α-^{32}P]dCTP to high specific activity.
7. 6*N* HCl.
8. Nensorb™ 20 cartridge (DuPont-NEN, Boston, MA). Since the capacity is approx 20 μg of nucleic acid, add contents of additional cartridges to process larger samples.
9. 50% Ethanol.
10. Siliconized 1.5-mL microfuge tubes.

2.4. Positive Selection and Generation of Double-Stranded cDNA

1. 10 μg of parental (or related) mRNA (twice selected on oligo[dT] cellulose).
2. 5X Hybridization buffer, 1% SDS, mineral oil.
3. Reagents for HAP chromatography (as in Section 2.3.).
4. 4X Second-strand buffer: 80 m*M* Tris-HCl, pH 7.4, 20 m*M* MgCl$_2$, 40 m*M* ammonium sulfate, 400 m*M* KCl, nuclease-free BSA at 0.2 μg/μL.
5. dNTP mixture, each at concentration of 2.5 m*M*.
6. [α-^{32}P]dCTP (3000 C$_i$/mmol).
7. RNase H, 2 U/μL.
8. DNA polymerase I, 5 U/μL.
9. 0.5*M* EDTA, pH 8.
10. Phenol.
11. CHCl$_3$.
12. TE buffer: 10 m*M* Tris-HCl, pH 7.4, 0.1 m*M* EDTA.
13. Sephadex G-50 column or spin column equilibrated with TE.
14. 20 μg of DNA-free RNA carrier, e.g., poly(A)- RNA.
15. 3*M* Sodium acetate, pH 6.
16. Ethanol.
17. Siliconized 1.5-mL microfuge tubes.

2.5. PCR Amplification of Subtractive cDNA

1. Thermal cycler, e.g., Cetus/Perkin-Elmer (Norwalk, CT).
2. 10X T4 polymerase buffer: 700 m*M* Tris-HCl, pH 7.7, 100 m*M* MgCl$_2$, 50 m*M* DTT.
3. dNTPs at 5 m*M* and 10 m*M* each.
4. 10 m*M* ATP.
5. T4 DNA polymerase, 5 U/μL.
6. T4 polynucleotide kinase, 5 U/μL.
7. 0.5*M* EDTA, pH 8.
8. Phenol.
9. CHCl$_3$.

10. 40 µg of DNA-free carrier RNA, e.g., poly(A)- RNA.
11. 3*M* Sodium acetate, pH 6.
12. Ethanol.
13. Amplification adapter at 500 ng/µL, e.g., mix 5 µg of 5'-AGCTAGAATTCGGTACCGTCGACC-3' with 5 µg of 5' phosphorylated-GGTCGACGGTACCGAATTCT-3'; dilute to 20 µL total vol with 2 µL 10X T4 polymerase buffer and H₂O, heat to 95°C for 5 min, and allow to cool gradually to room temperature. Stock solution can be stored at –20°C.
14. 5X T4 ligase buffer (e.g., BRL, Gaithersburg, MD).
15. T4 ligase (e.g., BRL).
16. TE buffer: 10 m*M* Tris-HCl, pH 7.4, 1 m*M* EDTA.
17. Sepharose CL-4B.
18. Sepharose 4B buffer: 0.1*M* NaCl, 20 m*M* Tris-HCl, pH 7.4, 1 m*M* EDTA.
19. 10X PCR amplification buffer: 100 m*M* Tris-HCl, pH 8.3, 500 m*M* KCl, 15 m*M* MgCl₂, 0.01% gelatin.
20. Amplification primer, e.g. 5'- AGCTAGAATTCGGTACCGTCGACC-3', 1 µg/µL.
21. *Taq* polymerase, 5 U/µL.
22. Mineral oil.
23. 1.5% Agarose gel and buffer (*see* Chapter 1).
24. φX174 *Hae* lll markers (e.g., BRL).
25. Acrylamide gel components (*see* Chapter 1).

3. Methods
3.1. First-Strand cDNA Synthesis

The choice of subtractive partners is the single most important consideration in using this approach. Subtractive cDNA methods work optimally for qualitative differences in mRNA expression, but are less certain of achieving enrichment of cDNA sequences corresponding to mRNAs that are expressed at a quantitatively higher level (this may be possible by using only a minimal excess of the subtractive partner mRNA for the subtractive hybridization). In principle, the highest degree of enrichment of unique parental cDNAs occurs when the subtractive partner differs only minimally from the parent. However, this highly enriched pool of subtractive cDNAs is likely to include a significant fraction of cDNAs representing very rare (<1 copy mRNA/cell) and clone-specific differences. To minimize the presence of these two categories, positive selection can be done to a relatively low Rot (*see* Notes 6 and 7) using mRNA from a cell line that shares critical properties with the parental cell.

1. Place 5 µg of parental mRNA in 16 µL of H_2O in a 1.5-mL microfuge tube (*see* Notes 1 and 2).
2. Incubate at 70°C for 3 min, and then place on ice.
3. Add in the following order:
   a. 10 µL of 5X RT buffer
   b. 2 µL of adapter-dT_{17} or random hexamer primer (*see* Note 8)
   c. 5 µL of solution containing each dNTP at 10 mM
   d. 1 µL of RNAsin, 30 U/µL
   e. 1 µL of nuclease-free BSA, 1 µg/µL
   f. 10 µL (100 µCi) of [α-^{32}P]dCTP (3000 C$_i$/mmol)
   g. 5 µL of MMLV reverse transcriptase, 200 U/µL.
4. Microfuge for 5 s. Mix gently and incubate at 37°C for 60 min.
5. Add 3 µL of 0.5M EDTA and 2 µL of 10N NaOH. Mix.
6. Incubate at 70°C for 30 min. Cool to room temperature.
7. Add 35 µL of 2M Tris-HCl, pH 7.4, and mix. Add 10 µL of 2N HCl and mix.
8. Extract with 100 µL of phenol:CHCl$_3$, and recover aqueous phase. Remove duplicate 1-µL aliquots and dilute each to 20 µL. Use a 2-µL aliquot from each dilution to determine total cpm. The remaining 36 µL can be analyzed on a denaturing acrylamide gel or an alkaline agarose gel to determine the size distribution of the first-strand cDNAs.
9. Remove labeled cDNA from unincorporated dNTPs by Sephadex G-50 chromatography (or by using a Sephadex G-50 spin column). Remove an aliquot from the excluded fraction to determine incorporated cpm.
10. The amount of first-strand cDNA in µg = (incorporated cpm/total cpm) × 66; this should be 0.5–2 µg (i.e., 10–40% yield). The specific activity of the first-strand cDNA is about 3000 cpm/ng (i.e., 1000–2000 Cerenkov cpm/ng or 1 ^{32}P/18,000 nucleotides). The material can be monitored by Cerenkov counts in all subsequent procedures.
11. Precipitate the excluded fractions by adding 0.1 vol of 3M Na-acetate and 2.5 vol of ethanol.
12. After 30 min on ice, microfuge the sample for 10 min at 4°C.
13. Discard supernatant. Add 1 mL ethanol, microfuge for 2 min at 4°C, and remove supernatant. Dry pellet in a vacuum centrifuge.

3.2. Hybridization of mRNA to cDNA (see Notes 6, 9, and 10)

1. Add 5 µg of poly(A)- and 50 µg of poly(A) + RNA from the subtractive partner (i.e., a 10-fold excess of mRNA compared to the parental mRNA used for first-strand cDNA synthesis) to the cDNA. Transfer sample to a 500-µL microfuge tube, and dry in a vacuum centrifuge.
2. Dissolve sample in 6 µL of H_2O. Then add 2 µL of 5X hybridization buffer plus 2 µL of 1% SDS.

3. Mix, microfuge for 5 s, and overlay with 100 μL of mineral oil.
4. Boil sample for 5 min, then incubate at 70°C to allow the reaction to achieve a Rot of 3000. At a concentration of 5 μg/μL of mRNA, this will require 20 h (*see* Note 6).

3.3. Separation of cDNA and cDNA:mRNA by Hydroxyapatite Chromatography

Single- and double-stranded polynucleotides can be conveniently separated from one another by chromatography on hydroxyapatite (HAP) columns *(1,16)*. Empirically it has been found that single-stranded and double-stranded nucleic acids are differentially bound to HAP in solutions containing low concentrations of phosphate. The best separations occur at 60°C, with single-stranded polynucleotides generally eluting quantitatively at approx 120 m*M*, pH 6.8 sodium phosphate buffer (PB), and double-stranded polynucleotides eluting quantitatively at 300–400 m*M* PB. Since the elution conditions may differ somewhat for different batches of HAP, it is recommended that the optimum PB concentrations for eluting single- and double-stranded polynucleotides be determined for each lot of HAP (*see* the following section).

3.3.1. Preparation of HAP

1. Mix 10 g of hydroxyapatite powder (Bio-Rad HTP) with 50 mL of 50 m*M* PB, 150 m*M* NaCl. Allow to settle for 10 min and remove fines.
2. Add 50 mL of 50 m*M* PB, 150 m*M* NaCl. Mix and heat 15 min in boiling water.
3. Mix and allow to settle for 10 min and remove fines. Note packed volume, which should be about 25 mL. Add 2 vol (approx 50 mL) of 50 m*M* PB, 150 m*M* NaCl to give a 3:1 slurry, which can be stored for several months at 4°C. The capacity is about 200 μg of double-stranded polynucleotide per milliliter of packed HAP.

3.3.2. HAP Fractionation of Single-Stranded cDNA from RNA-cDNA Duplexes (see Notes 4 and 5)

1. At the end of the hybridization, remove most of the mineral oil and dilute the aqueous phase with 250 μL of the 60°C HAP starting buffer. Transfer the sample to a 10-mL polypropylene tube, which contains about 0.5 mL of packed HAP and is maintained at 60°C. Rinse the hybridization tube with another 250 μL of the HAP starting buffer (the hybridization tube can be Cerenkov-counted to ensure quantitative transfer of the cDNA).

2. Mix the sample/HAP slurry and allow it to settle for 5 min. Remix the slurry, and allow it to settle for 5 min.
3. Mix the material a third time, pipet it into a column, and allow it to settle for 5 min.
4. Collect the flow-through in a microfuge tube.
5. Collect six 1-mL washes with the HAP starting buffer in six microfuge tubes.
6. Elute the column with six 1-mL aliquots of 450 mM PB, 150 mM NaCl into six microfuge tubes.
7. Determine which fractions contain single-stranded (flow-through and wash fractions) and double-stranded (450 mM PB) cDNA from Cerenkov counts. It is also possible to dissolve the HAP with 1 mL of 6N HCl to determine the portion of material that is not eluted; this is usually <5% of the total material applied to the column.
8. Pool the appropriate single-stranded fractions (e.g., following a subtractive hybridization step) and reapply to a second HAP column. It is not necessary to rechromatograph the double-stranded material following the positive selection hybridization step.
9. After addition of 10 µg of carrier RNA, quantitatively desalt and concentrate the appropriate single- and/or double-stranded fraction(s) by Nensorb 20 chromatography (according to the manufacturer's instructions), using 50% ethanol as an eluant. The sample is then dried by vacuum centrifugation (*see* Notes 11 and 12).

3.4. Positive Selection and Generation of Double-Stranded cDNA

The subtracted cDNA is positively selected by hybridization to a twofold excess of the parental mRNA or to mRNA derived from a cell that shares desired properties with the parental cell. Even if the positive selection is performed with the parental mRNA, it accomplishes several goals: (1) It selects against cDNA that has the same sense as mRNA (we find this to be a significant problem for random-primed first-strand cDNA but not for oligo(dT) primed cDNA); (2) it may provide some selection against contaminant DNAs; and (3) it provides a cDNA:mRNA duplex that serves as a good template for second-strand synthesis.

3.4.1. Positive Selection of First-Strand cDNA

1. Add 10 µg of poly(A)+ RNA from the parental line to the tube containing the subtracted cDNA. Dry the sample by vacuum centrifugation and then dissolve it in 3 µL of H$_2$O.
2. Add 1 µL of 5X hybridization buffer and 1 µL of SDS.

3. Mix, microfuge for 5 s, and overlay with 100 μL of mineral oil.
4. Place tube in boiling water bath for 5 min and incubate at 70°C to a Rot of 500 (*see* Note 7).
5. Isolate mRNA-cDNA duplex by HAP chromatography as described in Section 3.3. Add 10 μg of carrier RNA.
6. A rough estimate of the maximum enrichment of sequences in the subtractive cDNA = (100/ % ss cDNA from first subtractive HAP column) × (% ds cDNA from positive selection HAP column).
7. Desalt and concentrate sample by Nensorb chromatography.

3.4.2. Synthesis of Double-Stranded cDNA

1. Determine Cerenkov cpm of mRNA-cDNA duplex resuspended in 12 μL of H_2O.
2. Add the following:
   a. 10 μL of 4X second strand buffer
   b. 5 μL of mixture containing each dNTP at 2.5 mM
   c. 10 μL (100 μC_i) of [α-^{32}P]dCTP (3000 C_i/mmol)
   d. 1 μL RNase H (2 U/μL)
   e. 2 μl DNA polymerase I (5U/μL).
3. Microfuge for 5 s, mix gently, and incubate at 12°C for 2 h, and then at 22°C for 1 h.
4. Add 1 μL of 0.5M EDTA and 60 μL of H_2O. Mix and extract with 100 μL of phenol:$CHCl_3$ (1:1).
5. Remove 1 μL × 2 for total cpm. Separate the unincorporated nucleotides by G-50 Sephadex chromatography or a spin column. After addition of 10 μg of poly(A)- RNA as carrier and 0.1 vol of 3M sodium acetate, the excluded fractions are precipitated with ethanol, washed with 70% ethanol, and dried by vacuum centrifugation.
6. Resuspend the sample in 50 μL of TE buffer. The amount (ng) of second strand synthesis = (total Cerenkov cpm in excluded fraction – Cerenkov cpm in mRNA-cDNA duplex) × 16,500/total Cerenkov cpm in labeling reaction. The yield of second-strand should be comparable to the amount of first-strand cDNA present in the second-strand reaction.
7. The specific activity of second strand cDNA is approx 4000–8000 Cerenkov cpm/ng or about 1 ^{32}P/4000 nucleotides.

3.5. PCR Amplification of Subtractive cDNA

The double-stranded subtractive cDNA is treated to ensure blunt ends that contain 5' phosphates. A large excess of amplification adapters (containing *Sal* I, *Kpn* I, and *Eco* RI restriction sites) are then ligated to the subtractive cDNA. After removal of the excess amplification adapters and fragments of cDNA less than several hundred base pairs by Sepharose 4B chromatography, the subtractive cDNA is amplified by PCR. It appears that extensive PCR amplification of a heterogeneous population of subtractive cDNAs results in some terminal "sterile" cycles in which denaturation is not followed by quantitative annealing of the amplification oligonucleotide and subsequent polymerization. As a consequence, much of the amplified cDNA is at least partially single-stranded, perhaps owing to hairpin structures and heteroduplexes resulting from interactions of the terminal amplification sequences. A PCR chase reaction has been designed to permit denaturation of heteroduplexes and hairpin structures under conditions that will not denature homoduplexes, so that the partially single-stranded structures can be converted to homoduplexes. The PCR-chased subtractive cDNA is then size-fractioned on Sepharose 4B, resulting in a PCR-renewable source of subtractive cDNA that can be used to generate a probe or an insert for cloning in an appropriate vector.

3.5.1. Polishing and Kinasing the Ends of the Subtractive cDNA

1. To 50 μL of double-stranded cDNA, add the following:
   a. 7 μL of 10X T4 polymerase buffer
   b. 5 μL mixture of all four dNTPs at 5 mM each
   c. 5 μL of 10 mM ATP
   d. 2 μL of T4 DNA polymerase
   e. 2 μL of T4 polynucleotide kinase.
2. Microfuge for 5 s, mix gently, and incubate at 37°C for 30 min.
3. Add 2 μL of 0.5M EDTA. Extract with 70 μL phenol:CHCl$_3$ (1:1) and 70 μL CHCl$_3$.
4. Add 10 μg of poly(A)- RNA, 0.1 vol of 3M Na-acetate, and 2.5 vol of ethanol. Allow the sample to precipitate for 15 min on ice. Then collect precipitate by centrifugation at maximum speed in a microcentrifuge for 10 min, wash with ethanol, and dry by vacuum centrifugation. Resuspend sample in 5 μL of H$_2$O.

3.5.2. Amplification Adapter Ligation to Double-Stranded cDNA

1. To 5 μL of double-stranded cDNA, add:
   a. 1 μL of amplification adapter at 500 ng/μL (*see* Note 13)
   b. 1 μL of H$_2$O
   c. 2 μL of 5X T4 ligase buffer
   d. 1 μL of T4 ligase.
2. Incubate at 14°C overnight.

3. Add 90 µL of TE buffer, mix, and extract with 100 µL of phenol-CHCl$_3$ (1:1).
4. Chromatograph sample (*see* Note 14) on 4-mL column (i.e., 5-mL pipet) of Sepharose CL-4B, eluting with 200-µL aliquots. The eluted fractions can be analyzed by Cerenkov counting. The appropriate samples are pooled and precipitated by addition of 10 µg of poly(A)- RNA and 2.5 vol of ethanol. Resuspend in 30 µL of TE.

3.5.3. PCR Amplification of Subtractive cDNA

1. Transfer 5 µL of subtractive cDNA into the tube to be used for PCR amplification.
2. Add the following:
   a. 10 µL of 10X PCR amplification buffer
   b. 2 µL of a mix containing each dNTP at 10 m*M*
   c. 1 µL of the amplification oligonucleotide (1 µg/uL)
   d. 81 µL of H$_2$O
   e. 1 µL of *Taq* polymerase.
3. Mix gently and overlay with 100 µL of mineral oil.
4. Amplify by PCR using the following cycle profile:

30 main cycles	94°C,1 min (denaturation)
	50°C, 1 min (annealing)
	72°C, 1 min (extension)

   Remove 20-µL aliquots after 20, 25, and 30 cycles of amplification.
5. Analyze the three aliquots on a 1.5% agarose gel (*see* Chapter 1). Run ϕX174 *Hae* III size markers and 0.2 µg of the amplification oligonucleotide in separate lanes as controls. After the gel is stained with ethidium bromide, a heterogeneous smear of material with sizes up to 500 bp or more should be visible in one or more lanes. The presence of cDNA fragments with a unique size suggests the possibility of contaminants (*see* Note 15c).

3.5.4. PCR Chase of Amplified Subtractive cDNA

1. To 30 µL of amplified subtractive cDNA, add:
   a. 27 µL of 10X PCR buffer
   b. 6 µL of a mix of dNTPs at 10 m*M* each
   c. 1 µL of amplification oligonucleotide (1 µg/µL)
   d. 235 µL of H$_2$O
   e. 1 µL of *Taq* polymerase.
2. Mix gently and amplify by PCR using the following cycle profile:

2 cycles	94°C,1 min (denaturation)
	50°C, 1 min (annealing)
	72°C, 2 min (extension)

4 chase cycles	50°C, 1 min (annealing)
	72°C, 2 min (extension)

3. Extract the chased material with 300 µL of phenol:CHCl$_3$. After addition of 10 µg of carrier RNA and 30 µL of 3M Na-acetate, precipitate the aqueous phase with 750 µL of ethanol. Collect the precipitate by centrifugation, wash with 70% ethanol, and dry by vacuum centrifugation. Dissolve the pellet in 50 µL of TE buffer.

3.5.5. Size Fractionation of PCR Chased Subtractive cDNA (see Note 16)

1. Size-fractionate the PCR-chased cDNA by chromatography on a 4-mL Sepharose CL-4B column in a 5-mL pipet, collecting 200-µL fractions. Starting with fraction 6, subject 50-µL aliquots from each fraction to electrophoresis on a native 8% polyacrylamide gel (*see* Chapter 1), using φX174 *Hae* III fragments as size markers. Fractions containing essentially no material smaller than 350 bp are pooled, precipitated with ethanol after addition of 10 µg of carrier RNA, dried, and resuspended in 50 µL of TE buffer.

2. Estimate the approximate concentration of the size-fractionated, amplified subtractive cDNA by electrophoresis of a 5-µL aliquot on a native 8% acrylamide gel, by comparing the ethidium bromide staining of the heterogeneous subtractive cDNA with φX174 *Hae* III standards run on an adjacent lane. This stock of double-stranded subtractive cDNA can be used to prepare a radiolabeled probe or to prepare an insert suitable for cloning. A portion of it can also be PCR amplified and chased to renew the stock as desired.

See Notes 15 and 17–20.

4. Notes

1. Source of RNA. It is essential that trace contamination of the RNA (or any reagent used in the protocol for that matter) with DNA from any source be avoided since the contaminants can be amplified together with the very small quantities of cDNA that remain after subtraction, positive selection, and so forth. If possible, it is best to use cytoplasmic RNA for first-strand cDNA synthesis since this not only minimizes DNA contamination but also eliminates rare nuclear RNA species that may be enriched by subtraction. DNA can also be removed by one or more of the following procedures: precipitation of single-stranded polynucleotides (mainly RNA) with 2M LiCl, centrifugation of RNA through 5.7M CsCl, or treatment with DNase. Isolation of poly(A) + RNA by selection on oligo(dT) cellulose also removes trace DNA contaminants. We prefer to use two cycles of oligo(dT)-cellulose selection to isolate

mRNA for first strand cDNA synthesis, since one cyle of selection generates RNA that is substantially (approx 50%) contaminated with poly(A)- RNA, i.e., mostly rRNA. If RNA from only one cycle of selection is used for hybridization, a correction for the actual amount of mRNA should be made in estimating Rot values (*see* Note 6).

2. Amount of RNA. The amount of starting material can be decreased at least several fold although a minimum hybridization volume of about 4 μL, and adequate concentrations of RNA required to reach a high Rot are limiting. Also contaminant DNA is a more significant possibility as the amounts of RNA are decreased.

3. DNA contamination. This is a major problem to be avoided. Major sources include carrier RNA as well as the poly(A) + RNAs used for hybridization.

4. For a new lot of HAP, test binding/elution properties of single- and double-stranded DNA labeled to high specific activity. This can be done by applying each sample in 50 m*M* PB, 150 m*M* NaCl and eluting with duplicate 1-mL aliquots of 80, 100, 120, 140, 160, 180, 200, and 450 m*M* PB, 150 m*M* NaCl. The lowest concentration of PB that quantitatively elutes single-stranded DNA and little or no double-stranded DNA is then used to equilibrate the HAP slurry, to dilute samples for application to the column, and for washes to recover single-stranded cDNA. Some 10–15% of first-strand cDNAs (even if actinomycin D is present during synthesis) elute with double-stranded DNA; the single-stranded fraction elutes quantitatively in the single-stranded fraction on a second HAP column.

5. The columns used for HAP chromatography must be maintained at 60°C. Although a jacketed column can be used, we prefer a simple homemade system that uses disposable 3-mL plastic syringes as columns, essentially as described elsewhere *(16)*.

6. The kinetics of hybridization of cDNA to an excess of mRNA *(13–15)* is described by the pseudo-first-order rate equation:

$$C_t/C_o = e^{-k \times \text{Rot}} \qquad (1)$$

where C_o = the initial concentration of single-stranded cDNA, C_t = the amount of single-stranded cDNA remaining at time t, k = the rate constant in units liter. (mol nucleotide)$^{-1}$.s^{-1}, and Rot = the concentration of mRNA in mol of nucleotide/L multiplied by the time *(t)* in seconds. The rate constant, k, is determined from the following equation:

$$k = 10^6/C \qquad (2)$$

where C = the complexity of the mRNA in nucleotides. The constant k is based on hybridization at standard conditions, i.e., $T_m - 25°C$, cation concentration $= 0.18M$, and average size of single-stranded mRNA = 500 nucleotides. Since the complexity of mRNA in a typical mammalian cell line is approx 5×10^8 nucleotides (ca. 20 pg RNA or 0.3 pg mRNA/cell), $k = 2 \times 10^{-3}$. One can calculate that a cDNA derived from an mRNA present at 1 copy/cell is 50% double-stranded at a Rot of 345, 90% double-stranded at a Rot of 1150, 99% double-stranded at a Rot of 2300, and 99.9% double-stranded at a Rot of 3450. In contrast, cDNAs derived from mRNAs present at 10 and 0.1 copies/cell are 90% double-stranded at Rots of 115 and 11,500, respectively. Thus, during hybridization of cDNA to excess mRNA, a Rot of 3000 should be sufficient to ensure that all cDNAs derived from mRNAs present at least once per cell are present 99% or more as a double-stranded cDNA-mRNA duplex. For standard hybridization conditions, a 1 µg/µL concentration of mRNA gives a Rot of 11/h. For the hybridization conditions described below (i.e., cation concentration $= 0.60M$), mRNA at a concentration of 1 µg/µL gives a Rot (corrected to standard conditions) of 30/h *(15)*.

7. Rot for positive selection. By hybridizing only to a Rot of 500, 60% or more of sequences present ≥ 1 time/cell form a cDNA:mRNA duplex, whereas only 10% and 1% of sequences represented 0.1 or 0.01 time/cell, respectively, form a cDNA:mRNA duplex.

8. Choice of primer for cDNA synthesis. First-strand cDNA can be synthesized from the parental mRNA using either an adapter-dT_{17} primer (e.g. 5'-GGACTCGAGGTATCGATGCTTTTTTTTTTTTTTTTTT-3') or a random hexamer primer. Use of the adapter-dT_{17} primer results in the following advantages:

   a. First-strand cDNA is distinguished from contaminant DNAs (it should be noted that many subtractive cDNAs will lack adapter sequences as a result of fragmentation during the procedure);

   b. An adaptor probe can be used to detect clones that contain adaptor sequences;

   c. cDNA fragments containing the adapter-dT_{17} sequences can be enriched relative to other sequences by oligo-dT (or oligo dA) chromatography or by a PCR procedure (*see* Note 18);

   d. Single-stranded cDNA with the same sense as mRNA can be isolated on oligo-dT cellulose);

   e. Detection of cDNA clones that contain adapter sequences and are derived from the same mRNA (i.e., sibs) is efficient;

   f. Enriching for subtractive cDNA inserts that contain adapter-dT sequences decreases its complexity, thus enhancing the sensitivity of this material as a probe to identify clones in the subtractive library.

The random hexamer-primed cDNA has none of these advantages but does have three other potential advantages: (1) More mRNA sequences are represented so that inadvertant loss of representation of some mRNAs in the subtractive probe/library is minimized; (2) positive selection using mRNA from other species is more likely to be successful; (3) the subtractive cDNA library can be screened more reliably by conventional methods (e.g., antibodies), since clones in the library would not be biased toward the 3' ends of mRNA.

9. Correction of Rot to standard hybridization conditions. By definition, k varies as a function of hybridization conditions, whereas Rot is independent of these conditions. For convenience, however, we use a fixed k and correct Rot to standard conditions. Thus, for a given complexity of RNA, the fraction of cDNA in hybrid is calculated directly from the Rot corrected to standard hybridization conditions.

10. Hybridization conditions. To minimize fragmentation of cDNA, the hybridizations can be done at 52°C in 40% formamide, although the kinetics of hybridization are not as well worked out under these conditions *(9,13)*.

11. Desalting HAP fractions. The fractions can be concentrated by repeated n-butanol extractions, followed by G50 Sephadex chromatography to remove phosphate, and alcohol precipitation. We prefer to use Nensorb to optimize recoveries.

12. Alternative to HAP chromatography. Use of biotinylated driver mRNA or DNA has been used by others *(9,12)*. A recent report using oligo(dA) cellulose to remove single-stranded oligo(dT)-primed cDNA from cDNA:mRNA duplexes appears to provide another alternative *(10)*. Each alternative is alleged to cause less fragmentation of cDNA than HAP chromatography.

13. PCR adapter. Other potential amplification adapters may work but should be tested with a model substrate (e.g., PBR322 cut with *Sau*3A) to verify that all steps, including the blunting and kinasing reactions, work.

14. An identical Sepharose 4B column can be calibrated with blue dextran (void volume), phenol red or labeled dNTP (included volume), and ϕX174 *Hae* III markers. The first third of the samples between the void volume and the included volume contain the larger cDNAs, with amplification adapters ligated at each end, but should be devoid of unligated adapters and adapters that have self-ligated.

15. Characterization of amplified subtractive cDNA. Ideally, one would like to know the extent of enrichment of subtractive sequences, the number of independently amplified subtractive cDNA sequences, and whether or not a substantial fraction of contaminant DNAs have been amplified together with the cDNA.

a. Enrichment. A rough calculation of the maximum enrichment of subtractive sequences is given earlier. In the two cases tested (*11* and J. Battey, unpublished), this calculation was in agreement with a direct comparison of the frequency of clones in libraries prepared from subtractive cDNA and parental first-strand cDNA.

b. Number of independently amplified subtractive cDNA sequences. Although it is difficult to accurately assess this parameter, a minimum estimate can be made based on the number of amplification cycles required to generate 1 ug of subtractive probe, i.e. 20, 25, and 30 PCR amplification cycles would correspond to amplification of a minimum of 1.8×10^6, 5.8×10^4, and 1800 independent sequences having an average size of 500 bp.

c. Contaminant DNA. If the first-strand cDNA was synthesized with an adapter-dT_{17} primer and had an average size of 2000 bp at the time of synthesis and 400 bp after amplification, approx 20% of the amplified molecules of cDNA should contain an adapter sequence. This can be confirmed by cloning the insert into a convenient vector and probing with an adapter probe. If only a small fraction of clones with inserts contain adapter sequences, it is likely that contaminant DNAs have been coamplified with the subtractive cDNA. It is possible to enrich the amplified subtractive cDNA for molecules containing adapter sequences (*see* Note 18).

16. Size of subtractive cDNA. Subtractive cDNA prepared by the aforementioned classical protocol results in a substantial reduction in the size of first-strand cDNA, independent of any effect of PCR amplification. The reasons for this are poorly understood although thermal degradation during hybridization and some degradation during HAP chromatography may contribute to this result. The size selection of cDNA prior to amplification minimizes PCR selection of very short sequences, and the size selection after amplification further ensures that very short sequences are eliminated.

17. Yields of subtractive cDNA. If there is a theoretical 50-fold enrichment of subtractive cDNAs (Section 3.4.1.), the actual yield of positively selected first-strand cDNA in cDNA:mRNA duplex is several fold lower (e.g., 0.5% instead of 2.0%), because of sequential small losses from adsorption of cDNA to tubes, incomplete precipitation, incomplete recovery of material from HAP and Nensorb chromatography, and so forth. Although the total amount of cDNA should increase during the second-strand cDNA synthesis step, there may be an additional 90% loss of cDNA (i.e., fragments smaller than approx 200 bp) during the Sepharose 4B chromatography step following the ligation of amplification adapt-

ers. Thus, there is likely to be as little as 1–5ng of cDNA available for the PCR reaction.

18. Enrichment of subtractive cDNAs containing adapter-dT sequences. In addition to oligo(dA)-cellulose selection of subtractive cDNAs containing oligo(dT) sequences, it is possible to use PCR to enrich for subtractive cDNAs containing adapter sequences (cf *17*): Add dG to the 3' ends of the amplified subtractive cDNA using terminal deoxynucleotidyl transferase. The dG-tailed subtractive cDNA can then be amplified with three oligonucleotides: (1) the adapter portion of the adapter dT primer (the adapter-dT$_{17}$ primer *per se* does not seem to work as well for this PCR reaction); (2) a second adapter sequence, including a stretch of 12 dC residues at its 3' end, in limiting amounts; and (3) the second adapter sequence (without the 12 dC residues) at a higher level.

19. Screening libraries by hybridization with the subtractive cDNA probe. As noted in the introduction, a subtractive cDNA probe that is enriched 50-fold can detect clones containing sequences representing approx 0.002% of the mRNA sequences in a cell. Clones containing large inserts are more effectively detected than clones containing smaller inserts. Since subtractive cDNA libraries generally contain smaller inserts (e.g., 400 bp average) than conventional cDNA libraries (e.g., >2 kb), a subtractive probe will detect rare mRNAs more efficiently in a conventional cDNA library than in a subtractive cDNA library.

20. PCR-renewable cDNA to replace mRNA for hybridization. A recent publication *(18)* describes a method that uses PCR-renewable cDNA for both the parental cDNA and the biotinylated subtractive partner cDNA. This approach, which should decrease the need for large amounts of mRNA, is conceptually similar to the method described in this chapter. In principle, it could be used to further select the subtractive cDNA prepared by the method described in this chapter.

References

1. Alt, F. W., Kellems, R. E., Bertino, J. R., and Schimke, R. T. (1978) Selective multiplication of dihydrofolate reductase genes in methotrexate-resistant variant of cultured murine cells. *J. Biol. Chem.* **163,** 1357–1370.
2. Timberlake, W. E. (1980) Developmental gene regulation in Aspergillus nidulans. *Dev. Biol.* **78,** 497–510.
3. Sargent, T. D. and Dawid, I. B. (1983) Differential gene expression in the gastrula of Xenopus laevis. *Science* **222,** 135–139.
4. Hedrick, S. M., Cohen, D. I., Nielsen, E. A., and Davis, M. M. (1984) Isolation of cDNA clones encoding T cell-specific membrane-associated proteins. *Nature* **308,** 149–153.
5. Klickstein, L. B. (1987) Production of a subtractive cDNA library, in *Current*

Protocols in Molecular Biology, vol. 1 (Ausubel, F. M., Brent, R., Kingston, R. E., Moore, D. E., Seidman, J. G., Smith, J. A., and Struhl, K., eds.), Wiley, New York, pp. 5.8.6.–5.8.13.

6. Duguid, J. R., Rohwer, R. G., and Seed, B. (1988) Isolation of cDNAs of scrapie-modulated RNAs by subtractive hybridization of a cDNA library. *Proc. Natl. Acad. Sci. USA* **85,** 5738–5742.

7. Travis, G. H. and Sutcliffe, J. G. (1988) Phenol emulsion-enhanced DNA-driven subtractive cDNA cloning: isolation of low-abundance monkey cortex-specific mRNAs. *Proc. Natl. Acad. Sci. USA* **85,** 1696–1700.

8. Palazzolo, M. J. and Meyerowitz, E. M. (1987) A family of lambda phage cDNA cloning vectors, lambda SWAJ, allowing the amplification of RNA sequences. *Gene* **52,** 197–206.

9. Rubenstein, J. L. R., Brice, A. E. J., Ciaranello, R. D., Denney, D., Porteus, M. H., and Usdin, T. B. (1990) Subtractive hybridization system using single-stranded phagemids with directional inserts. *Nucleic Acids Res.* **18,** 4833–4842.

10. Batra, S. K., Metzgar, R. S., and Hollingsworth, M. A. (1991) A simple, effective method for the construction of subtracted cDNA libraries. *GATA* **8,** 129–133.

11. Timblin, C., Battey, J., and Kuehl, W. M. (1990) Application for PCR technology to subtractive cDNA cloning: identification of genes expressed specifically in murine playmacytoma cells. *Nucleic Acids Res.* **18,** 1587–1593.

12. Wieland, I., Bolger, G., Asouline, G., and Wigler, M. (1990) A method for difference cloning: Gene amplification following subtractive hybridization. *Proc. Natl. Acad. Sci. USA* **87,** 2720–2724.

13. Britten, R. J., Graham, D. E., and Neufeld, B. R. (1974) Analysis of repeating DNA sequences by reassociation. *Methods Enzymol.* **29,** 363–418.

14. Galau, G. A., Britten, R. J., and Davidson, E. H. (1977) Studies on nucleic acid reassociation kinetics: rate of hybridization of excess RNA with DNA, compared to the rate of DNA renaturation. *Proc. Natl. Acad. Sci. USA* **74,** 1020–1023.

15. Van Ness, J. and Hahn, W. E. (1982) Physical parameters affecting the rate and completion of RNA driven hybridization of DNA: new measurements relevant to quantitation based on kinetics. *Nucleic. Acids Res.* **10,** 8061–8077.

16. Sambrook, J., Fritsch, E. F., and Maniatis, T. (1989) Separation of single-stranded and double-stranded DNA by hydroxyapatite chromatography, in *Molecular Cloning: A Laboratory Manual,* vol. 3. Cold Spring Harbor Laboratory, Cold Spring Harbor, NY, pp. E.30–E.33.

17. Frohman, M. A., Dush, M. K., and Martin, G. R. (1988) Rapid production of full-length cDNAs from rare transcripts: Amplification using a single gene-specific oligonucleotide primer. *Proc. Natl. Acad. Sci. USA* **85,** 8998–9002.

18. Wang, Z. and Brown, D. D. (1991) A gene expression screen. *Proc. Natl. Acad. Sci. USA* **88,** 11,505–11,509.

CHAPTER 30

PCR-Based Full-Length cDNA Cloning Utilizing the Universal-Adaptor/Specific DOS Primer-Pair Strategy

David L. Cooper and Narayana R. Isola

1. Introduction

Biological systems are often influenced by molecules that are neither present in vast quantities or easily purified to homogeneity from other cellular constituents. The development of simple, efficient molecular cloning systems coupled with the relative ease of DNA sequence determination has made nucleic acid sequence determination the choice methodology when sequence determination of the complete biological molecule is indicated. However, one bottleneck that impedes direct access to sequence determination is the necessity to screen recombinant libraries containing a large number of clones for the low-abundance member. Advances in protein microsequencing techniques combined with application of the polymerase chain reaction (PCR) help simplify this process *(1–3)*. Utilizing only the *N*-terminal protein sequence information, we have developed a protocol for the selective amplification and subsequent cloning of specific full-length cDNAs, which in some instances may circumvent the time-consuming, tedious process of library screening. This approach, as illustrated in Fig. 1, employs the following steps:

1. First-strand cDNA synthesis initiated by use of a "universal" primer, which consists of a poly dT_{17} track and an "adaptor" deoxyoligonucleotide sequence *(4)* that ultimately introduces a directional cloning site into the 3' terminus of the cDNA;

From: *Methods in Molecular Biology, Vol. 15: PCR Protocols: Current Methods and Applications*
Edited by: B. A. White Copyright © 1993 Humana Press Inc., Totowa, NJ

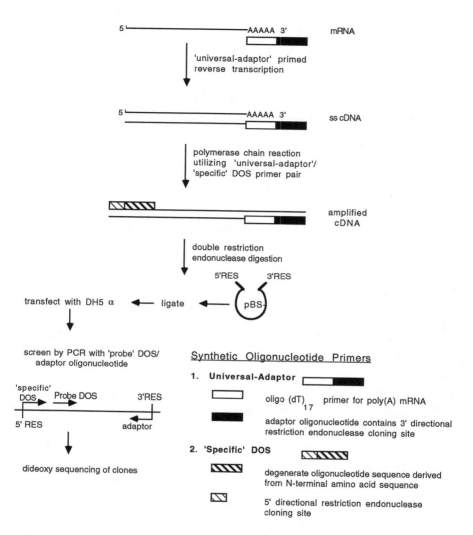

Fig. 1. Cloning strategy for cDNA amplification of low abundance mRNA.

2. Second-strand synthesis primed by annealing a "specific" deoxyoligonucleotide of limited degeneracy that represents the most likely codon
combinations following reverse translation *(5–7)* of a previously determined *N*-terminal amino sequence;

3. The PCR utilizing the specific degenerate oligonucleotide sequence
(DOS) and the unidirectional adaptor (but not the dT_{17} track) as priming sequences to amplify the appropriate cDNA sequence; and

4. Confirmation of appropriately amplified product is accomplished via a screening PCR utilizing a second "probe" DOS that is contiguous with the specific DOS *(8)*.

This technique originally utilized a poly $dT_{17}GGCC$ universal primer that ultimately introduced an *Xcy* I site into one terminus of the cDNA *(9)*. The second-strand cDNA and second PCR primer was a degenerate DOS that contained all possible codon combinations of the *N*-terminal amino acid sequence modified by addition of a synthetic *Eco* Rl linker. Subsequent amplification of cDNA sequence utilitizing this primer pair, although sufficiently effective to recover full-length cDNAs of low-abundance mRNAs, produced varying amounts of nonspecifically primed material when analyzed by agarose gel electrophoresis. Cloning of the specific cDNA via this PCR amplification with the universal/specific primer-pair approach was confirmed by screening with a second probe DOS located internal to the DOS employed to prime the PCR.

We have subsequently made three modifications that substantially improve the quality of the amplified product synthesized and the efficacy of the technique. First, the universal primer now contains an additional 18–22 bases of nucleotide sequence in addition to the mRNA-binding oligo (dT_{17}) track. This adaptor sequence is designed to match the basepair composition to the *N*-terminal reverse-translated DOS. It is constructed when possible to contain an internal rare restriction enzyme site (e.g., *Not* l) that functions as a unidirectional cloning site. Secondly, the specific DOS primer is now designed in such a way to accommodate only the most probable nucleotide sequences. Simply, this approach includes utilization of only the most prevalent two codons of an amino acid, compliance with dinucleotide codon frequencies *(10)*, and if possible a singular 3' most codon (i.e., methionine, tryptophan) or the most predominant codon of an amino acid *(11)*. The reduction in degeneracy is typically from a primer that previously averaged between 64-fold and 256-fold degenerate to primers currently in use which are rarely above 12- to 16-fold degenerate. This primer usually also contains a 5' synthetic restriction enzyme site(s) to simplify enzymatic manipulation in the subsequent cloning of PCR-generated product. The third modification replaces screening of recombinants by colony hybridization with an alternative PCR-based technique. In this approach, the probe DOS is used instead in a second round of PCR to confirm cloned recombinants contain inserts appropriate to the cDNA in question.

Table 1
Codon Utilization in Human Protein-Coding Sequences[a]

a.a.	codon	%	a.a.	codon	%	a.a.	codon	%
*F	UUU	0.35	P	CCU	0.24	*K	AAA	0.45
	UUU	*0.65*		*CCC*	*0.41*		*AAG*	*0.55*
				CCA	0.24			
L	UUA	0.05		CCG	0.11	*D	GAU	0.38
	UUG	0.09					*GAC*	*0.62*
	CUU	0.11	T	ACU	0.20			
	CUC	0.22		*ACC*	*0.47*	*E	GAA	0.40
	CUA	0.07		ACA	0.21		GAG	0.60
	CUG	*0.46*		ACG	0.12			
I	AUU	0.23	A	GCU	0.31	*C	UGU	0.30
	AUC	*0.64*		*GCC*	*0.40*		*UGC*	*0.70*
	AUA	0.13		GCA	0.17			
				GCG	0.12	*W	UGG	1.00
*M	AUG	1.00						
			*Y	UAU	0.47	R	CGU	0.09
V	GUU	0.13		*UAC*	*0.53*		CGC	0.19
	GUC	0.27					CGA	0.10
	GUA	0.09	*H	CAU	0.42		CGG	0.15
	GUG	*0.50*		CAC	0.58		*AGA*	*0.24*
							AGG	0.23
S	UCU	0.17	*Q	CAA	0.26			
	UCC	*0.26*		*CAG*	*0.74*	G	GGU	0.15
	UCA	0.11					*GGC*	*0.44*
	UCG	0.07	*N	AAU	0.34		GGA	0.17
	AGU	0.11		*AAC*	*0.66*		GGG	0.24
	AGC	*0.29*						

[a]Amino acids of limited codon degeneracy are marked by asterisk. Major fractional codon(s) utilized by an amino acid are boldfaced and italicized.

2. Materials

2.1. Design of "Specific DOS"

1. Codon utilization table for appropriate species (e.g., Table 1).
2. Dinucleotide frequency table for appropriate species (e.g., Table 2).

2.2. cDNA Synthesis and PCR

1. Total RNA or previously selected poly (A+) mRNA.
2. Universal-adaptor deoxyoligonucleotide, and specific DOS primer prepared with an appropriate automated synthesizer (*see* Notes 1 and 2).

Table 2
Optimum Codon Choice Influenced by Dinucleotide Frequencies
When Deducing a Probe Sequence from Human Amino Acid Sequence Data[a]

Amino acid	Optimum codon when first nucleotide of next codon is:	
	A or C or T	G
Methionine	ATG	nc
Tryptophan	TGG	nc
Tyrosine	TAC	TAT
Cysteine	TGC	TGT
Glutamine	CAG	nc
Phenylalanine	TTC	TTT
Aspartic acid	GAC	GAT
Asparagine	AAC	AAT
Histidine	CAC[b]	CAT
Glutamic acid	GAG	nc
Lysine	AAG	nc
Alanine	GCC	GCT
Isoleucine	ATC	ATT
Threonine	ACC	ACA[e]
Valine	GTG[c]	nc
Proline	CCC[d]	CCT
Glycine	GGC	nc[e]
Leucine	CTG	nc
Arginine	CGG	nc
Serine	TCC	TCT

[a]Optimum codon is the most frequent codon for all amino acids except arginine and serine.
[b]CAT when followed by C.
[c]GTC when followed by T.
[d]CCA when followed by T.
[e]Examples of cases where the "replace C by T" rule does not apply when the first nucleotide of the following codon is G.

3. MMLV reverse transcriptase (RT).
4. Annealing buffer: 10 mM Hepes, pH 6.9, 200 mM EDTA.
5. RT buffer: 50 mM Tris-HCl, pH 7.5, 75 mM KCL, 5 mM MgCl$_2$, 10 mM dithiothreitol, and 0.5 mM each of dATP, dCTP, dGTP, and dTTP.
6. Phenol:CHCl$_3$:isoamyl alchohol (PCIA): 25:25:1.
7. Nu-Clean D-50 spin columns (IBI, New Haven, CT).
8. *Thermus aquaticus (Taq)* DNA polymerase.
9. 10X PCR buffer: 100 mM Tris-HCl, pH 9.0 at 25°C, 500 mM KCl, 15 mM MgCl$_2$, 0.1% gelatin, and 1.0% Triton X-100.

10. 10X dNTP mix (for PCR): 2 m*M* each of dATP, dCTP, dGTP, and dTTP.
11. Mineral oil.

2.3. Size Selection and Amplified DNA Preparation

1. 20X TBE stock: 1*M* Tris base, 1*M* boric acid, 20 m*M* EDTA-disodium. Use 20X stock to prepare 1X TBE buffer.
2. Agarose, 0.8–1.2% (w/v) in 1X TBE buffer.
3. 1-cc Tuberculin syringe.
4. Ultrafree-MC filter, 0.45 µm (Millipore, Bedford, MA).
5. 1 m*M* Tris, pH 7.8, 0.1 m*M* EDTA buffer.
6. 3*M* Sodium acetate, pH 5.5.
7. Absolute ethanol.
8. 10 m*M* Tris-HCl, pH 8.5, 1 m*M* EDTA.
9. Appropriate restriction endonucleases and buffers.
10. PCIA (*see* item 6 in Section 2.2.).

2.4. Vector Preparation and Cloning

1. Appropriate restriction enzymes and buffers.
2. Supercoiled pBS- (Stratagene, La Jolla, CA) or a vector of choice.
3. Nu-Clean D50 spin columns (IBI, New Haven, CT).
4. Calf intestinal alkaline phosphatase and buffer.
5. 500 m*M* EDTA, pH 8.0.
6. TE buffer: 10 m*M* Tris-HCl, pH 7.4, 0.1 m*M* EDTA.
7. PCIA (*see* item 6 in Section 2.2.).
8. CHCl$_3$.
9. T4 DNA ligase and buffer.
10. Ultrapure, sterile water.
11. Transformation competent DH5α *E. coli* (BRL).
12. LB medium.
13. LB bacterial agar plates containing 100 µg/mL of ampicillin, 0.2 m*M* IPTG, and 40 µg/mL X-gal (or selection components appropriate for vector used).
14. Ultrafree-MC 30,000 NMWL (Millipore).
15. Horizontal agarase gel electrophoresis apparatus.

2.5. Screening Recombinant Clones

1. Probe DOS and universal/adaptor oligodeoxynucleotides.
2. Gel loading buffer.
3. Agarose gel and TBE buffer (*see* items 1 and 2 in Section 2.3.).
4. Ethidium bromide (10 mg/mL, store at 4°C in dark).
5. Components for dideoxy sequencing.

3. Methods

3.1. Design of "Specific DOS"

Owing to the inherent redundancy of the genetic code, amino acids may be encoded by from one to six codons. Synthesis of DOS primers of limited degeneracy based on reverse translation of protein sequences rich in amino acids with a restricted number of codons (Table 1) is a particularly helpful approach at controlling the loss of specificity characteristic of degenerate probes *(10)*. Further, it has been found that minimal homology requirements for PCR primers in most cases will be successful when limited to the preferred codon usage of the organism *(11)*. Specifically, we utilize the following approach to design our synthetic DOS primers *(12)*.

1. Choose primers of length >20 bases. Depending on amino acid sequence utilized, this length usually varies from between 18 and 25 nucleotides. Generally, the more inherent redundancy in an amino acid sequence, the longer the DOS synthesized needed to assure specificity.
2. Attempt to end the oligonucleotide at a Met or Trp residue, since this ensures the last three nucleotides will match completely. If this is impossible, synthesize all codons for that amino acid. Often terminating one nucleotide from the most highly degenerate third nucleotide position is a great help, ensuring the last two nucleotides are homologous. Alternatively, the final nucleotide can be synthesized as a T, since T has a low fidelity of basepairing (i.e., can basepair with G and C). This ensures that the 3'-most nucleotide is capable of promoting extension by *Taq* polymerase.
3. Utilize, at most, only the two most frequent codons for a particular amino acid. Attempt to limit the total degeneracy of the oligonucleotide to <32.
4. Design your DOS to the amino acid sequence that is encoded by the lowest possible number of codons from available protein sequence information.
5. Where design of the DOS has led to generation of intercodon C-G dinucleotides, consult the dinucleotide frequency table (Table 2) as a guide in the selection of optimum codons. The predicted increase in average homology may be small, although in sequences not including *Leu, Arg,* or *Ser*, predicted homology rises significantly *(10)*.
6. Design of the probe DOS also follows the aforementioned rules. To assure full-length cDNA cloning, the probe DOS sequence is derived from amino acid sequence just 3' to the specific DOS region. If this is impossible owing to a highly degenerate adjacent amino acid sequence, then adjacent, further 3' internal amino acid sequence, if available, may be utilized.

3.2. cDNA Synthesis and PCR Amplification Utilizing the Specific DOS/Universal Adaptor Oligonucleotide Primer Pair

1. Anneal the universal-adaptor oligonucleotide and mRNA (*see* Note 3) at an approx 1:1 (w/w) ratio at a mRNA concentration of 0.3 µg/µL in annealing buffer (typically 1–5 µg of each).
2. Heat the annealing mixture to 90°C for 2 min and then place immediately in ice water.
3. Perform first-strand cDNA synthesis in a final volume of 25–100 µL in RT buffer 125–500 U of cloned MMLV reverse transcriptase and incubate for 60 min at 37°C (*see* Note 4).
4. Extract the first-strand cDNA reaction with an equal volume of PCIA and apply to a Nu-Clean D-50 spin column (*see* Note 5) to remove RT and the universal adaptor oligonucleotide, respectively (*see* Note 6).
5. Take 1–2 µL of the first-strand cDNA, synthesized and purified free of first-strand primers as earlier, and add directly to a 50-µL (total volume) PCR reaction containing 100 n*M* synthetic oligonucleotide primers (adaptor molecule and unique DOS corresponding to *N*-terminal amino acid reverse translation product), 5 µL of dNTP mix, 5 µL of 10X PCR buffer and 2 U of *Taq* DNA polymerase.
6. Overlay reactions with 50–100 µL of mineral oil to prevent evaporation and amplify by PCR using the following cycle profile (*see* Notes 7–10):

Initially denaturation	95°C, 4 min
25–30 main cycles	95°C, 1 min (denaturation)
	55°C, 1 min (annealing)
	72°C, 2 min (extension)
Final extension	72°C, 5 min

3.3. Size Selection and Amplified cDNA Preparation (see Note 11)

1. Resolve the amplified products on a 1% agarose gel in TBE buffer (*see* Chapter 1) and cut fragments >1 kb, or the minimum cDNA coding requirement for the protein of interest, from the gel with a razor blade.
2. Freeze the agarose block containing DNA of interest at –20°C for 1 h.
3. Quick thaw at 37°C for 5 min.
4. Macerate the gel by passage through a 1-cc tuberculin syringe.
5. Transfer the macerated gel and accompanying liquid into a 0.45-µm Ultrafree-MC filter cup (Millipore).
6. Cap the tube and centrifuge for 10 min according to manufacturer's directions.

7. Resuspend agarose in 200 mL of 1 mM Tris-HCl, pH 7.8, 0.1 mM EDTA buffer and repeat centrifugation step. The PCR-generated cDNA is now in the filtrate tube, separated from contaminating gel fragments.

8. Extract the DNA with an equal volume of PCIA (approx 400 µL) and filter through a Nu-Clean D-50 spin column according to manufacturer's directions (IBI, New Haven, CT).

9. Precipitate the centrifugal membrane filtrate by addition of sodium acetate, pH 5.5, to 0.3M followed by 2 vol of cold 100% ethanol. Incubate on dry ice for 15 min, and pellet by centrifugation at 12,000g for 10 min in a microcentrifuge, and dissolve PCR product in 10–100 µL of 10 mM Tris-HCl, pH 8.5, 1 mM EDTA.

10. Generate first cloning site by adding an aliquot of the PCR product (typically 100 ng) to 2 µL of 10X restriction enzyme buffer, 1 µL of restriction enzyme of choice and sterile H$_2$O to a final volume of 20 µL.

11. Incubate at 37°C for 1 h. PCIA extract, purify by Nu-Clean D-50 column, and ethanol precipitate the PCR product as just described in steps 8 and 9.

12. Repeat this process for the second cloning site if directional cloning is to be attempted.

3.4. Vector Preparation and Cloning

1. Sequentially cleave selected vector, such as M13mp19 or pBS- (Stratagene) with the appropriate restriction endonucleases to generate the directional cloning site.

2. Eliminate the restriction endonuclease fragment by two successive Nu-Clean D50 spin columns.

3. In a 50-µL reaction containing 10 µg of double-digested plasmid vector, add 10 U CIAP to remove 5' phosphates. For protruding or flush ends incubate in CIAP buffer at 37°C, or at 60°C for recessed 5' ends, for 30 min.

4. Stop the reaction by adding 1 µL of 0.5M EDTA at pH 8. Add 100 µL of TE buffer and extract twice with an equal volume of PCIA and twice with CHCl$_3$.

5. Ethanol precipitate as described in step 9 in Section 3.4. and resuspend in TE buffer at 0.1 µg/µL.

6. Add together in a 400 µL microcentrifuge tube: 2 µL of supplied 10X ligase buffer, 1 µL (0.1 µg) of the prepared vector, 1 µL of T4 DNA ligase, and the insert DNA at a concentration of four times the molar concentration of the vector. Adjust volume with sterile H$_2$O to 20 µL, and incubate 4 h to overnight at 14°C.

7. Add 0.1 µg of the ligated material to 200 µL of competent DH5α *E. coli* and incubate on ice for 30 min.

8. Heat shock bacterial cells by incubating the tube in a 42°C water bath for 2 min.
9. Add 1 mL of LB medium and grow for 45–60 min at 37°C.
10. Spread 50, 100, and 200 µL of these cells with a bent-end glass rod over LB agar plates containing selection components. Allow medium to dry, then invert plates and grow overnight at 37°C.
11. Recombinant clones are visualized by the appearance of white colonies.

3.5. Screening Recombinant Colonies

1. Touch a micropipet tip to a colony *(12)* and rinse cells into a 50-µL PCR reaction mixture containing *Taq* polymerase, dNTPs, 1X PCR buffer (*see* step 5 in Section 3.2.) and the probe DOS and the adaptor oligonucleotide, each at a final concentration of 100 n*M*.
2. Amplify by PCR using 30–35 cycles and the cycle profile described in step 6 in Section 3.2.
3. Remove a 10-µL aliquot of the reaction product, and add 5 µL of gel loading buffer. Apply to a 0.8–1.2% agarose gel (w/v) in TBE buffer containing 10 µL of ethidium bromide (10 mg/mL) per 100 mL of gel volume.
4. Electrophorese at 70–100 V until dye front is at least three-quarters down the gel.
5. Following electrophoresis, visualize the product by UV illumination.
6. Prepare single-stranded or double-stranded plasmid DNA from colonies that amplify the appropriate size insert from the second screening PCR and submit to dideoxy sequencing (*see* Chapters 13 and 14).

4. Notes

1. Gel purification of the oligonucleotide primers is often of benefit only in reducing nonspecific product if the deoxyoligonucleotide synthesizer is not operating at its optimum coupling efficiency.
2. Design of the adaptor oligonucleotide should incorporate a closely matched G + C content, a rare restriction endonuclease recognition site for directional cloning, and lack of sequence homology with the unique oligonucleotide. Two computer software programs that are extremely useful in designing PCR oligonucleotide primers are OLIGO (National Biosciences Inc., Plymouth, MN) and Primer Detective (Clontech, Palo Alto, CA). Also, *see* Chapter 2.
3. Total RNA may be substituted without oligo (dT) cellulose selection when starting material is limiting.
4. Enzyme/buffer systems are available from a number of commercial vendors that allow for the oligo (dT) first-strand cDNA and second-strand PCR to proceed in one tube.

5. Removal of excess universal-adaptor oligonucleotide may be alternatively accomplished by centrifugal membrane filtration (Ultrafree-MC 30,000 NMWL, Millipore).

6. The product of first-strand cDNA synthesis is extremely stable. We routinely convert rare or difficult-to-prepare mRNA isolations to single-stranded cDNA for storage. We have successfully amplified specific cDNAs from such preparations more than one year later.

7. These are general initial conditions to be tested. Short amplification products, ≤1 kb, can often be amplified efficiently in 20–30 s if the PCR conditions are optimized. The optimal number of cycles of amplification is directly related to the number of appropriate target template molecules in the reaction. The PCR is so sensitive that even a transcript present at only 1 copy/cell can be amplified in 25–30 cycles if the mRNA from as little as 300,000–500,000 cells are present.

8. The nucleotide concentration in the PCR reaction mix is 0.5 mM. This has been utilized since, occasionally, raising the nucleotide concentration in such a manner has been found to limit the exonuclease activity of *Taq* polymerase and improve the specificity of sequence amplification.

9. It is suggested to titrate the concentration of oligonucleotide primers necessary for efficient amplification of product while limiting nonspecific artifacts. In most instances optimal primer concentration will fall between 50 nM and 3 μM. Oligonucleotide primers even of minimal degeneracy will sometimes falsely prime numerous transcripts. If possible, redesign your degenerate oligonucleotide and vary the annealing temperature accordingly. Remember that minimally 20–25% mismatch between oligonucleotide primer and template can be tolerated.

10. The annealing temperature as described in Section 3 is empirically derived. The optimum T_m for this step may be somewhat lower than predicted from the G + C content of the primer owing to the amount of mismatch between primer and template. Therefore, expect a lower optimum temperature of annealing from complex primer mixtures. Further, the optimum elongation temperature is similarly reduced if significant mismatch occurs between template and primer and may actually be closer to the annealing temperature than the optimum *Taq* polymerase elongation temperature (72°C).

11. Prior to proceeding with the cloning strategy outlined it is informative to gel-purify putative products and submit to a second round of PCR with the probe primer. Ultrafree-MC 0.45-μm centrifugal membranes may be used to recover the band of interest. Our laboratory routinely recovers between 70 and 95% of DNA cut from such an agarose block. Gel purification of appropriate sized products, followed by reamplification will often increase the cloning of the appropriate product by 5-

to 10-fold. Amplifications that do not result in generation of a specific product may be either cloned directly, then screened by conventional oligonucleotide filter hybridization methods utilizing the probe DOS, or by alternative PCR library screening techniques. Occasionally, direct PCR amplification of this heterogeneous product with the probe DOS will result in the production of the appropriate product.

References

1. Matsudaira, P. (1987) Sequence from picomole quantities of proteins electroblotted onto polyvinylidene difluoride membranes. *J. Biol. Chem.* **262,** 10,035–10,038.

2. Cooper, D. L., Baptist, E. W., Enghild, J., Lee, H., Isola, N. R., and Klintworth, G. K. (1990) Partial amino acid sequence determination of Bovine Corneal Protein 54K (BCP54). *Curr. Eye Res.* **9,** 781–786.

3. Saiki, R. K., Gelfand, D. H., Stoffel, S., Scharf, S. J., Higuchi, R., Horn, G. T., Mullis, K. B., and Erlich, H. A. (1988) Primer directed enzymatic amplification of DNA with a thermostable DNA polymerase. *Science* **239,** 487–491.

4. Welsh, J., Liu, J-P., and Efstratiadis, A. (1989) Cloning of PCR-amplified total cDNA: construction of a mouse oocyte cDNA library. *Genet. Anal. Techn. Appl.* **7,** 5–17.

5. Cooper, D. L. and Baptist, E. W. (1991) Degenerate oligonucleotide sequence-directed cross-species PCR cloning of the BCP54/ALDH 3 cDNA: priming from inverted repeats and formation of tandem primer arrays. *PCR Method. Applic.* **1,** 57–62.

6. Lee, C. C., Wu, X., Gibbs, R. A., Cook, R. G., Muzny, D. M., and Caskey, C. T. (1988) Generation of cDNA probes directed by amino acid sequence: cloning of urate oxidase. *Science* **239,** 1288–1291.

7. Lee, C. C. and Caskey, C. T. (1989) Generation of cDNA probes by reverse translation of amino acid sequence, in *Genetic Engineering: Principles and Methods* vol. 11, (Setlow, J. K. ed.), Plenum, New York, pp. 159–170.

8. Isola, N. R., Harn, H.-J., and Cooper D. L. (1991) Screening recombinant DNA libraries: a rapid and efficient method for isolating cDNA clones utilizing the PCR. *BioTechniques* **11,** 580–582.

9. Cooper, D. L. and Isola, N. R. (1990) Full-length cDNA cloning utilizing the polymerase chain reaction, a degenerate oligonucleotide sequence and a universal mRNA primer. *BioTechniques* **9,** 60–65.

10. Lathe, R. (1985) Synthetic oligonucleotide probes deduced from amino acid sequence data: theoretical and practical considerations. *J. Mol. Biol.* **183,** 1–12.

11. Sommer, R. and Tautz D. (1989) Minimal homology requirements for PCR primers. *Nucleic Acids Res.* **17,** 6749.

12. Cooper, D. L., Baptist, E. W., Enghild, J. J., Isola, N. R., and Klintworth, G. K. (1991) Bovine corneal protein 54K (BCP54) is a homologue of the tumor-associated (class 3) rat aldehyde dehydrogenase (RATALD). *Gene* **98,** 201–207.

CHAPTER 31

Use of Degenerate Oligonucleotide Primers and the Polymerase Chain Reaction to Clone Gene Family Members

Gregory M. Preston

1. Introduction
1.1. What Are Gene Families?

As more and more genes are cloned and sequenced, it is apparent that nearly all genes are related to other genes. Similar genes are grouped into families. Examples of gene families include the collagen, globin, and myosin gene families. There are also gene superfamilies. Gene superfamilies are composed of genes that have areas of high homology and areas of high divergence. Examples of gene superfamilies include the oncogenes, homeotic genes, and a newly recognized gene superfamily of transmembrane proteins related to the lens fiber cells *major intrinsic protein*, or the MIP gene superfamily *(1)*. In most cases, the different members of a gene family carry out related functions.

1.2. Advantages of PCR Cloning of Gene Family Members

There are several considerations that must be taken into account when determining the advantages of using polymerase chain reaction (PCR) to identify members of a gene family over conventional cloning methods of screening a library with a related cDNA, a degenerate primer, or an antibody. It is recommended that after a clone is obtained by PCR to use this template to isolate the corresponding clone from a

From: *Methods in Molecular Biology, Vol. 15: PCR Protocols: Current Methods and Applications*
Edited by: B. A. White Copyright © 1993 Humana Press Inc., Totowa, NJ

N-Terminal Sequence of 28kDa Protein.

(M) ASEFKKKLFWRAVVAEFLATTLFVFISIGSALGFK

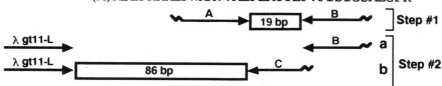

Fig. 1. Two-step strategy for cloning the 5' end of the CHIP28 gene. Listed on top is the *N*-terminal amino acid sequence of purified CHIP28 protein *(4)*, with the addition of the initiating methionine (in brackets) deduced from the sequence of the 5' clone, pPCR-2. Oligonucleotide primers are shown as lines, with arrow heads at their 3'-ends, and approximately representing 5' extensions containing restriction enzyme recognition sequences. In step 1, the sequence of the 19 bp separating degenerate primers 28k-A and -B was determined. The sequence was used to make the nondegenerate primer 28k-C. In step 2, a nested anchor amplification was made with primers 28k-B and -C, anchoring with primer λgt11-L. The 110-bp product obtained contained sequence corresponding to the *N*-terminal 22 amino acids, the initiating methionine, and 38 bp of 5'-untranslated sequence.

library, as mutations can often be introduced in PCR cloning. Alternatively, sequencing two or more PCR clones from independent reactions will also meet this objective. The following is a list of some of the advantages of cloning gene family members by PCR.

1. Either one or two degenerate primers can be used in PCR cloning. When only one of the primers is degenerate, the other primer must be homologous to sequences in the cloning vector, phage cloning site (Fig. 1, step 2), or to a synthetic linker sequence (a poly-G tail). The advantage to using only one degenerate primer is that the resulting clones contain all of the genetic sequence down from the primer (be it 5' or 3' sequence). The disadvantage to this anchor PCR approach is that one of the primers is recognized by every gene in the starting material, resulting in single-strand amplification of all sequences. This is particularly notable when attempting to clone genes that are not abundant in the starting material. This disadvantage can often be ameliorated in part by using a nested amplification approach to amplify desired sequences preferentially *(see* Section 3.3.).

2. It is possible to carry out a PCR reaction on first-strand cDNAs made from a small amount of RNA and, in theory, from a single cell. Several single-stranded "mini-libraries" can be rapidly prepared and analyzed

by PCR from a number of tissues or cell cultures under different hormonal or differentiation stages. Therefore, PCR cloning can potentially provide information about the timing of expression of an extremely rare gene family member, or messenger RNA splicing variants, which may not be present in a recombinant library.

3. Finally, the time and expense required to clone a gene should be considered. Relative to conventional cloning methods, PCR cloning can be more rapid, less expensive, and in some cases, the only feasible cloning strategy. It takes at least 4 d to screen 300,000 plaques from a λgt10 library. With PCR, an entire library containing 10^8 independent recombinants (~5.4 ng DNA) can be screened in one reaction. Again, to ensure authenticity of your PCR clones, you should either use the initial PCR clone to isolate recombinant clones from a library, or sequence at least two clones from independent PCR reactions.

1.3. Degenerate Oligonucleotide Theory

Because the genetic code is degenerate, primers targeted to particular amino acid sequences must also be degenerate to encode the possible permutations in that sequence. Thus, a primer to a six amino acid sequence with 64 possible permutations can potentially recognize 64 different nucleotide sequences, one of which is to the target gene. If two such primers are used in a PCR reaction, then there are 64×64 or 4096 possible permutations. The advantage of using highly degenerate primers such as this is that the target gene will be recognized by a small fraction (1/64) of both primers, and the amplification product from that gene will increase exponentially. The disadvantage is that some of the other 4095 possible permutations are likely to recognize other gene products. This disadvantage can be ameliorated by performing nested amplifications and by using "guessmer" primer. A guessmer primer is made by considering the preferential codon usage exhibited by many species and tissues (*see* Sections 3.1.), and therefore does not contain all the possible permutations in the amino acid sequence.

1.4. Codon Usage

In particular organisms, certain amino acid codons are "preferred" and are utilized more often than others *(2)*. For instance, the four codons for alanine begin with GC. In the third position of this codon, G is rarely used in humans (~10.3% of the time) or rats (~8.0%), but often used in *E. coli* (~35%). This characteristic of codon usage may be advantageously used when designing degenerate oligonucleotide primers (*see* Section 3.1.).

1.5. Strategy for Cloning the 5' End of a Red Cell CHIP28 Gene

In the Methods section, I describe how degenerate oligonucleotide primers were used to clone the 5' end of a gene that is expressed in the differentiating red blood cell. A cDNA for this 28-kDa protein (CHIP28, CHannel-forming Integral membrane Protein of M_r 28,000) was initially cloned from a human fetal liver λgt11 cDNA library, since the liver is the primary erythroid organ in the developing fetus. The protein encoded by this gene had been extensively characterized at a biochemical level and the first 35 amino acids from the *N*-terminus were determined prior to cloning *(3,4)*. The Methods section has been broken up into four parts.

1. In Section 3.1., the designing of the degenerate oligonucleotide primers is described.
2. In Section 3.2., a PCR amplification with degenerate primers (Fig. 1, step 1) is described to determine the authentic nucleotide sequence for the intervening amino acids.
3. In Section 3.3., the PCR cloning of the CHIP28 gene 5' end is described. Briefly, this was accomplished by probing the products of a nested anchor PCR amplification (Fig. 1, step 2) with the degenerate primer 28k-A.
4. Finally in Section 3.4., a brief description of how the 3' end of the CHIP28 gene was cloned is given.

2. Materials

2.1. Design of Degenerate Oligonucleotide Primers

No special materials are required here except the amino acid sequences to which the degenerate oligonucleotide primers are going to be designed and a codon usage table (Section 1.2.). A degenerate nucleotide alphabet (Table 1) provides a single-letter designation for any combination of nucleotides. Some investigators have successfully employed mixed primers containing inosine where degeneracy was maximal, assuming inosine is neutral with respect to base pairing, to amplify rare cDNAs using PCR *(5,6)*.

The primers used in this study are listed in Table 2. The degenerate primers 28k-A and B are made complementary to the indicated amino acid sequences. These primers are guessmers, e.g., they do not take

Table 1
The Degenerate Nucleotide Alphabet

Letter	Specification	Letter	Specification
A	Adenosine	C	Cytidine
G	Guanosine	T	Thymidine
R	puRine (A or G)	Y	pYrimidine (C or T)
K	Keto (G or T)	M	aMino (A or C)
S	Strong (G or C)	W	Weak (A or T)
B	Not A (G, C, or T)	D	Not C (A, G, or T)
H	Not G (A, C, or T)	V	Not T (A, C, or G)
N	aNy (A, G, C, or T)	I	Inosine[a]

[a]Although inosine is not a true nucleotide, it is included in this degenerate nucleotide list since many researchers have employed inosine containing oligonucleotide primers in cloning gene family members.

Table 2
Oligonucleotide Primers

Name	Amino acid sequence[a] and oligonucleotide sequence[b]
28k-A	[*Xba* I] F W R A V V A E (#9–16) 5'-ct*tctaga* TTC TGGAGG GCCGTS GTS GCNGA-3'
28k-B	F V F I S I G [*Cla* I] (#22–29) 3'-AAR CAS AAR TAR WSR TAR CCC *tagcta*at-5'
28k-C	E F L A T T L [*Xba* I] (#16–22) 3'-CTC AAGGACCGGTGC TGGGAGA*agatct*cg-5'
λgt11-L	5'-GGTGGCGACGACTCCTGGAGCCCG-3'

[a] The single letter abbreviations of amino acids are listed over the oligonucleotide sequences. The numbers refer to the amino acid sequence from the *N*-terminus of purified CHIP28 protein *(4)*. In brackets are the restriction endonuclease sites over the corresponding nucleotide sequence, which are italicized.

[b] Primer 28k-A and 28k-B are degenerate. Refer to Table 1 for a description of the degenerate nucleotide alphabet. Nucleotides corresponding to sequences in the CHIP28 gene or λgt11 are in caps. Restriction enzyme recognition sequences are italized.

into account all possible permutations in the amino acid sequence. For instance, only 256 out of the possible 3456 possible permutations in the amino acid sequence for primer 28k-B are taken into account. The sequence for primer 28k-C was deduced based on a PCR amplification with primers 28k-A and B (Section 3.2.). Primers B and C are complementary to the noncoding strand. The last primer, λgt11-L, is complementary to sequences flanking the *Eco* RI cloning site in the left arm of λgt11. It was used for anchor PCR cloning of the 5' end of the red cell CHIP28 cDNA (Section 3.3.).

2.2. PCR Step I: Degenerate Oligonucleotide PCR Cloning a Target DNA Sequence

1. Reagents and enzymes for the PCR, including *Taq* DNA polymerase, 10X PCR reaction buffer (Perkin Elmer Cetus, Norwalk, CT), dNTP stock solution (1.25 m*M* dATP, dCTP, dGTP, and dTTP), a programmable thermal cycler machine (available from a number of manufacturers, including Perkin Elmer Cetus and MJ Research, Inc.), and mineral oil, essentially as described in Chapter 1.

2. Degenerate oligonucleotide primers 28k-A and 28k-B (Table 2) at 20 pmol/µL in double-distilled 0.2 µm-filtered water (DDW) that has been autoclaved. The primers you choose will correspond to your amino acid sequence (*see* Sections 2.1. and 3.1.).

3. Heat denatured human fetal liver cDNA library DNA. The DNA template you use will depend on the available starting material. The DNA is heat-denatured at 99°C for 10 min and stored at 4°C or –20°C.

4. 7.5*M* Ammonium acetate, for precipitation of DNA. Ammonium acetate is preferred over sodium acetate because it inefficiently precipitates nucleotides and primers. Dissolve in DDW and filter through 0.2-µm membrane.

5. 100% Ethanol, stored at –20°C.

6. 70% Ethanol at room temperature.

7. Restriction endonucleases *Xba* I and *Cla* I. The enzymes you use will correspond to the sequences incorporated at the ends of your primers (Section 3.1.).

8. PC9: *See* Note 1. Mix equal volumes of buffer-saturated phenol (pH > 7.2) and chloroform; extract twice with an equal volume of 100 m*M* Tris-HCl, pH 9.0. Separate phases by centrifugation at room temperature for 5 min at 2000*g*. Store at 4°C or –20°C for up to 1 mo. Use polypropylene or glass tubes for preparation and storage only. **Do not use polystyrene tubes.** Buffer-saturated phenol is prepared using ultrapure redistilled crystalline phenol as recommended by the supplier (Gibco BRL, Gaithersburg, MD, product #5509).

9. 5X TBE: 0.45*M* Tris-HCl, pH 8.15, 0.45*M* boric acid, 10 m*M* EDTA. Dissolve in DDW and filter through 0.2-µm membrane. Store at room temperature.

10. Acrylamide gel apparatus.

11. 30% Acrylamide: *See* Note 1. Prepare 200 mL by dissolving 58 g of acrylamide plus 2 g of *N,N'*-methylenebisacrylamide in 150 mL of DDW.

Heat at 37°C and stir to dissolve. Adjust volume to 200 mL and filter into a dark bottle through Whatman No.1 paper. Store at 4°C. The pH should be ≤7.0; check before each use.

12. 10% AP: Dissolve 0.1 g of ammonium persulfate in 1 mL DDW. Store for up to 1 wk at 4°C.

13. TEMED.

14. *Hae* III digested φX174 DNA markers (Gibco BRL #5611SA).

15. 6X Gel loading buffer (GLOB): 0.25% bromophenol blue, 0.25% xylene cyanol FF, 30% glycerol in DDW. Store up to 4 mo at 4°C.

16. EtBr: *See* Note 1. Prepare 10 mg/mL stock of ethidium bromide in DDW and store at 4°C in a brown or foil-wrapped bottle. Use at 0.5–2.0 µg/ mL in DDW for staining nucleic acids in acrylamide and agarose gels.

17. 1-cc syringe barrels and plungers.

18. Siliconized glass wool.

19. Elution buffer: $0.5M$ ammonium acetate, 10 mM magnesium acetate, 2 mM EDTA, 0.05% sodium dodecyl sulfate (SDS). Prepare fresh from stock solutions of $7.5M$ ammonium acetate, $0.5M$ magnesium acetate, 200 mM EDTA, and 10% SDS.

20. TE: 10 mM Tris-HCl, pH 8.0, 1 mM EDTA. Dissolve in DDW and filter through 0.2-µm membrane. Store at room temperature.

21. pBLUESCRIPT II phagemid vector (Stratagene, La Jolla, CA). A number of comparable bacterial expression vectors are available from several companies.

22. T4 DNA Ligase (1 U/µL) and 5X T4 DNA Ligase Buffer (Gibco BRL).

23. Competent DH5α bacteria. This can be prepared *(7)* or purchased from a number of companies.

24. SOB media. Prepare from 20 g of bacto-tryptone, 5 g of bacto-yeast extract, and 0.5 g of NaCl dissolved in 950 mL DDW. Add 5 mL of $0.5M$ KCl. Adjust pH to 7.0 with NaOH. Adjust volume to 1000 mL with DDW and sterilize by autoclaving for 20 min at ≥15 lb/sq in. on liquid cycle.

25. SOB-Carb50 Plates. Prepare by adding 15 g of bacto-agar to 1000 mL SOB media prior to autoclaving. Add a stirring bar and sterilize by autoclaving 20 min on the liquid cycle. Gently swirl the media upon removing it from the autoclave to distribute the melted agar. **Be careful:** the fluid may be superheated and may boil over when swirled. Place the media on a stirrer and allow it to cool to 50°C. Add 1 mL of Carbenicillin (or ampicillin; 50 mg/mL in DDW, 0.2 µm filtered, store in 1-mL aliquots at –20°C), swirl to distribute, and pour 25–35 mL/90-mm plate. Carefully flame the surface of the media with a bunsen burner to remove air bubbles before the agar hardens.

26. IPTG. Dissolve 1 g of isopropylthiogalactoside in 4 mL of DDW, filter through 0.2-µm membrane, and store in aliquots at –20°C.

27. X-Gal. Dissolve 100 mg of 5-bromo-4-chloro-3-indolyl-β-D-galacto-pyranoside in 5 mL of dimethylformamide and stored at –20°C.

28. Plasmid DNA isolation kit, such as the QIAGEN>Plasmid<mini Kit (QIAGEN, Chatsworth, CA, #12123). These kits contain all reagent and instructions for plasmid DNA isolation form a small (1–5 mL) culture of bacteria.

29. DNA sequencing supplies or access to a DNA sequencing core facility.

2.3. PCR Step II: Anchor PCR to Clone the 5' End of the CHIP28 Gene

1. From Section 2.2., items 1, 3–6, 8–16, and 20–29.

2. Degenerate primer 28k-B, and nondegenerate primers 28k-C and λgt11-L (Table 2) at 20 pmol/μL in DDW that has been autoclaved.

3. Restriction enzymes *Xba* I and *Eco* RI.

4. Microfilter spin units (Millipore Ultrafree-MC #UFC3LTK, 30,000 mol-wt cut-off). *See* Note 2.

5. NEN Colony/plaque screen disc (Dupont, Boston, MA).

6. T4 Polynucleotide Kinase (T4 PNK; 20–50 U/μL).

7. 10X T4 PNK reaction buffer: $0.5M$ Tris-HCl, pH 7.6, 100 mM MgCl$_2$, 50 mM dithiothreitol, 1 mM spermidine HCl, 1 mM EDTA. Prepare fresh.

8. [γ-^{32}P]ATP (specific activity ~5000 Ci/mmol; 10 mCi/mL).

9. Glycogen (1 mg/mL), stored in small aliquots at –20°C.

10. 80% Ethanol, stored at –20°C.

11. 20X SSC: $3M$ NaCl, $0.3M$ sodium citrate, pH 7.0. Prepare in DDW and filter through 0.2-μm membrane. Store at room temperature.

12. 10% SDS. Prepare from 10 g of ultrapure SDS dissolved in 100 mL DDW at 37°C. Store at room temperature.

13. $0.5N$ NaOH. Prepare fresh in DDW.

14. $1M$ Tris-HCl, pH 7.5. Prepare in DDW and filter through 0.2-μm membrane. Store at room temperature.

15. Colony wash solution: 2X SSC, 0.1% SDS. For 200 mL, use 178 mL of DDW, 20 mL of 20X SSC, and 2 mL of 10% SDS.

16. 50X Denhardt's solution: 1% bovine serum albumin (BSA), 1% polyvinylpyrrolidone (PVP), 1% Ficoll. Prepare in DDW, 0.2-μm filter, and store in 10- to 20-mL aliquots at –20°C.

17. Prehybridization solution: 6X SSC, 5X Denhardt's, 0.1% SDS. For 25 mL, use 14.75 mL of DDW, 7.5 mL of 20X SSC, 2.5 mL of 50X Denhardt's, and 250 μL of 10% SDS.

18. Hybridization solution: same as the prehybridization solution, except without the SDS, and with 0.5–5.0 × 10⁶ dpm/mL end-labeled primer (≤10 ng/mL).

19. Heat-seal bags.

3. Methods

3.1. Design of Degenerate Oligonucleotide Primers

1. The first step in designing a degenerate primer is to write down the amino acid sequence, followed by the potential nucleotide sequence (or the complement of this sequence for a downstream primer), considering all possible permutations. For the amino acid sequence *Leu-Ile-Gly-Glu*, the degenerate nucleotide sequence (*see* Table 1) would be 5'-YTN-ATH-GGN-GAR-3'. If the amino acid sequence is relatively long, you can potentially design two or more degenerate primers. If only one is made, make it to sequences with a high GC content, as these primers can be annealed under more stringent conditions (e.g., higher temperatures).

2. The next step is to determine the number of permutations in the deduced nucleotide sequence. There are 192 permutations $[(2 \times 4) \times 3 \times 4 \times 2)]$ in the sequence 5'-YTN-ATH-GGN-GAR-3'. We can reduce the degeneracy by making educated guesses in the nucleotide sequence, i.e., by making a guessmer. The 3' end of a primer should contain all possible permutations in the amino acid sequence, since *Taq* DNA polymerase will not extend a prime with a mismatch at the extending (3') end. If the primer was to a human gene, a potential guessmer would be 5'-CTB-ATY-GGN-GAR-3', which only contains 64 permutations. This guessmer is proposed by taking into account the preferential codon usage for leucine and isoleucine in humans *(2)*.

3. The degeneracy of a primer can be reduced further by incorporating inosine residues in the place of N. This would result in the sequence 5'-CTB-ATY-GGI-GAR-3', which contains only 12 permutations. The advantages of using inosine-containing primers is that they have a reduced number of permutations, and the inosine presumably base pairs equally well with all four nucleotides, creating a single bond in all cases. The disadvantage is that inosines reduce the annealing temperature of the primer. Inosine-containing primers were therefore not used in this study.

4. It is often convenient to incorporate restriction endonuclease sites at the 5' ends of a primer to facilitate cloning into plasmid vectors. Different restriction sites can be added to the 5' ends of different primers so the products can be cloned directionally. However, not all restriction enzymes can recognize cognate sites at the ends of a double-stranded DNA molecule equally well (*see* Chapters 2 and 22). This difficulty can be ameliorated by adding a two to four nucleotide 5' overhang before the beginning of the restriction enzyme site (*see also* Note 3). The best restriction enzymes sites to use are *Eco* RI, *Bam* HI, and *Xba* I. Recent New England Biolabs Inc. (Beverly, MA) catalogs have a list of the

ability of different restriction enzymes to recognize short nucleotide sequences. A potential pitfall of this approach would be the occurrence of the same restriction site within the amplified product as used on the end of one of the primers. Therefore only part of the amplified product would be cloned.

5. The final consideration you should make is the identity of the 3'-most nucleotide. The nucleotide on the 3' end of a primer should not be N, I, or T, and preferably be G or C. The reason for this is that thymidine (and supposedly inosine) can nonspecifically prime on any sequence. Guanosines and cytidine are preferred since they form three H-bonds at the end of the primer, a degree stronger than an A:T basepair.

3.2. PCR Step I: Degenerate Oligonucleotide PCR Cloning a Target DNA Sequence

The first step in cloning the CHIP28 gene was to determine the nucleotide sequence between the degenerate primers, 28k-A and B (Fig. 1, step 1). The template for these reactions was a human fetal liver library in λgt11. When first amplified, the titer of this library was 5×10^9 PFU/mL. A 5-μL aliquot would therefore contain ~2.5×10^7 PFU, or ~1.5 ng of DNA. Prior to PCR amplification, the DNA was heat denatured at 99°C for 10 min. I have found no significant difference in the PCR amplification of crude phage lysates and DNA isolated from the bacteriophage capsids. In all cases, nonrecombinant λ DNA is also PCR amplified with the same primers as a negative control.

3.2.1. PCR Reaction (see Notes 4 and 5)

1. Pipet into 0.5-mL microcentrifuge tubes in the following order:
   a. 58.5 µL of autoclaved DDW
   b. 5.0 µL of heat denatured library DNA
   c. 5.0 µL of Primer 28k-A
   d. 5.0 µL of Primer 28k-B
   e. 10 µL of 10X PCR reaction buffer
   f. 16 µL of 1.25 mM dNTP stock solution
   g. 0.5 µL of Amplitaq DNA polymerase (Perkin Elmer Cetus).

   If several reactions are being set up concurrently, a master reaction mix can be made up, consisting of all the reagents used in all of the reactions, such as the DDW, primers, reaction buffer, dNTPs, and the polymerase. This reaction mix should be added last.
2. Briefly vortex each sample, and spin for 10 s in a microfuge.
3. Overlay each sample with 2–3 drops of mineral oil.

4. Amplify using the following cycle profile.

   24 main cycles   94°C, 60 s (denaturation)
                                       50°C, 90 s (annealing; *see* Note 6)
                                       72°C, 60 s (extension)
   Final extension   72°C, 4 min
                                       10°C (hold)

5. Remove the reaction tubes from the thermal cycler and add 200 μL of $CHCl_3$. Spin for 10 s in a microfuge to separate oil-$CHCl_3$ layer from the aqueous layer.

6. Carefully transfer the aqueous layer to a clean microfuge tube. If any of the oil-$CHCl_3$ layer is also transferred, it must be removed by extracting the sample again with another 100 μL $CHCl_3$.

7. If performing a nested anchor PCR amplification (as in Section 3.3.), it is necessary to remove all primers from the primary PCR reaction first. If this applies to your experiment, *see* Note 2 at this time. Otherwise continue.

8. Set up four secondary PCR amplifications with 10-μL aliquots of the first reactions, using the same procedure and cycling parameters. Again extract all of the samples with $CHCl_3$ to get rid of the oil, and pool all four reactions into a 1.5-mL microfuge tube.

3.2.2. DNA Recovery and Restriction Enzyme Digestion

1. Add 150 μL of 7.5*M* ammonium acetate (50% vol) to 300 μL of the PCR reaction. Vortex briefly to mix. Precipitate the DNA with 1 mL of 100% ethanol, vortex the samples for 5–10 s and ice for 15 min.

2. Spin down the DNA at 12,000*g* for 10 min at 4°C in a microfuge. Decant the aqueous waste. Add 500 μL of 70% ethanol. Vortex briefly and spin another 5 min at 4°C. Decant the ethanol and allow the pellets to dry inverted at room temperature, or dry in a SpeedVac (Savant Instruments, Inc., Farmingdale, NY) for 2–10 min.

3. The DNA pellets are resuspended in 100 μL of DDW. This DNA is then digested with 100 U of the restriction enzymes *Xba* I and *Cla* I. The enzymes you use will correspond to the restriction sites incorporated at the ends of your primers. Add 115 μL of DDW, 90 μL of the PCR-amplified DNA, 25 μL of 10X restriction enzyme reaction buffer, 10 μL of *Xba* I (10 U/μL), and 10 μL of *Cla* I (10 U/μL). Incubate at 37°C for 3 h.

4. Extract the DNA with 250 μL of PC9. Vortex for 15 s and spin at 12,000*g* for 2 min at room temperature.

5. Transfer the aqueous phase to a clean 1.5-mL microfuge tube. Add 125 μL of 7.5*M* ammonium acetate, and vortex briefly to mix.

6. Precipitate the DNA with 800 µL of 100% ethanol. Vortex the samples for 5–10 s and ice for 15 min. Repeat step 2 in this section and resuspend the DNA in 10–20 µL of DDW.

In the next two sections, the PCR amplified DNA is run on a polyacrylamide gel for the optimal separation of PCR product sizes, followed by the elution of specific sized PCR products from that gel. If you do not know the expected size of your target sequence, an alternate approach will have to be devised. If your amplified product is expected to be greater that 400 bp, you could run the PCR products on an agarose gel and GeneClean (Bio 101, Inc., La Jolla, CA) or electroelute the DNA of the expected size (*see* Notes 2 and 7).

3.2.3. Polyacrylamide Gel Electrophoresis

1. Run the PCR products on a 15% polyacrylamide gel (19:1 acrylamide/ *bis*) in 1X TBE. Prepare this gel by adding 11.6 mL of DDW, 20 mL of 30% acrylamide, and 8 mL of 5X TBE to a vacuum flask. The polymerization of the acrylamide is catalyzed by adding 320 µL of 10% AP and 50 µL of Temed. Quickly pour the solution between the glass plates and insert the comb. When polymerization is complete, carefully remove the comb and wash the wells out with 1X TBE.
2. Prepare the samples for electrophoresis by adding 1 µL of 6X GLOB to every 5 µL of sample. In one tube, add 1 µL of *Hae* III-digested ϕX174 DNA, 9 µL of DDW, and 2 µL of 6X GLOB.
3. Prerun the gel at 100 V for 5–15 min. Load the samples and run at 100 V until the bromophenol blue dye is at the bottom of the gel.
4. Turn the power off, remove the glass plate-gel sandwich, and separate the glass plates. Notch one corner of the gel for orientation purposes and transfer the gel into a tank containing 200–500 mL of DDW. Add EtBr to 1 µg/ml and gently shake at room temperature for 5–15 min to stain the DNA. Photograph the gel under a UV lamp.

3.2.4. Crush and Soak Extraction of DNA from Polyacrylamide Gels

1. Excise the products in the EtBr-stained gel from about 70–90 bp with a clean razor blade. Transfer the gel slice into a microfuge tube that has been weighed. Reweigh the tube to determine the weight of the gel.
2. Crush the gel with a pipet tip and add 2 vols of elution buffer (300 µL of buffer to 150 mg of gel). Close the tube, wrap the top with parafilm, and shake at 37°C for 2 h on a shaker platform. Using a large pipet tip, pipet the suspension up and down to help break up the acrylamide gel. Incubate another hour at 37°C with rapid shaking.

3. Transfer the entire suspension into a 1-cc syringe barrel containing a siliconized glass wool plug. Slowly pass the supernatant through the column by pushing with the syringe plunger, collecting the supernatant into a clean microfuge tube.

4. Add an additional 200 µL of elution buffer (at 37°C) to the top of the column and pass it through the column to wash off any remaining DNA.

5. Precipitate the DNA by adding 2 vol of 100% ethanol, vortex the samples for 5–10 s and ice for 60 min. Spin down the DNA at 12,000g for 15 min at 4°C. Decant the aqueous waste and allow the pellets to dry inverted at room temperature, or dry in a SpeedVac for 2–10 min.

6. Resuspend the DNA in 100 µL of TE plus 50 µL of 7.5*M* AmAc, and precipitate with 350 µL of 100% ethanol as just described in step 5.

7. Resuspend the DNA in 10 µL of DDW.

3.2.5. DNA Ligation, Bacterial Transformation, Plasmid DNA Minipreps, and DNA Sequencing

1. Digest 1 µg of pBluescript II KS phagemid vector DNA (Stratagene) with 10 U of *Cla* I and *Xba* I in a 50-µL vol. Incubate at 37°C for 2 h. Extract once with 50 µL of PC9. Vortex 15 s, and spin 2 min at 12,000g in a microfuge.

2. Transfer the aqueous phase to a clean tube, add 25 µL of 7.5*M* ammonium acetate, vortex 5 s, and precipitate with 200 µL of 100% ethanol. After 15 min on ice, collect the DNA by centrifugation at 12,000g for 10 min at 4°C.

3. Decant and wash the pellet in 200 µL of 70% ethanol, vortex briefly, and spin another 5 min at 4°C. After drying the pellets, resuspend the DNA in 10 µL DDW.

4. Set up the ligation reactions on ice with the *Xba/Cla*-digested vector and insert as follows:

   Reaction 1:   1 µL vector (100 ng; vector control)
   Reaction 2:   1 µL vector + 1 µL insert
   Reaction 3:   1 µL vector + 4 µL insert

5. Add 2 µL of 5X T4 DNA ligase buffer and DDW to 9.5 µL. Add 0.5 µL of T4 DNA Ligase, gently vortex, and spin 5 s in a microfuge.

6. Incubate at 14°C for 4–20 h. Stop the reaction with an equal volume of TE and store the samples at −20°C.

7. Set up a bacterial transformation with competent DH5α bacteria, or a comparable strain of bacteria. Be sure to include a positive control (10 ng undigested vector DNA) and a negative control (DDW). To 1.5-mL microfuge tubes, add half of the ligation mix (10 µL) or 10 µL of control DNA or DDW, and 100 µL of competent bacteria (thawed slowly on ice), and incubate on ice for 45 min. Heat shock at 37°C for 3 min.

Return to room temperature for 2 min. Add 400 µL of SOB media containing 10% glycerol. Mix end-over-end and allow bacteria to recover and express the ampicillin resistance gene by growing at 37°C for 1 h.

8. Prewarm SOB-Carb50 plates at 37°C for 45 min. About 30 min before plating the bacteria on the plates, add 40 µL of X-Gal and 4 µL of IPTG, and quickly spread over the entire surface of the plate using a sterile glass spreader. Spread 20–200 µL of the transformation reactions on these plates. Allow the inoculum to absorb into the agar and incubate the plates inverted at 37°C for 12–24 h. Afterwards, placing the plates at 4°C for 2–4 h will help enhance the blue color development.

9. Colonies that contain active β-galactosidase will appear blue, whereas those containing a disrupted LacZ gene will be white. Set up minipreparations of plasmid DNA by inoculating 2 mL of SOB media, containing ampicillin (or carbenicillin) at 50 µg/mL, with 12–24 individual white colonies. After growing at 37°C overnight, isolate the plasmid DNA using a QIAGEN plasmid mini kit (or another available kit). Resuspend the DNA in 50 µL of DDW.

10. Digest 20 µL of the DNA with *Xba* I and *Cla* I, PC9 extract, and ethanol precipitate as described in Section 3.2.2. Run the digested DNA on a 15% polyacrylamide gel along with *Hae* III digested ϕX174 DNA markers, as described in Section 3.2.3. Perform double-stranded DNA sequencing on the recombinants containing an insert of about the correct size. One out of five of my clones (pPCR-1) contained an insert corresponding to the intervening amino acids, FLATTL. A nondegenerate antisense primer, 28k-C, was then made corresponding to this sequence (Fig. 1, step 2; Table 2) for anchor PCR cloning the 5' end of the CHIP28 gene.

3.3. PCR Step II: Anchored PCR to Clone the 5' End of the CHIP28 Gene

3.3.1. PCR Reaction I

See Notes 4 and 5. As in Section 3.2., nonrecombinant λ DNA is also PCR amplified with the same primers as a negative control. The first step in cloning the 5' end of the CHIP28 gene was to perform an anchored PCR reaction with primers 28k-B and λgt11-L. A second set of reactions was performed using primer 28k-B and an oligo to the right arm of λgt11. However, Southern blot analysis, probing with primer 28k-A, suggested only a single product in the λgt11-L primed reaction and nothing in the λgt11-R reaction. The first PCR reaction with the degenerate primer 28k-B was included to enrich for PCR products corresponding to the CHIP28 gene.

1. Pipet into 0.5-mL microcentrifuge tubes in the following order:
   a. 53.5 µL of DDW that has been autoclaved
   b. 10 µL of heat-denatured library DNA
   c. 5.0 µL of Primer 28k-B
   d. 5.0 µL of Primer λgt11-L
   e. 10 µL of 10X PCR reaction buffer
   f. 16 µL of 1.25 mM dNTP stock solution
   g. 0.5 µL of Amplitaq DNA polymerase.
2. Briefly vortex each sample, then spin for 10 s in a microfuge.
3. Overlay each sample with 2–3 drops of mineral oil.
4. Amplify using the following cycle profile:

31 main cycles	94°C, 60 s (denaturation)
	54°C, 60 s (annealing; *see* Note 6)
	72°C, 60 s (extension)
Final extension	72°C, 4 min
	10°C (hold)

5. Remove the reaction tubes from the thermal cycler and add 200 µL of CHCl$_3$. Spin for 10 s in a microfuge to separate oil-CHCl$_3$ layer from the aqueous layer.
6. Carefully transfer the aqueous layer to a clean microfuge tube. If any of the oil-CHCl$_3$ layer is also transferred, it must be removed by extracting the sample again with another 100 µL of CHCl$_3$.
7. To remove all unincorporated nucleotides and primers, spin the sample through a microfilter spin unit (e.g., Millipore, Bedford, MA, Ultrafree MC), as recommended by the manufacturer. The DNA will remain on top of the membrane, whereas nucleotides and primes will pass through. Wash the sample with an additional 400 µL of TE and respin to concentrate. Repeat wash one more time. Add TE to the remaining sample to bring the final volume of retained fluid to 50 µL. Transfer to a clean sterile tube.

3.3.2. PCR Reaction II

Repeat steps 1–6 from Section 3.3.1. except in step 1, amplify a 5-µL aliquot of the primary PCR reaction using primers 28k-C and λgt11-L.

3.3.3. Restriction Enzyme Digestion and DNA Clean-Up

1. Digest the secondary PCR reaction with 100 U of the restriction enzymes *Eco* RI (recognition site in the phage) and *Xba* I. The enzymes you use will correspond to the restriction sites incorporated at the ends of your primers and the starting material. Assemble the reaction consisting of 115 µL of DDW, 90 µL of the PCR amplified DNA, 25 µL of 10X restriction enzyme reaction buffer, 10 µL of *Xba* I (10 U/µL), and 10 µL of *Eco* RI (10 U/µL). Incubate at 37°C for 3 h.

2. Remove the low-mol-wt oligonucleotides, nucleotides, and PCR restriction ends by diluting the sample with 500 µL of TE buffer and concentrating in a microfilter spin unit (Millipore, Ultrafree MC). After concentrating, wash the sample four times with 300 µL of TE.
3. Adjust the volume of the sample on the top of the microfilter spin membrane to 100 µL and transfer it to a clean microfuge tube. Extract the DNA with 100 µL of PC9. Vortex for 15 s. Spin at 12,000*g* for 2 min at room temperature. Transfer the aqueous phase to a clean microfuge tube.
4. Add 50 µL of 7.5*M* ammonium acetate, vortex, and precipitate with 350 µL of 100% ethanol. After 15 min on ice, spin down the DNA at 12,000*g* for 10 min at 4°C in a microfuge. Decant the aqueous waste. Add 300 µL 70% ethanol. Vortex briefly and spin another 5 min at 4°C. Decant the ethanol and allow the pellets to dry inverted at room temperature, or dry in a SpeedVac for 2–10 min. Resuspend the DNA in 10 µL DDW.

3.3.4. DNA Ligation and Bacterial Transformation

Since the size of the 5' end of the gene is not known, the entire sample is ligated into pBluescript II phagemid vector (Stratagene), transformed into DH5α competent bacteria, and plated on SOB-Carb50 plates as described in steps 1–8 in Section 3.2.5., with the exception that the vector is digested with *Eco* RI and *Xba* I.

To enrich for the identification of recombinants containing an insert of interest, bacterial colony lifts are made on plates containing 100–400 colonies, and probed with ^{32}P-end-labeled nonoverlapping degenerate primers. In the next two sections, the protocols for ^{32}P-end-labeling of oligonucleotide primers and bacterial colony lift/hybridization are described.

3.3.5. End-Labeling Oligonucleotide Primers with T4 Polynucleotide Kinase

1. Set up a reaction in a 1.5-mL microfuge tube as follows (at room temperature):

Primer 28k-A (or your primer)	1.0 µL (15–30 pmol)
10X T4PNK reaction buffer	2.0 µL
DDW (autoclaved)	11.0 µL
[γ-^{32}P]ATP	
(5000 Ci/mmol; 10 mCi/mL)	5.0 µL (10 pmol)
T4 Polynucleotide Kinase	
(10 U/µL)	1.0 µL

   Gently vortex and spin 10 s in microfuge. Incubate the reaction at 37°C for 45 min. *See* Note 8.

2. Terminate the reaction by heat inactivation of the enzyme at 68°C for 10 min.

3. Add 2 µL of glycogen (1 mg/mL), 3 µL of TE buffer, and 50 µL of 7.5M ammonium acetate. Vortex 5 s and spin 10 s. Remove 1 µL into 99 µL of TE buffer for determining the percent incorporation of ^{32}P into the primer and for calculating the specific activity. Set this tube aside.

4. Precipitate the primers with 185 µL of 100% ethanol. Vortex 10 s and ice for 30 min. Spin down the primers by centrifugation at 12,000g for 30 min at 4°C. Carefully transfer the radioactive aqueous waste into appropriate containers using disposable pipet tips.

5. Add 400 µL of 80% ethanol. Vortex briefly and spin another 10 min at 4°C. Decant the radioactive wash. If desired, repeat 80% ethanol wash once more, and allow the pellets to dry inverted at room temperature behind a plexiglass screen.

6. Dissolve the radioactive primers in 75 µL of DDW. Remove 1 µL into 99 µL of TE for percent incorporation and specific activity calculations. Store the labeled primers at –20°C and use within 7 d.

7. Determine the percent incorporation and specific activity of the primer by counting 10 µL of the pre- and postprecipitation aliquots in a scintillation counter.

3.3.6. Bacterial Colony Lift-Hybridizations

1. Lift colonies from plates containing 100–400 bacterial colonies onto NEN colony/plaque screen disc (Dupont) as recommended by the manufacturer with the following modifications. Denature and neutralize the colonies on Whatman 3MM paper that has been saturated with the appropriate solutions.

2. Before prehybridization, wash the filters in 2X SSC, 0.1% SDS at 37°C (25–100 mL/filter), and gently rub the filters with a Kimwipe soaked in this solution to remove bacterial debris. After 20 min, add fresh SSC/SDS solution and incubate for 2 h at 55°C.

3. Blot dry the filters, transfer them into heat-seal bags, and prehybridized at 65°C for 2–4 h in prehybridization solution (2–5 mL/filter).

4. Remove the prehybridization solution and add equivalent volume of the hybridization solution containing 3.7 × 10^7 cpm/mL of ^{32}P-end-labeled 28k-A. Hybridize overnight at 45°C.

5. The hybridization solution can be saved for future uses. Wash the filters in 6X SSC, 0.1% SDS (25–100 mL/filter) for 30 min with increasing temperature. The final wash was at 65°C. By increasing the temperature and decreasing the salt concentration, the stringency will increase. You will have to determine the optimal conditions for every primer employed.

I have washed colony lifts with a 25-mer primer (with a high GC content) at 50°C in 1X SSC and was able to detect preferentially hybridizing colonies with a 6-h exposure.

3.3.7. Plasmid DNA Minipreps, Restriction Enzyme Digestion, and DNA Sequencing

1. Set up minipreparations of plasmid DNA by inoculating 2 mL of SOB media, containing ampicillin (or carbenicillin) at 50 µg/mL, on 12–24 individual white colonies that correspond to colonies that preferentially hybridize with the ^{32}P-labeled primer. After growing at 37°C overnight, isolate the plasmid DNA. Resuspend the DNA in 50 µL of DDW.
2. Digest 10 µL of the DNA with *Xba* I and *Eco* RI, PC9 extract, and ethanol precipitate as described in Section 3.2.2. Run the digested DNA on a 15% polyacrylamide gel along with *Hae* III-digested φX174 DNA markers, as described in Section 3.2.3.
3. Perform double-stranded DNA sequencing on recombinants containing different sizes of inserts. In this experiment, the insert sizes ranged from about 60–1200 bp. A plasmid with a 118-bp insert contained a nucleotide sequence corresponding to the first 22 amino acids from the *N*-terminus of purified CHIP28 protein, proceeded by an initiating methionine and 38 bp of 5'-untranslated nucleotides. The sequence around the initiating methionine agreed well with the consensus for translational initiation as determined by Kozak *(8)*. The clone was named pPCR-2.

3.4. PCR Step III: Anchor PCR to Clone the 3' End of the CHIP28 Gene

Briefly, the rest of the CHIP28 cDNA was cloned by designing partially overlapping sense primers to this 5' region, and performing another nested anchor amplification of the DNA in the human fetal liver library, this time anchoring with a primer complementary to the right arm of λgt11. Between sequential amplifications, a GeneClean purification step was performed (*see* Note 2) to remove unincorporated primers, nucleotides, and small amplification products. The amplified DNA was run on a 2.6% agarose gel in 1X TAE *(7)*, stained with ethidium bromide, and transferred to nylon by alkaline transfer essentially as described elsewhere *(9)*. A single 850-bp product hybridized with ^{32}P-labeled primer 28k-C. This product (pPCR-3) was cloned and sequenced, and used to screen an adult human bone marrow library for full-length recombinants *(10)*.

4. Notes

1. Carcinogens cautions. Contact your hazardous waste department for proper disposal procedures in your area.
2. After a PCR reaction, it is often advantageous to remove the unincorporated nucleotides and primers from the reaction. This step is essential when performing a nested PCR amplification. I have used two different methods to purify PCR products. Spin dialysis columns are ideal for purifying PCR products from 50–500 bp in length, and larger. These columns are available from a number of manufacturers (e.g., Millipore and Centricon). I prefer Millipore Ultrafree MC micro filter units (30,000 mol-wt cut-off) because of the convenience of being able to spin the samples in a tabletop microfuge. Spin columns will also work with PCR products >500 bp, but for several reasons, I would recommend using the GeneClean II kit (Bio 101 Inc.) for these applications. The Glassmilk suspension in these kits will bind to DNA molecules >500 bp, but binds poorly to smaller molecules. Therefore, smaller PCR products can be effectively removed. This kit is also useful for gel-purifying specific products from agarose gels. The instructions supplied with the kit are quite simple and explicit, and therefore require no further explanation.
3. When designing primers with restriction enzyme sites and 5' overhangs, note that this 5' overhang should not contain sequence complementary to the sequence just 3' of the restriction site, as this would facilitate the production of primer dimers. Consider the primer 5'-ggg.*aagctt*. CCCAGCTAGCTAGCT-3', which has a *Hind* III site proceeded by a 5'-ggg and followed by a CCC-3'. These 12 nucleotides on the 5' end are palindromic, and can therefore easily dimerize with another like primer. A better 5' overhang would be 5'-cac.
4. All PCR reactions should be set up in sterile laminar flow hoods using either positive displacement pipetors or pipet tips containing filters to prevent the contamination of samples, primers, nucleotides, and reaction buffers by DNA. Similarly, all primers, nucleotides, and reaction buffers for PCR should be made up and aliquoted using similar precautions.
5. When cloning a gene from a recombinant library by PCR, remember that not all genes are created equally. In particular, genes with high G:C contents have proven more difficult than most. Several researchers have made contributions in a search for factors to enhance the specificity of PCR reactions. A nonionic detergent, such as Nonident P-40, can be incorporated in rapid sample preparations for PCR analysis without significantly affecting *Taq* polymerase activity *(11)*. In some cases, such

detergents are absolutely required in order to reproducibly detect a specific product *(12)* presumably attributable to inter- and intrastrand secondary structure. More recently, tetramethylammonium chloride has been shown to enhance the specificity of PCR reactions by reducing nonspecific priming events *(13)*.

6. A critical parameter when attempting to clone by PCR is the selection of a primer-annealing temperature. This is especially true when using degenerate primers. The primer melting temperature (T_m) is calculated by adding 2° for A:T base pairs, 3° for G:C base pairs; 2° for N:N base pairs, and 1° for I:N base pairs. The T_m for primer 28k-B would be 47°C, excluding the restriction enzyme site. Most PCR chapters suggest you calculate the T_m and set the primer annealing temperature to 5–10°C below the lowest T_m. Distantly related gene superfamily members have been cloned using this rationale *(14)*. However, I have found that higher annealing temperatures are helpful in reducing nonspecific priming, which can significantly affect reactions containing degenerate primers. In reaction containing primer 28k-B, 50°C and 54°C primer-annealing steps were selected.

7. An effective alternative to GeneCleaning of PCR products from agarose gels is to elute the DNA from an excised piece of gel with an electrical field. The procedure for electroelution is described elsewhere *(7)*.

8. In labeling the oligonucleotide primers with T4 kinase, the primers concentration is in excess relative to the [γ-^{32}P]ATP. This will result in significant incorporation of the nucleotides onto the primers, while reducing the specific activity of the probe. For higher specific activity probes, use 5–10 pmol of primer and 20 pmol on [γ-^{32}P]ATP.

9. When cloning a gene from a recombinant library by PCR, remember that not all inserts are created equally. Attempts to clone the CHIP28 gene from a bone marrow or fetal liver library by immunological reactivity with an anti-28 kDa antibody were unsuccessful. Also attempts to express the full-length CHIP28 cDNA in bacteria have failed. It is thought that the hydrophobic CHIP28 product is toxic to the cells.

Acknowledgments

I am grateful to my colleagues for their support and helpful discussions. I especially thank Eric Fearon for technical assistance, and Peter Agre for his consent to report these procedures in detailed form and for his generous support. This work was supported in part by NIH grant HL33991 to Peter Agre. I am the recipient of Postdoctoral Fellowships T32HL07525 ('90–'91) and American Heart Association Maryland Affiliate ('91–'93).

References

1. Pao, G. M., Wu, L-F., Johnson, K. D., Höfte, H., Chrispeels, M. J., Sweet, G., Sandal, N. N., and Saier Jr., M. H. (1991) Evolution of the MIP family of integral membrane transport proteins. *Mol. Microbiol.* **5,** 33–37.
2. Wada, K.-N., Aota, S.-I., Tsuchiya, R., Ishibashi, F., Gojobori, T., and Ikemura, T. (1990) Codon usage tabulated from the GenBank genetic sequence data. *Nucleic Acids Res.* **18,** 2367–2411.
3. Denker, B. M., Smith, B. L., Kuhajda, F. P., and Agre, P. (1988) Identification, purification, and partial characterization of a novel Mr 28,000 integral membrane protein from erythrocytes and renal tubules. *J. Biol. Chem.* **263,** 15,634–15,642.
4. Smith, B. L. and Agre, P. (1991) Erythrocyte Mr 28,000 transmembrane protein exists as a multisubunit oligomer similar to channel proteins. *J. Biol. Chem.* **266,** 6407–6415.
5. Knoth, K., Roberds, S., Poteet, C., and Tamkun, M. (1988) Highly degenerate, inosine-containing primers specifically amplify rare cDNA using the polymerase chain reaction. *Nucleic Acids Res.* **16,** 10932.
6. Chérif-Zahar, B., Bloy, C., Kim, C. L. V., Blanchard, D., Bailly, P., Hermand, P., Salmon, C., Cartron, J-P., and Colin, Y. (1990) Molecular cloning and protein structure of a human blood group Rh polypeptide. *Proc. Natl. Acad. Sci. USA* **87,** 6243–6247.
7. Sambrook, J., Fritsch, E. F., and Maniatis T. (1989) *Molecular Cloning: A Laboratory Manual.* Cold Spring Harbor Laboratory, Cold Spring Harbor, NY.
8. Kozak, M. (1987) An analysis of 5'-noncoding sequences from 699 vertebrate messenger RNAs. *Nucleic Acids Res.* **15,** 8125–8132.
9. Reed, K. C. and Mann, D. A. (1985) Rapid transfer of DNA from agarose gels to nylon membranes. *Nucleic Acids Res.* **13,** 7207–7221.
10. Preston, G. M. and Agre, P. (1991) Isolation of the cDNA for erythrocyte integral membrane protein of 28 kilodaltons: member of an ancient channel family. *Proc. Natl. Acad. Sci. USA* **88,** 11,110–11,114.
11. Weyant, R. S., Edmonds, P., and Swaminathan, B. (1990) Effects of ionic and nonionic detergents on the Taq polymerase. *BioTechniques* **9,** 308–309.
12. Bookstein, R., Lai, C-C., To, H., and Lee, W-H. (1990) PCR-based detection of a polymorphic Bam HI site in intron 1 of the human retinoblastoma (RB) gene. *Nucleic Acids Res.* **18,** 1666.
13. Hung, T., Mak, K., and Fong, K. (1990) A specificity enhancer for polymerase chain reaction. *Nucleic Acids Res.* **18,** 4953.
14. Zhao, Z-Y. and Joho, R. H. (1990) Isolation of distantly related members in a multigene family using the polymerase chain reaction technique. *Biochem. Biophys. Res. Comm.* **167,** 174–182.

CHAPTER 32

Single Specific Primer-Polymerase Chain Reaction (SSP-PCR) and Genome Walking

*Venkatakrishna Shyamala
and Giovanna Ferro-Luzzi Ames*

1. Introduction

The polymerase chain reaction (PCR) is used for selective amplification of DNA fragments from both prokaryotes and eukaryotes (1–3). The only requirement for amplification is that the sequence of the extremities of the DNA fragment to be amplified be known *(4)*. This places a limitation on the use of PCR in the amplification of adjacent unknown regions. We have developed a method that allows the amplification of double-stranded DNA even when the sequence information is available at one end only *(5)*. This method, the single specific primer-PCR (SSP-PCR), permits amplification of genes for which only a partial sequence information is available, and allows unidirectional genome walking from known into unknown regions of the chromosome.

The basic principle of the SSP-PCR procedure is schematically described in Fig. 1. Briefly, (1) the chromosomal DNA is digested with one or two restriction enzymes; (2) the unknown end of the restricted chromosomal DNA is ligated to a suitable oligomer (Generic Oligomer) of known sequence, sufficiently long to serve as a PCR primer (Generic Primer); or the unknown end can be ligated to a vector, in which case the vector sequence can be used to design a PCR primer; (3) the ligation mixture is then amplified with the generic primer annealing to the generic oligomer or to the vector and a specific primer annealing

From: *Methods in Molecular Biology, Vol. 15: PCR Protocols: Current Methods and Applications*
Edited by: B. A. White Copyright © 1993 Humana Press Inc., Totowa, NJ

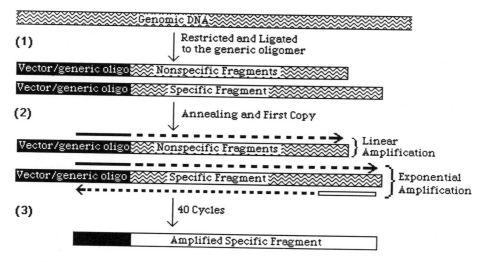

Fig. 1. Schematic representation of single specific primer-PCR (SSP-PCR). Shaded boxes represent nonspecific DNA (essentially all genomic DNA) and the fragments derived by digestion; the blackened boxes represent the vector or generic oligomer ligated to the unknown end of the DNA fragments; blackened bar, generic primer; open bar, fragment-specific primer; dashed arrows, amplified DNA. In step 1, the genomic DNA is digested with a combination of enzymes and the unknown end is ligated to an oligomer of known sequence. In steps 2 and 3, the ligation reaction mixture is amplified with a primer specific for the known end of the fragment and a generic primer complementary to the vector or generic oligomer. Although the generic primer anneals to the unknown ends of all fragments, the resulting products increase only linearly. However, simultaneous annealing of the fragment-specific primer and the generic primer to the specific product results in exponential amplification of the specific product.

to the known sequence. Even though a variety of fragments would be ligated to the generic oligomer and during PCR, the generic primer would anneal to all these ends, specificity is imparted by the specific primer. The exponential accumulation of the product resulting from the combination of the generic primer and the specific primer will dominate over the linear accumulation of nonspecific products from the generic primer alone (*see* Fig. 1).

The most obvious use of SSP-PCR is for unidirectional movement into unknown regions for routine cloning and sequencing and this has been successfully demonstrated by taking two consecutive steps of genome walking in *S. typhimurium (5)* and for the amplification of ribosomal RNA gene promoters in *Thiobacillus ferroxidans* (T. White, personal communication). SSP-PCR is also ideal for situations when the existing sequence information cannot be utilized; for example, cloning of genes from *E. coli* strains other than K12 for which no restriction map is available *(6)*; or cloning genes from a different genus of a related organism such as *S. typhimurium*. This feature has been successfully tested in *S. typhimurium* for amplifying the intergenic region between *argA* and *recD* utilizing the available *E. coli* sequence *(5,7)*.

Another use of SSP-PCR is in the case of deletion *(7a)*, duplication, insertion, and recombination mutations. For example, the sites of transposon or viral insertion can be identified by SSP-PCR, by using a primer specific for the transposon or virus of interest. This feature has been utilized to analyze genetic duplication mutations in the histidine biosynthetic operon in *S. typhimurium (8)*. SSP-PCR has been used successfully in eukaryotes to analyze translocations in the human X chromosome *(9)* and transgenic mice *(10)*. SSP-PCR should also be useful for the analysis of intron-exon junctions in eukaryotic genomic DNA.

The feature of SSP-PCR that makes it unique as compared to other published procedures for amplifying unknown regions is its ability to generate a genomic library. Since in most cases restriction data in the unknown region may not be available, several combinations of restriction digestion has to be performed, one of which is likely to yield the amplification product. Even though the product may not be realized with other combinations, these can be used for the next step of genome walking. In other words, these ligation reaction mixes are the equivalent of genomic libraries, since they can be used repeatedly. For each new step of gene walking, a new primer is designed on the basis of the newly obtained sequence and used as specific primer in SSP-PCR, with an aliquot of all the ligation mixes (except for the one generating the last sequenced fragment). This feature offers a supply of template for continuous genome walking by PCR.

2. Materials

2.1. Isolation of Bacterial Chromosomal DNA

1. LB medium *(11)*. Prepare by dissolving 10 g of bactotryptone, 5 g of yeast extract and 10 g of NaCl into 1 L of double-distilled water. Adjust pH to 7.0 with 1*N* NaOH and autoclave for 20 min on liquid cycle.
2. STE: 150 m*M* NaCl, 100 m*M* Tris-HCl, pH 8.0, 100 m*M* EDTA.

3. SE: 150 mM NaCl, 100 mM EDTA
4. Lysozyme: 20 mg/mL in 0.25M Tris-HCl, pH 8.0.
5. 10% SDS (w/v) in water.
6. Pronase: 20 mg/mL in 50 mM Tris-HCl, pH 7.4, self-activated at 37°C for 2 h.
7. TE buffer: 10 mM Tris-HCl, pH 7.5, 1 mM EDTA.
8. RNase solution: Prepare from RNase A (10 mg/mL) in 100 mM sodium acetate, pH 5.5; and RNase T1 (10,000 U/mL). Heat-treat both in boiling water for 10 min.

2.2. Restriction and Ligation of DNA

1. A range of restriction enzymes and suitable buffers.
2. TE buffer (*see* Section 2.1., step 7).
3. Commercial cloning vector or suitable oligomer (*see* Note 1).
4. T4 DNA ligase and 10X ligase buffer as supplied by manufacturer.

2.3. Polymerase Chain Reaction

1. 10X PCR buffer: 670 µM Tris-HCl, pH 8.8 at 25°C, 67 mM MgCl$_2$, 160 mM ammonium sulfate.
2. dNTPs: 25 mM of each (by mixing 100 mM dNTPs; Pharmacia).
3. Primers: 30 µM.
4. *Taq* polymerase.
5. Mineral oil.

2.4. Analysis of the PCR Product and Testing for Specificity

1. Agarose.
2. TBE: 100 mM Tris-HCl, 100 mM borate, 1mM EDTA.
3. Ethidium bromide: 10 mg/mL in water.
4. Denaturation solution: 400 mM NaOH, 800 mM NaCl.
5. 5X SSPE: 50 mM sodium phosphate, pH 7.0, 900 mM NaCl, 5 mM EDTA.
6. 10% SDS (w/v) in water.
7. Calf thymus DNA (1 mg/mL).
8. Primers: Internal primer (1 µM).
9. Polynucleotide kinase and kinase buffer supplied by manufacturer.
10. [γ-^{32}P]ATP (>5000 Ci/mmol).

2.5. Sequencing of PCR DNA Following Asymmetric Amplification

1. Limiting primer: 0.3–0.6 µM.
2. Excess primer: 30 µM.
3. Sephacryl S-300 (Pharmacia).

4. Glass wool (Corning), siliconize by soaking in Sigmacote and air-dry in fume hood.
5. Sequencing paraphernalia.

3.1. Isolation of Chromosomal DNA (5)

1. Grow a 40-mL culture of *S. typhimurium* in LB overnight.
2. Harvest cells by centrifugation and wash in 10 mL of STE.
3. Centrifuge and resuspend in 0.5 mL of SE, and add 50 µL of freshly prepared (20 mg/ml) lysozyme.
4. Incubate at 37°C for 20 min. Add 200 µL of 10% SDS, incubate at 60°C for 10 min.
5. Treat cell lysate with 0.5 mL of self-activated pronase by shaking at 37°C for 2 h.
6. Phenol/chloroform extract and ethanol precipitate the DNA.
7. Centrifuge and solubilize the pellet in 5.0 mL of TE.
8. Digest with 100 µL of RNase solution for 1 h at 37°C.
9. Ethanol precipitate the DNA and dissolve in 0.1 mL of TE. Generally the yield is in the range of 0.5 mg.

3.2. Restriction and Ligation (5,8)

The following protocol is for DNA for which no restriction data are available. If restriction information is available for the unknown region, the DNA should be digested only with the necessary enzyme(s). *See* Note 2.

1. Digest in a 50-µL final vol 20 µg of test-chromosomal DNA (for a final total of 10 different enzyme digests) with a restriction enzyme that cuts in the known region. Phenol/chloroform extract and ethanol precipitate according to standard procedures.
2. Solubilize the precipitated test-chromosomal DNA in 80 µL. Use 8 µL (2 µg total) per digest for the second restriction enzyme (10 µL final volume including 1 µL of enzyme and 1 µL of 10X buffer).
3. Digest 50 µg of vector DNA (mp18/mp19) with the first restriction enzyme (enough for 10 ligations). Phenol/CHCl₃ extract and ethanol precipitate according to standard procedures.
4. Solubilize vector DNA precipitate in 80 µL. Depending on the requirement, use 8 µL or its multiples for the second restriction enzyme digestion. For example, *Sal* I is compatible with *Sal* I, *Xho* I, *Ava* I, and so forth. So digest 24 µL (3 × 8 µL) with *Sal* I. Following digestion, carry out phenol/CHCl₃ extraction and ethanol precipitation. Solubilize the DNA in 6 µL of TE buffer per reaction or its multiples if intended for more reactions.

5. Dilute 1 µL (0.2 µg) of digested test chromosomal DNA to 30 µl with TE to obtain a concentration of 6.6 ng/µL. Perform ligation reaction in a total volume of 10 µL. Each ligation contains 2 µL (13.2 ng) of chromosomal DNA, 6 µL (5 µg) of vector DNA, 1 µL of 10X ligation buffer and 1 U of ligase (*see* Note 3). Incubate at 15°C overnight (*see* Note 4).

3.3. Polymerase Chain Reaction (5)

1. For a final volume of 25 µL/PCR, prepare a reaction mix for 12 samples containing 30 µL of 10X PCR buffer, 8 µL of 30 µ*M* specific primer, 8 µL of 30 µ*M* vector specific primer, 8 µL of 25 m*M* dNTPs and 3.0 U of *Taq* polymerase.
2. Aliquot 24 µL of reaction mix per tube. Add 1 µL of ligation mix and a drop of mineral oil.
3. Amplify by PCR using the following cycle profile (*see* Note 5):
   40 main cycles   94°C, 1 min (denaturation)
                             55°C, 1 min (annealing)
                             72°C, 2 min (extension)

3.4. Testing for the Product and Specificity (12)

1. Electrophorese 2 µL of the PCR reaction mix in 1% agarose in TBE buffer.
2. Visualize the DNA with ethidium bromide staining (*see* Chapter 1).
3. Perform Southern analysis for specificity on the agarose gel. Denature the DNA by treating the gel with 400 m*M* NaOH, 800 m*M* NaCl for 30 min. Dry the gel on to Whatman 3MM paper.
4. Remove the dried gel from the paper by wetting the paper in water.
5. Soak the released gel in 5X SSPE, 0.3% SDS containing 10 µg/mL sonicated calf thymus DNA for 30 min at 50°C.
6. Kinase an appropriate internal primer in a total volume of 10 µL including 5 µL of 1 µ*M* primer, 3 µL of [γ-^{32}P]ATP, 1 µL of polynucleotide kinase (10 U) in 1X polynucleotide kinase buffer for 1 h at 37°C. Stop the reaction with 1 µL of 10% SDS and increase the volume to 50 µL with water.
7. Prepare a spin column by layering siliconized glass wool in a 1.0-mL syringe barrel and packing it with Sephadex G-50 by centrifugation at 1600*g* for 4 min. Wash the packed matrix once with 50 µL of TE buffer by centrifuging at 1600*g* for 4 min. Discard flow-through. Layer the 50-µL solution of kinased oligo on top of the matrix and centrifuge as in the washing step. Collect the primer as flow-through and use a 20-µL aliquot for probing. Incubate the gel with probe for 4 h at 55°C.
8. Wash the gel twice with 2X SSPE, 0.1% SDS at RT for 15 min and then twice with 5X SSPE, 0.1% SDS at 52°C for 15 min each and expose to film.

3.5. Single Primer Method for Testing Specificity (5,8)

1. Electrophorese 2 µL of the PCR on 1% agarose gel. After ethidium bromide staining, cut out the DNA band. Depending on the concentration of DNA (5–40 ng), 100–500 µL of TE is added to the agarose piece in an Eppendorf tube.
2. Incubate the Eppendorf tube in a boiling water bath to melt the agarose. If the agarose solidifies owing to too small a volume of TE, reheat the tube prior to use.
3. Use a 5-µL aliquot of this for amplification in a volume of 100 µL.
4. Set up three reactions, one with generic primer alone, second with the gene-specific primer alone, and a third one with both primers.
5. PCR is performed as described in Section 3.3. with one modification of going through 20 cycles in the place of 40 cycles.
6. If the product is genuine, only the third tube should give the original size product. The first and second tubes might give some nonspecific products, but they would be of different size.

3.6. Asymmetric PCR for Direct Sequencing of Products <500 bp (3,5)

See Note 6.

1. Use a 5-µL aliquot of the agarose solubilized DNA described in Section 3.5. as the starting material.
2. The PCR is carried out in a total volume of 100 µL with limiting primer at 0.006–0.012 µ*M* (2 µL of 0.3 µ*M* stock); and the excess primer at 0.6 µ*M* (2 µL of 30 µ*M* stock). Generally the reaction is set up in two tubes with the primer concentrations reversed to permit sequencing of both strands.
3. Perform PCR as described in Section 3.3., with the modification that the extension time at 72°C is reduced to 1 min, since the product is <500 bp.
4. Following amplification, extract the PCR reaction mix with 300 µL of chloroform to solubilize the mineral oil.
5. Prepare a spin column by layering siliconized glass wool in a 1.0-mL syringe barrel and packing it with Sephacryl S-300 by centrifugation at 1600g for 4 min. The packed column is washed once with 100 µL of TE buffer by centrifuging at the same speed.
6. Layer the chloroform-extracted aqueous PCR product on top of the Sephacryl S-300 matrix and centrifuge at the same speed as in the washing step.
7. Collect the flow-through and ethanol-precipitate the DNA.
8. Solubilize the DNA pellet in 15 µL of TE. Use a 7-µL aliquot for sequencing with 1 µL of the limiting primer.

4. Notes

1. The main emphasis in SSP-PCR is to introduce a known sequence at the end of the unknown sequence. At the simplest level this could be an oligomer of known sequence and sufficient length (20 bp), with complementarity to the restricted unknown end. However, we have preferred to use M13 or similar vectors (SK, pGem, and so forth) for the following reasons (1) commercial availability of the vectors and the generic primers, (2) multiple cloning sites for ligating a variety of ends, (3) use of a single generic primer for all amplifications.

2. In contrast to IPCR *(14)*, SSP-PCR does not require circularization of the DNA template and therefore does not require ligation of both ends of the chromosomal DNA fragment. Only the unknown end of the restricted chromosomal DNA has to be ligated to the generic oligomer (vector) to make it contiguous with the generic primer-binding site. Although ligation following a single restriction enzyme digestion of both the test DNA and the vector is sufficient to generate an amplification product, digestion of the vector with a second enzyme is desirable to decrease vector-vector ligation and to minimize the presence of generic-primer binding sites at the two ends of the vector dimers.

3. One of the critical parameters in SSP-PCR is the ratio of generic oligomer (vector) to the test DNA during ligation. Ideally, the ligation reaction should result in the joining of the monomeric form of the unknown end of the test DNA fragment to the vector. For this purpose, restricted chromosomal DNA is used at very low concentration, as suggested by Collins and Weissman *(13)*. This should minimize the ligation of restricted DNA fragments to each other, thus decreasing the number of fragments appropriately ligated to the vector. The vector DNA is used at a high concentration to saturate the system and seek out individual chromosomal DNA fragments for ligation. The ratio of vector DNA to the test DNA fragment for ligation could vary for each situation.

4. It is useful to have a positive control for restriction-ligation and PCR. Any DNA fragment of known sequence with a suitable restriction site can be carried through the whole procedure and will need only one additional specific primer for PCR.

5. Although PCR supports the amplification of products up to several kilobases, the largest fragment obtained by SSP-PCR reported to date is 1.8–2.0 kb *(9)*. The absence of a product with most combinations of the restriction-ligation mix suggests that either the restriction site was farther away or that the extension time was insufficient to generate larger products. However, this is not a serious limitation, since, it is necessary in any case to synthesize multiple sequencing primers to sequence a

product of 2.0 kb. The same primer(s) can be used to perform the next step of genome walking.

6. For products >500 bp, sequencing is preferably performed after cloning into any of the commercially available vectors. If SSP-PCR is performed by ligating the unknown end to the multiple cloning site in a vector like M13, SK, or pGEM, it offers a choice of amplified restriction sites for cloning of the SSP-PCR product *(5)*.

Acknowledgment

We wish to thank Anil Joshi for many helpful discussions. This work was supported by: National Institutes of Health Grant GM39415 to G.F.-L.A.

References

1. Shyamala, V. and Ames, G. F.-L. (1992) Genome walking by single specific primer-polymerase chain reaction (SSP-PCR). *Methods Enzymol.* (in press).
2. White, T. J., Arnheim, N., and Erlich, H. A. (1989) The polymerase chain reaction. *Trends Genet.* **5,** 185–189.
3. Shyamala, V. and Ames, G. F.-L. (1989) Amplification of bacterial genomic DNA by the polymerase chain reaction and direct sequencing after asymmetric amplification: application to the study of periplasmic permeases. *J. Bacteriol.* **171,** 1602–1608.
4. Mullis, K. B. and Faloona, F. A. (1987) Specific synthesis of DNA *in vitro* via a polymerase catalysed chain reaction. *Methods Enzymol.* **155,** 335–350.
5. Shyamala, V. and Ames, G. F.-L. (1990) Genome walking by single-specific-primer polymerase chain reaction; SSP-PCR. *Gene* **84,** 1–8.
6. Kohara, Y., Akiyama, K., and Isono, K. (1987) The physical map of the whole *E. coli* chromosome; application of a new strategy for rapid analysis and sorting of a large genomic library. *Cell* **50,** 495–508.
7. Brown, K., Finch, P. W., Hickison, I. D., and Emmerson, P. T. (1987) Complete nucleotide sequence of the *Escherichia coli argA* gene. *Nucleic Acids Res.* **15,** 10586.
7a. Shyamala, V. and Ames, G. F.-L. (1992) Rapid genome walking by single specific primer-polymerase chain reaction (SSP-PCR): Analysis of deletion mutants with unkown ends, in *PCR Protocols*. Lavoisier, France.
8. Shyamala, V., Schneider, E., and Ames, G. F.-L. (1990) Tandem chromosomal duplications; role of REP sequences in the recombination event at the join-point. *EMBO J.* **9,** 939–946.
9. Bodrug, S. E., Holden, J. J. A., Ray, P. N., and Worton, R. G. (1991) Molecular analysis of X-autosome translocations in females with Duchenne muscular dystrophy. *EMBO J.* **10,** 3931–3939.
10. MacGregor, G. R. and Overbeek, P. A. (1992) Use of a simplified single-site PCR to facilitate cloniong of genomic DNA sequences flanking a transgene integration site. *PCR Meth. Appl.* **1,** 129–135.

11. Miller, J. H. (1972) *Experiments in Molecular Genetics.* Cold Spring Harbor Laboratory, Cold Spring Harbor, NY.
12. Bos, J. L., Vries, M. V., Jansen, A. M., Veeneman, G. H., Van Boon, J. H., and Van der Eb, A. J. (1984) Three different mutations in codon 61 of the human N-ras gene detected by synthetic oligonucleotide hybridization. *Nucleic Acids Res.* **12,** 9155–9163.
13. Collins, F. S. D. and Weissman, S. A. (1984) Directional cloning of DNA fragments at a large distance from an initial probe: a circularization method. *Proc. Natl. Acad. Sci. USA* **81,** 6812–6816.
14. Triglia, T., Peterson, M. G., and Kemp, D. J. (1988) A procedure for in vitro amplification of DNA segments that lie outside the boundaries of known sequences. *Nucleic Acids Res.* **16,** 8186.

CHAPTER 33

cDNA Cloning by Inverse
Polymerase Chain Reaction

Sheng-He Huang, Chun-Hua Wu, Bing Cai,
and John Holcenberg

1. Introduction

Since the first report on cDNA cloning in 1972 *(1)*, this technology has been developed into a powerful and universal tool in the isolation, characterization, and analysis of both eukaryotic and prokaryotic genes. But the conventional methods of cDNA cloning require much effort to generate a library that is packaged in phage or plasmid and then to survey a large number of recombinant phages or plasmids. There are three major limitations of those methods. First, a substantial amount (at least 1 µg) of purified mRNA is needed as starting material to generate libraries of sufficient diversity *(2)*. Second, the intrinsic difficulty of multiple sequential enzymatic reactions required for cDNA cloning often leads to low yields and truncated clones *(3)*. Finally, screening of a library with hybridization technique is time-consuming.

Polymerase chain reaction (PCR) technology can simplify and improve cDNA cloning. Using PCR with two gene-specific primers, a piece of known sequence cDNA can be specifically and efficiently amplified and isolated from very small numbers ($<10^4$) of cells *(4)*. However, it is often difficult to isolate full-length cDNA copies of mRNA on the basis of very limited sequence information. The unknown sequence flanking a small stretch of the known sequence of DNA cannot be amplified by the conventional PCR. Recently, anchored-

From: *Methods in Molecular Biology, Vol. 15: PCR Protocols: Current Methods and Applications*
Edited by: B. A. White Copyright © 1993 Humana Press Inc., Totowa, NJ

PCR *(5,6* and Chapter 31) and inverse PCR *(7–9)* have been developed to resolve this problem. Anchored PCR techniques have the common point: DNA cloning goes from a small stretch of known DNA sequence to the flanking unknown sequence region with the aid of a gene-specific primer at one end and a universal primer at the other end. Because of only one gene-specific primer in the anchored PCR, it is easier to get a high level of nonspecific amplification than by PCR with two gene-specific primers *(9,10)*. The major advantage of inverse PCR (IPCR) is to amplify the flanking unknown sequence by using two gene-specific primers.

At first IPCR was successfully used in the amplification of genomic DNA segments that lie outside the boundaries of known sequence *(7,8)*. We have a new procedure that extends this technique to the cloning of unknown cDNA sequence from total RNA *(9)*. Double-stranded cDNA is synthesized from RNA and ligated end to end (Fig. 1). Circularized cDNA is nicked by heating in boiling water or selected restriction enzyme. The reopened circular cDNA is then amplified by two gene-specific primers. The following protocol was used to amplify cDNA ends for the human stress-related protein ERp72 *(9)* (Fig. 2).

2. Materials
2.1. First-Strand cDNA Synthesis

1. Total RNA prepared from human CCRF/CEM T-cell lymphoblastoid cells *(11). See* Chapters 1, 17, 18, and 19 for procedures on RNA isolation.
2. dNTP mix (10 mM of each dNTP).
3. Random primers (Boehringer-Mannheim, Indianapolis, IN). Prepare in sterile water at 1 µg/µL. Store at –20°C.
4. RNasin (Promega, Madison, WI).
5. Actinomycin D (1 mg/mL). Actinomycin D is light sensitive and toxic. It should be stored in a foil-wrapped tube at –20°C.
6. MMLV reverse transcriptase.
7. 5X first-strand buffer: 250 mM Tris-HCl, pH 8.3, 375 mM KCl, 50 mM MgCl$_2$, 50 mM DTT, and 2.5 mM spermidine. The solution is stable at –20°C for more than 6 mo.

2.2. Second-Strand Synthesis

1. 10X second-strand buffer: 400 mM Tris-HCl, pH 7.6, 750 mM KCl, 30 mM MgCl$_2$, 100 mM (NH$_4$)$_2$SO$_4$, 30 mM DTT, and 0.5 mg/mL of BSA. The solution is stable at –20°C for at least 6 mo.
2. NAD (1 mM).

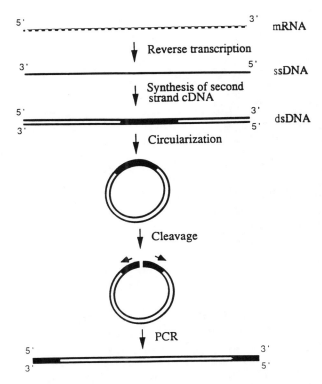

Fig. 1. Diagram of IPCR for cDNA Cloning. The procedure consists of five steps: reverse transcription, synthesis of second-strand cDNA, circularization of double-strand cDNA, reopen the circular DNA, and amplification of reverse DNA fragment. The black and open bars represent the known and unknown sequence regions of double-stranded cDNA, respectively.

3. RNase H (2 U/µL).
4. *E. coli* DNA polymerase I (5 U/µL).
5. *E. coli* DNA ligase (1 U/µL).
6. Nuclease-free H_2O.
7. T4 DNA polymerase.
8. EDTA (200 mM), pH 8.0.
9. GeneClean (Bio 101 Inc., La Jolla, CA).
10. TE buffer: 10 mM Tris-HCl, pH 7.6, 1 mM EDTA. Sterile filter.
11. DNA standards. Prepare 1-mL aliquots of a purified DNA sample at 1, 2.5, 5, 10, and 20 µg/mL in TE buffer. Store at –20°C for up to 6 mo.
12. TE/ethidium bromide: 2 µg/mL of ethidium bromide in TE buffer. Store at 4°C for up to 6 mo in a dark container.

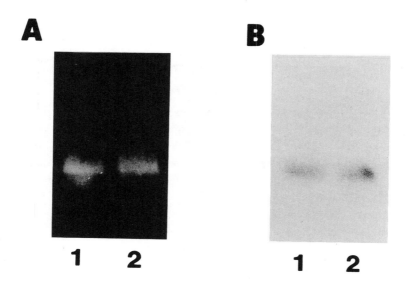

Fig. 2. Amplification of cDNA Ends of Human ERp72. Application of IPCR to amplifying the joining region (280 bp) from 5' (160 bp) and 3' (120 bp) sequences of human ERp72 cDNA. Amplified cDNAs from CCRF/CEM cells sensitive (lane 1) and resistant (lane 2) to cytosine arabinoside stained by ethidium bromide (**A**) or hybridized with ^{32}P-labeled ERp72 cDNA (**B**). *See* text for the sequences of the primers and the parameters of IPCR.

2.3. Circularization and Cleavage

1. 5X ligation buffer (supplied with T4 DNA ligase).
2. T4 DNA ligase (1 U/μL).
3. T4 RNA ligase (4 μg/μL).
4. Hexaminecobalt chloride (15 μ*M*).
5. GeneClean.

2.4. Inverse PCR

1. PCR reagents (*see* Chapter 1).
2. Deoxyoligonucleotides were synthesized on an Applied Biosystems (Foster City, CA) 380B DNA synthesizer and purified by OPEC column from the same company. Primer pairs were selected from the 5' and 3' sequence of the cDNA coding for human ERp72 stress-related protein (5'-primer: 5'-TTCCTCCTCCTCCTCCTCTT-3'; 3'-primer: 5'-ATCTAAATGTCTAGT-3')*(9)*.

3. Methods

3.1. First-Strand cDNA Synthesis (12)

Perform reverse transcription in a 25-μL reaction mixture, adding the following components:

5X first-strand buffer	5.0 μL
dNTP mix	2.5 μL
random primers	2.5 μL
RNasin	1.0 U
actinomycin D	1.25 μL
MMLV reverse transcriptase	250 U
RNA	15–25 μg of total RNA

(Heat denature RNA at 65°C for 3 min prior to adding to reaction)

Nuclease-free H$_2$O — to 25 μL final vol

3.2. Second-Strand Synthesis (13)

1. Add components to the first-strand tube on ice in the following order:

10X second-strand buffer	12.5 μL
NAD	12.5 μL
RNase H	0.5 μL
E. coli DNA polymerase I	5.75 μL
E. coli ligase	1.25 μL
Nuclease-free water	67.5 μL

2. Incubation at 14°C for 2 h.
3. Heat the reaction mix to 70°C for 10 min, spin in a microfuge for a few seconds, and then put in ice.
4. Add 4 U of T4 DNA polymerase and incubate at 37°C for 10 min to blunt the ends of double-stranded cDNA.
5. Stop the reaction with 12.5 μL of EDTA and 200 μL of sterile H$_2$O.
6. Concentrate and purify the sample with GeneClean. Resuspend the DNA in 100–200 μL of TE buffer.
7. Estimate the DNA concentration by comparing the ethidium bromide fluorescent intensity of the sample with that of a series of DNA standards on a sheet of plastic wrap (14). Dot 1–5 μL of sample onto plastic wrap on a UV transilluminator. Also dot with 5 μL of DNA standards. Add an equal volume of TE buffer containing 2 μg/mL of ethidium bromide, mix by repipetting up and down. Use proper UV shielding for exposed skin and eyes.

3.3. Circularization and Cleavage (see Notes 1–3)

1. Set up the circularization reaction mix containing the following components: 100 μL (100 ng DNA) of purified sample, 25 μL of 5X liga-

tion buffer, and 6 µL of T4 DNA ligase. Finally, add 2 µL of T4 RNA ligase or 15 µL of 15 µ*M* hexaminecobalt chloride (*see* Note 4).
2. Incubate at 18°C for 16 h.
3. Boil the ligated circular DNA for 2–3 min in distilled water or digest with an appropriate restriction enzyme to reopen circularized DNA.
4. Purify the DNA sample with GeneClean as described in step 6 in Section 3.2.

3.4. Inverse PCR (see Note 5)

1. Add 1/10 of the purified cDNA to a 100 µl amplification mix *(15)* containing 100 pmol of each gene specific primer and 1 U of *Taq* DNA polymerase.
2. Amplify by PCR using the following cycle profile:

    25 main cycles   94°C, 1 min (denaturation)

    65°C, 2 min (annealing)

    72°C, 4 min (elongation)

4. Notes

1. For maximum efficiency of intramolecular ligation, low concentration of cDNA should be used in the ligation mix. High density of cDNA may enhance the level of heterogeneous ligation, which creates nonspecific amplification.
2. Cleavage of circularized cDNA at a selected known sequence region is important since circular DNA tends to form supercoil and is a poor template for PCR *(16)*. Circularized DNA is only good for amplification of a short DNA fragment (<300 bp).
3. The following three ways can be considered to introduce nicks in circularized DNA. Boiling is a simple and common way. But sometimes this method is not sufficient to make nicks in circular DNA. A second method is selected restriction enzyme digestion. The ideal restriction site is located in the known sequence region of cDNA. In most cases, it is difficult to make the right choice of a restriction enzyme because the restriction pattern in the unidentified region of cDNA is unknown. If an appropriate enzyme is not available, EDTA-oligonucleotide-directed specific cleavage may be tried *(17,18)*. Oligonucleotides linked to EDTA-Fe(II) at T can bind specifically to double-stranded DNA by triple-helix formation and produce double-stranded cleavage at the binding site.
4. Inclusion of T4 RNA ligase or hexaminecobalt chloride can enhance the efficiency of blunt-end ligation of double-stranded DNA catalyzed by T4 DNA ligase *(14,19)*.
5. IPCR can be used to efficiently and rapidly amplify regions of unknown sequence flanking any identified segment of cDNA or genomic DNA. This technique does not need construction and screening of DNA libraries to

obtain additional unidentified DNA sequence information. Some recombinant phage or plasmid may be unstable in bacteria and amplified libraries tend to lose them *(16)*. IPCR eliminates this problem.

References

1. Verma, I. M., Temple, G. F., Fan, H., and Baltimore, D. (1972) In vitro synthesis of double-stranded DNA complimentary to rabbit reticulocyte 10S RNA. *Nature* **235,** 163–169.
2. Akowitz, A. and Mamuelidis, L. (1989) A novel cDNA/PCR strategy for efficient cloning of small amounts of undefined RNA. *Gene* **81,** 295–306.
3. Okayama, H., Kawaichi, M., Brownstein, M., Lee, F., Yokota, T., and Arai, K. (1987) High-efficiency cloning of full-length cDNA; Construction and screening of cDNA expression libraries for mammalian cells. *Methods Enzymol.* **154,** 3–28.
4. Brenner, C. A., Tam, A. W., Nelson, P. A., Engleman, E. G., Suzuki, N., Fry, K. E., and Larrick, J. W. (1989) Message amplification phenotyping (MAPPing): a technique to simultaneously measure multiple mRNAs from small numbers of cells. *BioTechniques* **7,** 1096–1103.
5. Frohman, M. A. (1990) RACE: Rapid amplification of cDNA ends, in *PCR Protocols: A Guide to Methods and Applications* (Innis, M. A., Gelfand, D. H., Sninsky, J. J., and White, T. J., eds.), Academic, San Diego,CA, pp. 28–38.
6. Shyamala, V. and Ames, G. F.-L. (1989) Genome walking by single-specific-primer polymerase chain reaction: SSP-PCR. *Gene* **84,** 1–8.
7. Ochman, H., Gerber, A. S., and Hartl, D. L. (1988) Genetic applications of an inverse polymerase chain reaction. *Genetics* **120,** 621–625.
8. Triglia, T., Peterson, M. G., and Kemp, D. J. (1988) A procedure for in vitro amplification of DNA segments that lie outside the boundaries of known sequences. *Nucleic Acids Res.* **16,** 8186.
9. Huang, S.-H., Hu, Y. Y., Wu, C.-H., and Holcenberg, J. (1990) A simple method for direct cloning cDNA sequence that flanks a region of known sequence from total RNA by applying the inverse polymerase chain reaction. *Nucleic Acids Res.* **18,** 1922.
10. Delort, J., Dumas, J. B., Darmon, M. C., and Mallet, J. (1989) An efficient strategy for cloning 5' extremities of rare transcrips permits isolation of multiple 5'-untranslated regions of rat tryptophan hydroxylase mRNA. *Nucleic Acids Res.* **17,** 6439–6448.
11. Davis, L. G., Dibner, M. D., and Battey, J. F. (1986) *Basic Methods in Molecular Biology*, Elsevier, New York.
12. Krug, M. S. and Berger, S. L. (1987) First strand cDNA synthesis primed by oligo(dT). *Methods Enzymol.* **152,** 316–325.
13. Promega (1989) Protocols and Applications, pp. 167–180.
14. Sambrook, J., Fritsch, E. F., and Maniatis, T. (1989) *Molecular Cloning* (2nd ed.), Cold Spring Harbor Laboratory, Cold Spring Harbor, NY.
15. Saiki, R. K., Gelfand, D. H., Stoffel, S., Scharf, S. J., Higuchi, R., Horn, G. T., Mullis, K. B., and Erlich, H. A. (1988) Primer-directed enzymatic amplification of DNA with a thermostable DNA polymerase. *Science* **239,** 487–491.

16. Moon, I. S. and Krause, M. O. (1991) Common RNA polymerase I,II, and III upstream elements in mouse 7SK gene locus revealed by the inverse polymerase chain reaction. *DNA Cell Biol.* **10,** 23–32.
17. Strobel, S. A. and Dervan, P. B. (1990) Site-specific cleavage of a yeast chromosome by oligonucleotide-directed triple-helix formation. *Science* **249,** 73–75.
18. Dreyer, G. B. and Dervan, P. B. (1985) Sequence-specific cleavage of single-stranded DNA: Oligodeoxynucleotide-EDTA.Fe(II). *Proc. Natl. Acad. Sci. USA* **82,** 968–972.
19. Sugino, A., Goodman, H. M., Heynecker, H. L., Shine, J., Boyer, H. W., and Cozzarelli, N. R. (1977) Interaction of bacteriophage T4 RNA and DNA ligases in joining of duplex DNA at base-paired ends. *J. Biol. Chem.* **252,** 3987.

CHAPTER 34

Amplification of Gene Ends from Gene Libraries by Polymerase Chain Reaction with Single-Sided Specificity

Sheng-He Huang, Ambrose Y. Jong, Wu Yang, and John Holcenberg

1. Introduction

Isolation of a full-length gene on the basis of a limited sequence information is often troublesome and challenging. Tremendous effort is needed to isolate a specific gene by screening cDNA or genomic libraries by oligonucleotide or nucleic acid probes. In those methods, basically nucleic acid probes are used in a screening process to check whether or not a plaque or a colony contains the sequence of interest. There have been attempts to isolate specific DNA fragments using immobilized DNA, in which particular DNA fragments were enriched by hybrid selection and then the concentrated library was screened by a specific DNA probe *(1,2)*. Recently, polymerase chain reaction (PCR) has been applied to the cloning of genes. Friedmann et al. *(3)* first used PCR to screen λgt11 library with two gene-specific primers. This protocol can be effectively used to isolate a particular DNA fragment between two specific primers or to generate nucleic acid probe from cDNA libraries. The unknown sequences flanking the fragment between the two specific primers cannot be amplified by this method.

Anchored PCR or single-specific-primer PCR *(4)* and inverse PCR *(5)* have been adapted to cloning of full-length cDNAs with the knowledge of a small stretch of sequence within the gene. Both methods start from mRNA and are good for cDNA cloning when a cDNA library is

From: *Methods in Molecular Biology, Vol. 15: PCR Protocols: Current Methods and Applications*
Edited by: B. A. White Copyright © 1993 Humana Press Inc., Totowa, NJ

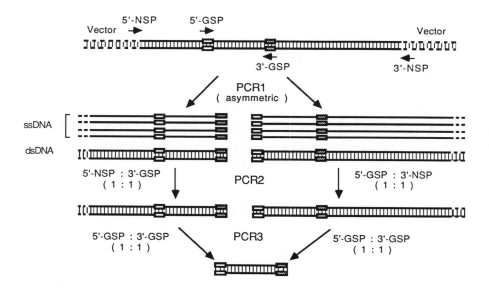

Fig. 1. The scheme for rapid amplification of gene ends (RAGE). 5'-NSP is λgt11 forward primer (5'-GACTCCTGGAGCCCG-3'). 3'-NSP: λgt11 reverse primer (5'-GGTAGCGACCGGCGC-3'). 5'-GSP: 5ASPR with *Eco*R1 restriction site (5'-AGACTGAATTCGGTACCGGCGGTACTATCGCTTCC-3'). 3'-GSP: 3ASPB containing *Bam*H1 site (5'-CTGATGGATCCTGGCAGTGGCTGGACGC-3').

not available. Cloning of full-length cDNA is usually far more diffi-cult than any other recombinant DNA work because the multiple sequen-tial enzymatic reactions often result in low yield and incomplete clones *(6).* Shyamala and Ames *(7)* extended the use of anchored PCR to amplify unknown DNA sequences from genome on the basis of using a short stretch of known sequence for designing a gene-specific primer (*see* Chapter 32). We developed a much simpler method to isolate full-length cDNA and flanking genomic DNA from gene libraries by anchored PCR. We used the yeast gene coding for asparaginase II *(8)* as a model to verify this method. Recently, we amplified and isolated 5'-cDNA fragment of 5-hydroxytryptamine 2 (5-HT2) receptor from λ SWAJ-2 mouse brain cDNA library by this technique *(9).* The principle of this tech-nique is schematically depicted in Fig. 1. Briefly, two ends of a gene were amplified by two nonspecific primers (NSP) complementary to the vector sequences flanking the polylinker region and two gene-specific primers (GSP) complementary to 5' and 3' parts of the known gene

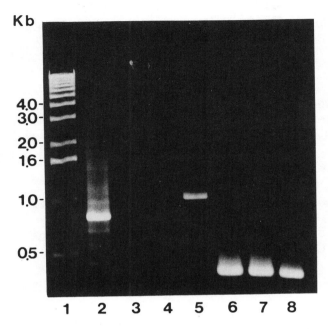

Fig. 2. Gel pattern of PCR products. Fragments amplified from the purified 5' (lane 6) and 3' (lane 7) gene ends and λgt11 yeast genomic library (lane 2, 3, 4, 5, and 8) were resolved in 1.5% agarose gel and stained with ethidium bromide. Lane 1 was 1-kb ladder DNA marker (BRL). There were five pairs of primers used in PCR: (1) 5'-NSP and 3'-GSP (lane 2); (2) 3'-NSP and 3'-GSP (lane 3); (3) 5'-NSP and 5'-GSP (lane 4); (4) 3'-NSP and 5'-GSP (lane 5) and (5) 5'-GSP and 3'-GSP (lane 6, 7, and 8).

sequence. The 5'-cDNA region of 5-HT2 receptor was successfully amplified from λ SWAJ-2 mouse brain cDNA library with a GSP 5'-TTCTGCCTGAGACTAAAAAGGGTTAAGCCCTTATGATGGCA-3' and an NSP (*Xba* I-T15 adaptor primer: 5'-GTCGACTCTTAGAT-3') *(9)*. Two ends of the yeast asparaginase II gene *(8)* were amplified from λgt11 yeast genomic library by this method. The size of the full-length gene is 1.7 kb and there is 353-bp overlapping sequence between the two GSPs (5ASPR and 3ASPB). The 3' gene end from 5ASPR to the extreme 3' end of the gene and the 5' gene end from 3ASPB to the very beginning of 5' part of the gene are 1.15 and 0.9 kb, respectively. Figure 2 shows that there was only one orientation for the asparaginase II gene clone and the expected sizes of the two amplified gene ends were obtained. The same size fragment was amplified from the gene library and the two purified gene ends by using the two GSPs as primers. The gel-purified PCR DNA fragments were sequenced and the sequence data were consistent with the literature *(8)*.

2. Materials

1. PCR reagents (*see* Chapter 1).
2. 10X *Taq* polymerase buffer supplied by New England Biolabs (Beverly, MA).
3. Yeast genomic library, lambda SWAJ-2 mouse brain cDNA library and 5' and 3' λgt11 primers were obtained from ClonTech (Palo Alto, CA) (*see* Note 1).
4. GeneClean (Bio 101 Inc., La Jolla, CA).
5. Sequenase II (United States Biochemical Corporation, Cleveland, OH).
6. Dimethylsulfoxide (DMSO).
7. Deoxyoligonucleotides were synthesized on an Applied Biosystems (Foster City, CA) 380B DNA synthesizer and purified by an OPEC column from the same company.

3. Methods

3.1. PCR 1 (see Notes 2 and 3)

1. In order to enhance the specificity of amplification, carry out asymmetric PCR in a 50-µL reaction containing a 1-µL aliquot of λgt11 yeast genomic library (about 1×10^7 PFU), 2.5 pmol nonspecific primer (NSP), 50 pmol gene-specific primer (GSP), 5 µL of DMSO, 5 µL of 10X *Taq* DNA polymerase buffer, 1.5 m*M* of each dNTP and 2.5 U of *Taq* DNA polymerase. Before adding the enzyme, heat the PCR cocktail to 94°C for 3 min to disrupt the phage particles.
2. Because there may be two separate orientations for each insert in the gene library, set up four reactions for the amplification of two gene ends in the first round of PCR. Thus, each gene-specific primer is matched with each nonspecific primer (*see* Fig. 2).
3. Amplify by PCR using the following cycle profile:
     35 main cycles   94°C, 1 min (denaturation)
                          48°C, 30 s (annealing)
                          72°C, 8 min (extension)
     The major product, single-stranded DNA, can be used for sequencing and Southern blotting analyses (*see* Chapters 1, 13, and 14).

3.2. PCR 2

1. Dilute each of the first PCR products at 1:10 in H_2O.
2. Amplify 1-µL aliquots as described in PCR 1, except use equal amounts (50 pmol) of the NSP and GSP. The selected dsDNA fragments are

available for cloning, which can be facilitated by incorporating restriction sites in the primers (*see* Chapters 2 and 22, and *see* Note 4).

3.3. PCR 3

PCR 3 is used to test the products of PCR 2 since amplification of the two GSPs should produce the same fragment from a gene library and the two purified gene ends.

1. Using two GSPs (50 pmol each) and 1 µL of 1:10 diluted PCR 2 product or 1-µL aliquot of gene library, amplify by PCR using the following cycle profile:

35 main cycles	94°C, 30 s
	48°C, 30 s
	72°C, 4 min

2. Run DNA fragments from PCR 1, 2, and 3 on an agarose gel (*see* Chapter 1).
3. Excise specific bands and purify by GeneClean.
4. Sequence by the dideoxy chain termination method *(10)* with Sequenase with the aid of NP-40 *(11). See* Chapters 13 and 14.

4. Notes

1. The successful isolation of clones from cDNA or genomic libraries is dependent on the quality of the library. There is a big difference between the primary and the amplified libraries because different recombinant clones may grow at very different rates, resulting in unequal distribution of the recombinants in the amplified library *(12)*. It may be better to use the primary library for PCR amplification. If a short stretch of known DNA sequence (more than 100 bp) is available, it is easy to test the quality of the library by PCR amplification with two GSPs. In case the library is not good, an alternative strategy is to use inverse or anchored PCR to isolate the clone from genomic DNA or self-made cDNA.
2. This method does not appear to be suitable for DNA amplification with degenerate primers based on the highly conserved regions of a protein from other species or limited amino acid sequence data because degeneracy of primers can create more problems with nonspecific amplification. This limit may be reduced by incorporation of deoxyinosine into wobble positions of degenerate oligonucleotides *(13)*.
3. The most obvious and common problem for PCR with single-sided specificity is nonspecific amplification of DNA fragments without significant homology with the gene of interest *(14,15)*. In this study, we found two things that improve specific amplification. We found that *Taq* DNA polymerase buffer from New England Biolab works well for PCR with single-specific primer. Second, asymmetric PCR with a relatively large

amount of GSP was performed in the first round of PCR in order to enhance the specificity of amplification.

4. The two gene ends with overlapping sequence can be simply linked by two ways. First, ssDNAs of the two gene ends from PCR 1 can be annealed and end-filled by Klenow in presence of random primers. Second, when the sequence information is obtained from the two gene ends, a full-length cDNA or gene can be amplified by two specific primers that represent the sequences at the extreme 3' and 5' ends of the gene or cDNA.

References

1. Schott, H. and Bayer, E. (1979) Template chromatography. *Adv. Chrom.* **17,** 181–229.
2. Tsurui, H., Hara, E., Oda, K. Suyama, A., Nakada, S., and Wada, A. (1990) A rapid and efficient cloning method with a solid-phase DNA probe: application for cloning the 5'-flanking region of the gene encoding human fibronectin. *Gene* **88,** 233–239.
3. Friedmann, K. D., Rosen, N. L., Newman, P. J., and Montgomery, R. R. (1988) Enzymatic amplification of specific cDNA inserts from λgt11 libraries. *Nucleic Acids Res.* **16,** 8718.
4. Frohman, M. A., Dush, M. K., and Martin, G. R. (1988) Rapid production of full-length cDNAs from rare transcripts:amplification using a single gene-specific oligonucleotide primer. *Proc. Natl. Acad. Sci. USA* **85,** 8998–9002.
5. Huang, S. H., Hu, Y. Y., Wu, C. H., and Holcenberg, J. (1990) A simple method for direct cloning cDNA sequence that flanks a region of known sequence from total RNA by applying the inverse polymerase chain reaction. *Nucleic Acids Res.* **18,** 1922.
6. Okayama, H., Kawaichi, M., Brownstein, M., Lee, F., Yokota, T., and Arai, K. (1987) High-efficiency cloning of full-length cDNA; construction and screening of cDNA expression libraries for mammalian cells. *Methods Enzymol.* **154,** 3–28.
7. Shyamala, V. and Ames, G.F.-L. (1989) Genome walking by single-specific-primer polymerase chain reaction: SSP-PCR. *Gene* **84,** 1–8.
8. Kim, K. W., Kamerud, J. Q., Livingston, D. M., and Roon, R. J.(1988) Asparaginase II of saccharomyces cerevisiae: characterization of the ASP3 gene. *J. Biol. Chem.* **263,** 11,948–11,953.
9. Yang, W., Chen, K., Lan, N.C., Huang, S.-H., and Shih, J. C. (1992) Cloning of the gene and cDNA for the mouse 5-HT2 receptor. Submitted.
10. Sanger, F., Nickler, S., and Coulson, A. R. (1977) DNA sequencing with chain-terminating inhibitors. *Proc. Natl. Acad. Sci. USA* **74,** 5463–5467.
11. Bachmann, B., Lucke, W., and Hunsmann, G. (1990) Improvement of PCR amplified DNA sequencing with the aid of detergents. *Nucleic Acids Res.* **18,** 1309.
12. Frischauf, A.-M. (1987) Construction and characterization of a genomic library in λ. *Methods Enzymol.* **152,** 190–199.

13. Patil, R. V. and Dekker, E. E. (1990) PCR amplification of an Escherichia coli gene using mixed primers containing deoxyinosine at ambiguous positions in degenerate amino acid codons. *Nucleic Acids Res.* **18,** 3080.

14. Loh, E. Y., Elliott, J. F., Cwirla, S., Lanier, L. L., and Davis, M. M. (1989) Polymerase chain reaction with single-sided specificity: analysis of T cell receptor ∂ chain. *Science* **243,** 217–220.

15. Frohman, M. A. (1990) RACE: Rapid Amplification of cDNA Ends, in *PCR Protocols: A Guide to Methods and Applications* (Innis, M. A., Gelfand, D. H., Sninsky, J. J., and White, T. J., eds.), Academic, San Diego, CA, pp. 28–38.

CHAPTER 35

Anchoring a Defined Sequence to the 5' Ends of mRNAs

The Bolt to Clone Rare Full Length mRNAs and Generate cDNA Libraries from a Few Cells

Jean Baptiste Dumas Milne Edwards, Jacques Delort, and Jacques Mallet

1. Introduction

1.1. Amplification of Low Amounts of mRNA

Among numerous applications, the polymerase chain reaction (PCR) *(1,2)* provides a convenient means to clone 5' ends of rare mRNAs and to generate cDNA libraries from tissue available in amounts too low to be processed by conventional methods. Basically, the amplification of cDNAs by the PCR requires the availability of the sequences of two stretches of the molecule to be amplified. A sequence can easily be imposed at the 5' end of the first-strand cDNAs (corresponding to the 3' end of the mRNAs) by priming the reverse transcription with a specific primer (for cloning the 5' end of rare messenger) or with an oligonucleotide tailored with a poly (dT) stretch (for cDNA library construction), taking advantage of the poly (A) sequence that is located at the 3' end of mRNAs. Several strategies have been devised to tag the 3' end of the ss-cDNAs (corresponding to the 5' end of the mRNAs). We *(3)* and others have described strategies based on the addition of a homopolymeric dG *(4,5)* or dA *(6,7)* tail using terminal deoxyribonucleotide transferase (TdT) ("anchor-PCR" *[4]*). However, this strategy has important limita-

From: *Methods in Molecular Biology, Vol. 15: PCR Protocols: Current Methods and Applications*
Edited by: B. A. White Copyright © 1993 Humana Press Inc., Totowa, NJ

tions. The TdT reaction is difficult to control and has a low efficiency (unpublished observations). But most importantly, the return primers containing a homopolymeric (dC or dT) tail generate nonspecific amplifications, a phenomenon that prevents the isolation of low abundance mRNA species and/or interferes with the relative abundance of primary clones in the library. To circumvent these drawbacks, we have used two approaches. First, we devised a strategy based on a cRNA enrichment procedure, which has been useful to eliminate nonspecific-PCR products and to allow detection and cloning of cDNAs of low abundance *(3)*. More recently, to avoid the nonspecific amplification resulting from the annealing of the homopolymeric tail oligonucleotide, we have developed a novel anchoring strategy that is based on the ligation of an oligonucleotide to the 3' end of ss-cDNAs. This strategy is referred to as SLIC for *s*ingle-strand *li*gation to ss-*c*DNA *(8)*.

These methods have been applied to the cloning of the 5' ends of the rat tryptophan hydroxylase (TPH). In the rat, the corresponding mRNAs display a diversity both at 3' *(9)* and 5' *(3)* noncoding regions. Two populations of TPH mRNAs with distinct 5' ends, designated TPH-α and TPH-β, have been characterized by S1 mapping experiments and molecular cloning *(3)*. TPH-α, the sequence of which is included in TPH-β (Fig. 1A), is the most abundant species and it accounts for about 0.5% of total mRNAs in the pineal gland. TPH-β is about 100-fold less abundant than TPH-α. These characteristics made this system ideal for comparing the efficiencies of cloning strategies.

Finally, the potential of the SLIC method to generate a cDNA library will also be addressed.

1.2. Single-Strand Ligation Mediated Anchor-PCR: The SLIC Strategy

As the strategy described in this chapter requires many oligonucleotides, location and sequences of all the oligonucleotides used in our studies are given in Fig. 1. The strategy is illustrated in Fig. 2. The ss-cDNA is primed from the poly A tail (oligonucleotide BM 3', Fig. 1E,F) or of a known sequence within the RNA (oligonucleotide PEX, Fig. 1A,B). After RNAs hydrolysis (Sections 2.6. and 3.5.) and removal of the primers (Sections 2.5. and 3.4.), a modified oligonucleotide, BM 5' (Fig. 1C,D), is ligated to the 3' end of the ss-cDNA in the presence of T4 RNA ligase (Sections 2.7. and 3.6.), which has also been shown to ligate short single-stranded DNA fragments *(10,11)*.

Three precautions are needed to target the ligation of the BM 5' oligonucleotide specifically to the 3' end of the ss-cDNA and to avoid self-ligation and/or circularization. First, the oligonucleotide that serves to prime the ss-cDNA synthesis must not contain a phosphate at its 5' end. Second, the 3' end of the BM 5' oligonucleotides must be blocked; this was performed by adding a dideoxyribonucleotide to the 3' end of the primer (*see* Section 2.7.1. and 3.6.1.). Finally, the oligonucleotide used to prime the ss-cDNA synthesis must be removed (Sections 2.5.1. and 3.4.) before carrying out the ligation experiment since it can compete with the ss-cDNA for ligation to BM 5'.

This method has been tested in two experimental situations: (1) to assess the usefulness of the strategy in cloning 5' ends of rare messengers we have recloned the TPH-β 5' end, and (2) a cDNA library has been generated starting from the total RNAs from only 10^4 cells.

1.3. Cloning of TPH-β Species Using SLIC Methodology (8)

Total RNAs were extracted from one pineal gland and primer extension was carried out with the PEX oligonucleotide (Fig. 1A,B) using one fifth (approx 200 ng) of the RNAs. After removal of PEX, the single-stranded cDNA was ligated to the modified oligonucleotide BM 5' with T4 RNA ligase, as described in Section 3. A 45-cycle amplification was then performed with the two oligonucleotides BM 1-5' (Fig. 1C,D) and Ba (Fig. 1A,B).

At this stage, the presence of TPH-β was tested by Southern blotting (Flying Southern, Section 2.9. and 3.9.), using an oligonucleotide, Fs, which is specific to TPH-β. As no signal was obtained, we performed a further set of 30 cycles of amplifications with primer Ca and BM 2-5'. A specific TPH-β signal could then be detected. Blunt-end cloning of these PCR products led to 60% of the recombinants being TPH clones, of which approx 1% were TPH-β clones. The TPH-β clones that were sequenced contained the 5 nucleotide sequence (TGCCC) in addition to the TPH-ß sequence that had been determined after dG tailing-mediated PCR.

The failure to find these 5 bases previously is probably owing to the inherent limitations of tailing-mediated anchor-PCR, although we cannot exclude the fact that it merely reflects a difference in RNA preparations. In the anchored PCR, cDNAs are 3'-tailed with a homopolymeric (dA) or (dG) stretch, the return primers contain a homopolymeric (dT) or (dC) tail, respectively. These homopolymeric sequences,

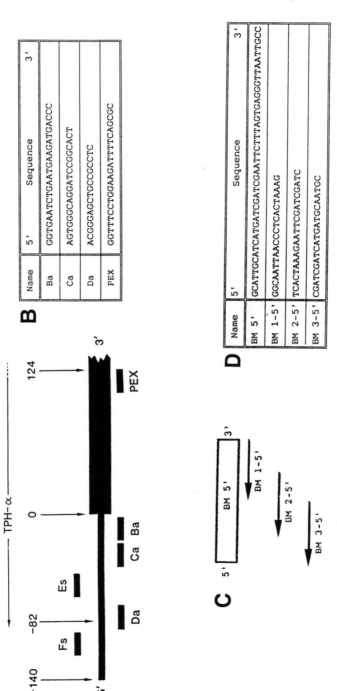

B

Name	5'	Sequence	3'
Ba		GGTGAATCTGAATGAAGATGACCC	
Ca		AGTGGGCAGGATCCGGCACT	
Da		ACGGGAGCTGCCGCCTC	
PEX		GGTTTCCTGGAAGATTTTCAGCGC	

D

Name	5'	Sequence	3'
BM 5'		GCATTGCATCATGATCGATCGAATTCTTTAGTGAGGGTTAATTGCC	
BM 1-5'		GGCAATTAACCCTCACTAAAG	
BM 2-5'		TCACTAAAGAATTCGATCGATC	
BM 3-5'		CGATCGATCATGATGCAATGC	

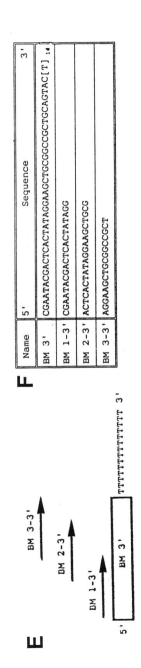

F

Name	5' Sequence 3'
BM 3'	CGAATACGACTCACTATAGGAAGCTGGGCCGCTGCAGTAC$[T]_{14}$
BM 1-3'	CGAATACGACTCACTATAGG
BM 2-3'	ACTCACTATAGGAAGCTGCG
BM 3-3'	AGGAAGCTGCGGGCCGCT

Fig. 1. Designation and sequences of the oligonucleotides used in this study. (**A**) Oligonucleotides relative to the 5' ends of tryptophan hydroxylase mRNAs; thick line: 5' end of coding region; thin line: 5' untranslated region. Sense and antisense oligonucleotides are represented above and below, respectively. (**B**) Sequences of the oligonucleotides described in **A**. (**C**). The modified 46-nucleotide oligonucleotide BM 5' is ligated to the single-stranded cDNA. Following ligation, nested amplifications were carried out with the anticomplementary oligonucleotides BM 1-5', BM 2-5', BM 3-5', which overlap over 9–10 bases. (**D**) Sequences of the oligonucleotides described in C. (**E**) Priming of cDNA synthesis from PC12 cells RNAs was done with BM 3', a 40-base oligonucleotide, which has a 14-base dT tail at its 3' end. The nested amplifications were carried out with three overlapping oligonucleotides that are parts of BM 3': BM 1-3', BM 2-3', BM 3-3'. (**F**) Sequence of the oligonucleotides described in E.

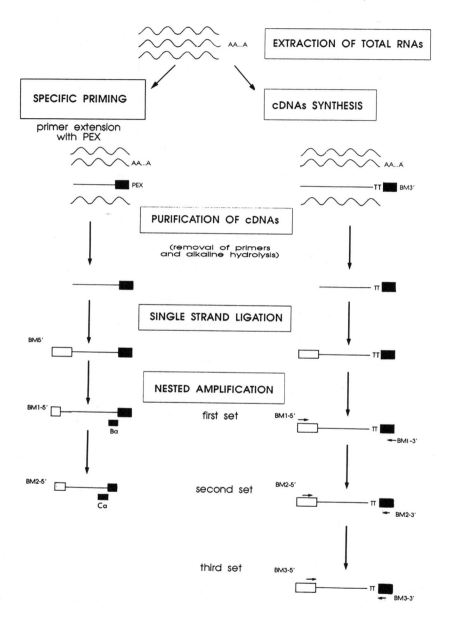

especially dC tails, are likely to hybridize not only to dA or dG tails but also to dA- or dG-rich sequences, respectively. Such a phenomenon may have accounted in our previous cloning experiments *(3)* for the absence of the $^{3'}$ACGGG$^{5'}$ sequence at the 3' end of the fully extended TPH-β cDNAs, although the length of the return primer has been purposely reduced to six residues in order to limit nonspecific hybridization. Therefore, in contrast to the SLIC strategy, the use of dG tailing strategies not only yields high levels of background but may also generate double-stranded cDNAs shorter than their corresponding single-stranded cDNAs.

1.4. Construction of a cDNA Library Based on the SLIC Method (8)

We tested the potential of the SLIC method to generate a cDNA library from a limited number of cells. To allow us to control that the library would be representative, we used a PC12 cell line where it has been shown that the tyrosine hydroxylase (TH) mRNA represents approx 0.05% of the total mRNAs.

Total RNA was prepared from a pellet containing 10^4 cells. Half of the material (about 10 ng) was used to synthesize single-stranded cDNAs primed with BM 3' (Fig. 1E,F). Following removal of the primer, the ss-cDNA was ligated with the modified oligonucleotide BM 5' (Fig. 1C,D). These cDNAs were PCR amplified, using BM 1-5' and BM 1-3' (Fig. 1C–F). S-400 chromatography (Section 3.8.) was used to eliminate short-sized DNAs (mainly primer dimers and BM 5'/BM 3' ligation). Two successive nested PCR were then carried out with the pair of primers, respectively BM 2-5', BM 2-3' and BM 3-5', BM 3-3'.

Fig. 2. *(opposite page)* SLIC strategy. Following the extraction of total RNAs, the synthesis of the single-stranded cDNA is carried out using a specific primer PEX (cloning of 5' ends of messengers) or a primer containing a homopolymeric dT tail (library construction). These primers *do not* contain phosphate at their 5' end. The primers are removed by AcA34 chromatography and the RNAs are hydrolyzed in 0.3M NaOH. Single-stranded ligation is carried out with BM 5'. This oligonucleotide has been modified so as to have no hydroxyl radical at its 3' end and a phosphate group at its 5' end . The ss-cDNAs are suitable for amplification experiments. Oligonucleotide BM 5' has been designed to permit the use of three oligonucleotides (BM 1-5', BM 2-5', BM 3-5') in the further amplification experiments. In conjunction with the use of three specific oligonucleotides it allows nested-PCR experiments, thereby increasing the specificity of the amplification products. BM 3' has also been designed to permit the choice of three oligonucleotides (BM 1-3', BM 2-3', BM 3-3') usable in nested PCR experiments with the 5' oligonucleotides.

After an additional purification by S-400 chromatography, the products generated after the BM 3-3'/BM 3-5' amplification were cloned in λZAP II vector, yielding about 3×10^6 clones. Analysis of 24 clones chosen at random indicated that the average size of the library was about 400 bases. Finally, 30,000 primary clones were screened with a TH cDNA probe, yielding 10 (0.03%) positive clones, shown by partial sequencing to be TH clones, thus confirming the representative nature of the library. Thus, no major distortion had been introduced in the abundance of TH clones through the SLIC strategy.

1.5. Conclusion

Anchoring a defined sequence at the 3' end of the ss-cDNAs is a crucial issue not only in cloning 5' ends of mRNAs but also in generating cDNAs libraries starting from a low number of cells. The nonspecific amplification associated with the use of homopolymeric tails prevents the isolation of low abundant mRNAs and decreases the representativity of cDNA libraries. In contrast, single-strand ligation-mediated anchor-PCR (the SLIC strategy) yields single-stranded molecules with two defined ends containing specific sequences usable as targets for PCR primers. Thus, the problems encountered using the SLIC strategy are only those characteristic of classical PCR amplifications.

2. Materials

2.1. Oligonucleotides

2.1.1. Choice of the Oligonucleotides

All of the oligonucleotides used in this study have been synthesized by GENSET (France). *See* Notes 1 and 2.

1. As regards oligonucleotides BM 3', BM 5' and the related oligonucleotides (BM 1,2 and 3-3' and BM 1,2 and 3-5'), the sequences have been arbitrarily chosen. In an attempt to facilitate the cloning of the amplified cDNAs, we have synthesized oligonucleotides (BM 5' and BM 3') with sequences recognized by restriction enzymes (*Not*I and *Pst*I for BM 5' and β*Sp* H1 and *Sp*H 1 for BM 3'). Unfortunately, cutting of PCR products with these enzymes were highly inefficient and we have found that blunt-ended cloning of the cDNAs originating from the 5' ends of messengers and addition of linkers to the ends of the PCR products before cloning of the amplified cDNAs (to construct cDNA library) was the most efficient method.
2. Oligonucleotides related to BM 3' and BM 5' are respectively part of BM 3' or BM 5', overlapping each other by approx 10 bases. The oli-

gonucleotides related to BM 3' are in the same orientation as BM 3', whereas the oligonucleotides related to BM 5' are anticomplementary to BM 5'.

3. For cloning of the 5' ends of mRNAs, the sequence of the oligonucleotides is dictated by the mRNA that has to be cloned. At least four oligonucleotides need to be used:

    a. One oligonucleotide for primer extension experiment.
    b. Two (at least) for PCR experiments. The compatibility of these oligonucleotides and their BM 5' partners must be checked.
    c. One as a probe for Southern blotting experiments. This oligonucleotide must be chosen close to the known 5' end of the messenger. It is preferable to choose an oligonucleotide antisense. Thus, this oligonucleotide could be used with one of two oligonucleotides previously chosen to check the length of the primer extension product prior to carrying out the SLIC procedure (*see* Note 3).

2.2. Extraction of RNAs (see Note 3)

1. Starting material: pelleted cells stored at –80°C upon collection or pieces of organs frozen in liquid nitrogen and stored at –80°C. Ice.
2. 5*M* Guanidium isothiocyanate (GITC), stored at 4°C.
3. 5% *N*-lauryl sarcosyl. Prepare in pasteurized or dd H_2O and store at room temperature.
4. 1*M* Sodium citrate, pH 7.0. Store at –20°C.
5. β-mercaptoethanol. Store at 4°C.
6. Lysis buffer: 4*M* GITC, 25 m*M* sodium citrate, pH 7.0, 0.5% *N*-lauryl sarcosine, 0.1*M* β-mercaptoethanol. Prepare fresh.
7. 1*M* Sodium acetate, pH 4.0. Store at –20°C.
8. Phenol saturated with water, stored at 4°C.
9. $CHCl_3$.
10. DT40 (Dextran T 40 Pharmacia, Saint Quenten en Yvelines, Cedex, France, #17-0270-01). Prepare a stock solution at 5 g/L in dd H_2O, and store at 4°C.
11. 100% Ethanol. Aliquot and store at –20°C.
12. 70% Ethanol. Store at 4°C.
13. DEP-treated H_2O (*see* Chapter 1) or dd H_2O.

2.3. Synthesis of the cDNAs (see Note 3)

1. Water bath at 90°C and then at 42°C.
2. Dry ice powder.
3. Acetylated bovine serum albumin.
4. RNAsin (Promega, Madison, WI, #N2111).
5. β-mercaptoethanol.
6. 1*M* Tris-HCl, pH 8.3.
7. 1*M* KCl.
8. dNTPs: 100 m*M*.

9. 20 m*M* Na Pyrophosphate.
10. [α-^{32}P] dCTP, 400 Ci/mmol (Amersham International, #PB10165).
11. AMV reverse transcriptase (Promega #M5101).
12. The primer that you have chosen (BM 3' if generation of a cDNA library is needed or a specific primer if experiment is carried out to clone the 5' end of a messenger) in solution in dd H$_2$O at 100 mg/L.

2.4. Desalting Samples with Spun Columns

1. Trisacryl GF: 0.5*M* (IBF, Villeneuve la Garenne, France, #259112). Suspend 1 vol of resin in 1 vol of 1 m*M* Tris-HCl, pH 7.4, 0.2 m*M* EDTA (TE 0.1). Centrifuge at 50*g* for 7 min. Remove the supernatant and repeat three times. Resuspend the resin in a small volume of TE 0.1 and store at 4°C up to 5 mo.
2. Microfuge tubes (0.5, 1.5, and 2 mL).
3. Needle syringe.
4. A glass bead 3-mm diameter (Ateliers Cloup, Champigny sur Marne, France, #00 065.03). A bench-top centrifuge.

2.5. Removal of the Primers

1. PrimeErase Quick Push Columns (Stratagene, La Jolla, CA, #400705). Material and buffers are provided ready to use.
2. Bench centrifuge for Eppendorf tubes.
3. 4*M* Ammonium acetate.
4. 2-propanol.
5. 75% Ethanol.
6. AcA 34 Ultrogel chromatography (IBF: #230141).
7. 2-mL disposable pipet (length = 25 cm)
8. Sterilized glass wool or a glass bead of 3-mm diameter.
9. GEB: 10 m*M* Tris-HCl, pH 8.0, 300 m*M* NaCl, 1 mM EDTA, 0.05% SDS.
10. Loading buffer. Prepare 5 mg of bromophenol blue, add 10 mL glycerol 60% (v/v), filter through a 0.45-μm filter, aliquot, and store –20°C.
11. ^{32}P-labeled primer. Assemble a 10-μL reaction containing: 50 m*M* Tris-HCl, pH 7.5, 10 m*M* MgCl$_2$, 0.1 m*M* spermidine, 0.1 m*M* EDTA, 5 m*M* DTT, 100 ng of primer, 2.5 μL of [γ^{32}P]-ATP (3000 Ci/mmol, Amersham International, #PB10168), and 1 U of terminal deoxyribonucleotide transferase. Incubate 30 min at 37°C. Remove unincorporated nucleotides using Tris-acryl GF 0.5*M* spun columns. Typically half of the radioactivity is incorporated.

2.6. RNAs Hydrolysis and Precipitation of ss-cDNA

1. A water bath at 50°C.
2. 2*N* NaOH.
3. 2.2*N* Acetic acid.

4. Dextran T 40 (Pharmacia).
5. 10 mM LiCl.
6. 100% Ethanol.
7. 75% Ethanol.

2.7. Single-Strand Ligation

2.7.1. Functionalization of the Ligated Oligonucleotide

1. Terminal transferase with supplied buffer and 25 mM CoCl$_2$ (Boehringer Mannheim, Germany, #220582).
2. [α-^{32}P] ddATP, 3000 Ci/mmol (Amersham International, #PB10233).
3. 8% Polyacrylamide gel.
4. ^{32}P-kinased BM 5' (*see* Section 2.5.).
5. Spin-X cartridge (Costar, Brumath, France, #8160).
6. Dry ice powder.

2.7.2. Ligation of the ss-cDNAs to the Modified Oligonucleotide

1. Purified ss-cDNAs.
2. 10X RNA ligase buffer: 500 mM Tris-HCl, pH 8.0, 100 mM MgCl$_2$, 100 µg/1 µg of bovine serum albumin, 500 µM ATP, 10 mM hexamine cobalt chloride. Sonicate to disrupt possible DNA contaminant that could be amplified in further PCR experiments. Filter through 0.22-µm membrane. Store at –20°C.
3. 40% PEG 6000 (Appligene, Illkirch, France). Prepare in dd H$_2$O and stored at –20°C.
4. Modified oligonucleotide BM 5'.
5. T4 RNA ligase (New England Biolabs, Beverly, MA, #204L).

2.8. PCR Amplification

PCR reagents (*see* Chapter 1), including primers.

2.9. Flying Southern and Hybridization Procedure

1. 0.4N NaOH.
2. Filter paper sheets (at least 1-cm thickness) larger than the gel.
3. Hybond N$^+$ (Amersham) or other positively charged membrane. (The type of membrane is important since the fixation of DNAs is performed in alkaline conditions.) Cut at the size of the agarose minigel.
4. Four sheets of Whatman 3MM cut at the size of the gel.
5. A centrifuge allowing the centrifugation of microtiter plates and the appropriate gondola.
6. 50X Denhardt's solution (*see* Chapter 1).
7. 20X SSC (*see* Chapter 1).
8. 0.5 mM Sodium phosphate, pH 7.0.

9. 20% SDS.
10. 0.5*M* EDTA.
11. Denatured herring sperm DNA (4 mg/mL).
12. The ^{32}P-kinased oligonucleotide probe (*see* Section 2.5.).
13. An incubator at 42°C.

2.10. Cloning of PCR the Products

As cloning strategies, screening of the libraries, and isolation of clones of interest are the basic background knowledge of all molecular biologists and are well described in classical literature (*12,13*), they are not detailed here and we limit discussion to the generation of clonable PCR products.

The complexity of the PCR product mixture is the most important criterion in the choice of cloning strategy. If the PCR products are amplified 5' ends of messengers (PCR product exhibiting a weak complexity with the SLIC strategy), they can be cloned in blunt-ended vectors after polishing the 3' ends with T4 DNA polymerase. If PCR products correspond to a cDNA population that is highly complex, the use of linkers or adaptators is recommended to generate sticky ends compatible with the appropriate cloning vector. In our ends, these strategies were more efficient than using PCR primers containing sequences recognized by restriction enzymes. The choice of the strategy depends on the number of clones necessary to be screened to obtain the desired sequence.

1. Restriction enzyme buffer A (Boehringer Mannheim).
2. 2 m*M* dNTP.
3. T4 DNA polymerase (Boehringer Mannheim, Germany, #1004786).
4. 16°C water bath.
5. A spin column (Section 2.4. and 3.3.).
6. Blunt-ended PCR products.
7. 10X Ligation buffer: 0.5*M* Tris-HCl, pH 7.5, 0.1*M* MgCl$_2$, 1 m*M* spermidine, 1 m*M* EDTA, 0.5*M* DTT, 100 m*M* ATP.
8. T4 DNA ligase.
9. Phosphorylated linkers of your choice. We have successfully used *Eco* RI linkers purchased from New England Biolabs.
10. Cloning vectors. To clone the 5' ends of mRNAs, we have successfully used *Eco* RV-cut, phosphatased pBluescript II (Stratagene). To clone the cDNA library we have used *Eco* RI-cut, phosphatased λZAP II (Stratagene) following the supplier's instructions.

3. Methods

3.1. Extraction of Total RNAs

All manipulations must be carried out in an RNase free environment (*see* ref. *14*).

Pelleted cells or organs or fragments of organs can be used as starting material. Upon collection they must be stored at –80°C as quickly as possible.

1. For 10^4 cells, add 40 µL of lysis buffer and 0.5 µg of DT 40 to the sample.
2. If cells are used, vortex 10 min to shear DNA. If organs are used, a plastic micropotter is recommended to homogenize the tissue.
3. Sequentially add: 2.5 µL of $1M$ sodium acetate, pH 4.0, 50 µL of water-saturated phenol, and 10 µL of $CHCl_3$. Mix after each addition. *See* Note 4.
4. Leave 10 min on ice.
5. Spin 10 min at 4°C.
6. Collect the aqueous phase in a clean tube.
7. Add 280 µL of 100% ethanol.
8. Precipitate (2×5 min) in liquid nitrogen.
9. Spin $12,000g$ for 30 min at 4°C.
10. Wash one or two times with 70% ethanol.
11. Vacuum dry (no more than 3 min, longer drying times stick the RNAs to the tube).
12. Dissolve the pellets in 10–20 µL of DEP-treated (or dd) H_2O. Store at –80°C.

3.2. Synthesis of the ss-cDNA

3.2.1. Synthesis of ss-cDNAs Primed by BM 3' (cDNA Library)

1. RNAs are diluted in 18 µL of DEP-treated H_2O.
2. Heat the tubes at 90°C for 5 min.
3. Spin and freeze the tubes in dry ice powder.
4. Warm up the tube in ice.
5. Assemble a reaction of 47 µL final vol containing: 0.1 mg/mL of bovine serum albumin, 0.1 U/µL of RNAsin, 70 mM β-mercaptoethanol, 10 mM Tris-HCl, pH 8.3 (at 42°C), 8 mM KCl, 1.6 mM $MgCl_2$, 1 mM each of dATP, dTTp, and dGTP, 0.05 mM dCTP, 3 pmol of BM 3', and 4 mM sodium pyrophosphate.
6. Incubate at 4°C for 15 min.
7. Add 2 µL of [α-^{32}P] dCTP (400 Ci/mmol) and approx 15 U of AMV reverse transcriptase.
8. Incubate for 45 min at 42°C.
9. Add dCTP to 1 mM.
10. Incubate for 30 min at 42°C.
11. Store at –20°C until use (*see* Note 5).

3.2.2. Synthesize of cDNAs Primed
with Specific Primers (see Note 6)

1. In a final volume of 18 μL, dilute the RNAs and add 6 pmol of primers.
2. Heat the tubes at 90°C for 5 min.
3. Cool slowly to the annealing temperature of the primer.
4. To a final volume of 50 μL, add 0.1 mg/mL of bovine serum albumin, 0.1 U/μL of RNAsin, 70 mM β-mercaptoethanol, 10 mM Tris-HCl, pH 8.3 (42°C), 8 mM KCl, 1.6 mM MgCl$_2$, 1 mM of each dNTP, 4 mM sodium pyrophosphate, and 15 U of AMV reverse transcriptase.
5. Incubate for 45 min at 42°C.

3.3. Desalting Samples with a Spin Column

1. Pierce the 0.5-mL microfuge tube with a needle.
2. Put the glass bead in the bottom of the tube.
3. Place the 0.5-mL tube in the 2-mL microfuge tube.
4. Fill the 0.5-mL microfuge tube with resin.
5. Spin in a swinging bucket rotor at 700g for exactly 2 min.
6. Transfer the microcolumn in a 1.5-mL microfuge tube.
7. Load the microcolumn with the sample that has to be desalted (do not fill with vol >30 μl).
8. Spin in a swinging bucket rotor at 700g for exactly 2 min.
9. Remove the column and store the eluate until use (*see* Note 7).

3.4. Removal of the Primers

To avoid ligation of the primers used in the synthesis of the cDNAs to BM 5', these primers need to be removed prior to carrying out ligation. The length of primer used to perform the primer extension is the critical criteria to choose the method of elimination. If the length is <25 nucleotides, PrimeErase Quick Push Columns (Stratagene) or ammonium acetate/2-propanol differential precipitation are efficient to remove primers quickly. If primers are longer, use AcA 34 (IBF).

3.4.1. PrimeErase Quick Push Columns

Follow the supplier's instructions.

3.4.2. Ammonium Acetate / 2-propanol
Differential Precipitation

1. Add 1 vol of ammonium acetate and 2 vol of 2-propanol. Mix.
2. Incubate at room temperature at least 10 min.
3. Spin in microfuge at maximum speed for 10 min at room temperature.
4. Remove supernatant, add 5 vol of 75% ethanol, and resuspend the pellets.

5. Spin as in step 3.
6. Remove supernatant and dry the pellets.
7. Resuspend in dd H_2O.

3.4.3. AcA 34 Chromatography

1. Equilibrate 10 mL of gel in 50 mL of GEB.
2. Allow the gel to sediment and discard the supernatant.
3. Resuspend the gel in 50 mL of GEB.
4. Repeat steps 2 and 3 at least three times.
5. Plug a 2-mL disposable pipet (length = 25 cm) with glass wool or with a glass bead of 3-mm diameter.
6. Load a homogeneous Ultrogel suspension into the pipet until the packed gel reaches 1 cm from the top.
7. Equilibrate the column with 5–10 mL of GEB.
8. Load the sample (100 μL of DNA, 4 μL of 0.5M EDTA, 2 μL of bromophenol blue loading buffer, and 50,000 cpm of ^{32}P-kinased BM 3').
9. Collect 100-μL fractions until the bromophenol blue reached the bottom of the column.
10. Count the Cerenkhov radiation of the fractions (*see* Note 8).
11. Pool the cDNA fractions, avoiding contamination with the primer fraction.

3.5. Alkaline Hydrolysis of RNA and Precipitation of ss-cDNA

1. Add 0.15 vol of 2N NaOH.
2. Incubate 30 min at 50°C.
3. Add 0.15 vol of 2.2N acetic acid.
4. Add 0.03 vol of 0.5 μg/μL of Dextran T40, 0.07 vol of 10M LiCl, 2.8 vol of 100% ethanol, and mix well.
5. Freeze thaw twice for 5 min in liquid nitrogen.
6. Spin 30 min at 12,000g at 4°C.
7. Wash with 70% ethanol.
8. Dry the pellets under vacuum no longer than 3 min.
9. Dissolve the pellets in 5 μL of dd H_2O and store the purified ss-cDNAs at −80°C.

3.6. Single-Strand Ligation

3.6.1. Protection of the Oligonucleotide (see Note 9)

The oligonucleotide used in the ligation has to be synthesized with a 5'-phosphate end and its 3' hydroxyl must be removed. One way to do this is to tail the 3' end of the oligonucleotide with a dideoxyribonucleotide.

1. In a 25-µL reaction, add 500 ng of oligonucleotide, 5 µL of 5X terminal transferase buffer, 1 µL of CoCl$_2$, 100 mM ddATP, and 2.5 µCi of [α-^{32}P] ddATP (3000 Ci/mM).
2. Add 25 U of terminal transferase.
3. Incubate 1 h at 37°C.
4. Heat 10 min at 75°C.
5. Load on an 8% polyacrylamide gel. In a neighboring lane, load a sample of the nontailed but ^{32}P-kinased oligonucleotide.
6. Excise the correct band (after autoradiography of the wet gel).
7. Transfer the band into SPIN-X cartridge (Costar #8160), and add 200 µL of water.
8. Freeze in dry ice powder. Thaw.
9. Spin for 5 min at 12,000 rpm.
10. Precipitate following steps 4–8 of Section 3.4.
11. Dissolve the pellets in 50 µL of dd H$_2$O (approx 5 ng/µL).

3.6.2. Ligation of the ss-cDNAs to the Modified Oligonucleotide (see Note 10)

1. Mix the following:
   1 µL of purified ss-cDNAs
   1 µL of 10X RNA ligase buffer
   0.5 µL of the modified oligonucleotide
   6.25 µL of 40% PEG 6000
   1 µL of T4 RNA ligase.
2. Incubate 48 h at 22°C.
3. Remove 5 µL of each sample and store at –20°C until use.
4. Incubate the remaining mixture for another 48-h period.
5. Store the samples at –20°C until use.

3.7. PCR Amplification

Conditions of PCR amplification depends on the T_m of the oligonucleotides used. The primers used in this study have been successfully used with annealing temperatures of 51°C or 55°C. For the two cloning experiments, we have performed nested PCR (*see* Notes 11 and 12).

3.7.1. First PCR

1. In a 500-µL microfuge tube, mix the following:
   a. Half of the ligation product.
   b. 10 µL of 10X PCR buffer, 200 nmol of dNTP, primers for the first amplification (BM 5'-1/BM 3'-1 or the specific primer of your choice). Adjust volume to 100 µL with dd H$_2$O.
   c. 1 U of *Taq* polymerase.

d. Overlay with 100 µL of mineral oil.
2. Perform PCR using the following cycle profile:

Initial denaturation   93°C, 3 min
35 main cycles   55°C, 30 s (annealing)
94°C, 30 s (denaturation)   72°C, 1 min (extension)

Cool down the reaction tubes to 4°C.

3.7.2. Second PCR

1. Use 1 µL of the first PCR amplification in a 50-µL final vol reaction similar to the reaction mixture of the first PCR except that the primers of the second amplification are BM 5'-2/BM 3'-2 or the second specific primer of your choice.
2. Perform PCR as in the first PCR reaction.

3.7.3. Analysis of the PCR Products

Analysis of the PCR products can be performed with 1% agarose electro-phoresis as described elsewhere (*13; see* Chapter 1). It is recommended to check the specificity of the PCR with Southern blotting experiments. A fast and efficient protocol, named "Flying Southern" is given in Section 3.8.

3.8. Flying Southern and Hybridization Procedure

This protocol is derived from ref. *15*. It allows quick and accurate analysis of the specificity of PCR products using agarose minigels (7.5 × 10.5 cm). This protocol requires the use of a centrifuge that accepts microtitre plates.

1. After UV visualization of the nucleic acids, soak the agarose gel in 0.4*N* NaOH.
2. Soak a Hybond N⁺ membrane (Amersham International) in sterile H_2O and then in 6X SSC.
3. Prepare in a plastic box:
   Filter paper sheets (1-cm total thickness)
   Hybond N⁺ membrane
   Agarose gel (avoiding bubbles between the membrane and the gel)
   Four sheets of Whatman 3MM paper soaked in 0.4*N* NaOH.
4. Spin in microtitre plate gondola at 1000*g* for 20–30 min (agarose gels from 0.8–2%).
5. Remove the Whatman paper.
6. Mark the position of the wells on the membrane.
7. Remove the membrane from the gel and rinse twice for 5 min each in 6X SSC.
8. Prehybridize in 1 mL (7.5 × 10.5 cm membrane) of buffer (1X

Denhardt's, 6X SSC, 25 mM sodium phosphate, pH 7, 25 mM EDTA, 250 µg/mL of denatured herring sperm DNA, 1% SDS). Load the buffer in a Petri dish, then put the membrane DNA face toward the bottom of the dish above the buffer. Avoid air bubbles. Close the Petri dish.

9. Incubate 20 min at 42°C.
10. Remove the prehybridization buffer and replace it by 300 µL of hybridization buffer containing 2–3 × 10^6 cpm of ^{32}P-kinased oligonucleotide. Place the membrane in a Petri dish as described earlier.
11. Incubate for 30 min at 42°C.
12. Wash from 6X SSC to 1X SSC with 1% SDS, at 42°C. Check radioactivity with a bench monitor.

3.9. Cloning PCR Products

3.9.1. Blunt-Ending of PCR Products (see Note 13)

1. After having removed primers (Section 3.4.2.), assemble a 20-µL reaction containing 2 µL of buffer A (Boehringer), 1 µL of 2 mM dNTPs, and 2 µL of T4 DNA polymerase (Boehringer). To limit the activity of the enzyme, keep at 4°C.
2. Incubate 15 min at 16°C. Warm to 75°C for 10 min to denature the enzyme.
3. Remove nucleotides and salts using a Tris-acryl GF 0.5M spun column.

3.9.2. Ligation of Linkers to the PCR Products

1. To approx 500 ng of blunt-ended PCR products, add 1 µg of phosphorylated *Eco* R1 linkers, 5 µL of 10X ligation buffer, and dd H$_2$O to 46 µL. Finally, add 4 µL of T4 DNA ligase. Incubate overnight at 16°C.
2. Heat 5 min at 75°C.
3. Add 9 µL of buffer H (Boehringer), 2 µL of 100 mM spermidine, 33 µL of dd H$_2$O, and 4 µL of *Eco* RI (Boehringer 60 U/µL). Incubate 2 h at 37°C.
4. Precipitate as described in Section 3.4.2.
5. Resuspend in 10–20 µL of dd H$_2$O and use to ligate to *Eco* RI-cut cloning vector following classical protocols (*14*) or supplier's instructions.

4. Notes

1. For all the oligonucleotides used, secondary structures and self-complementary structures should be avoided. Compatibility in T_m and nonannealing of each pair of oligonucleotides should be checked. Softwares facilitating this analysis are now available.
2. All the oligonucleotides except BM 3' and BM 5' can be used as crude preparations, after ammonia deprotection and lyophilization. As BM 3' and BM 5' are long oligonucleotides, they must be purified (using polyacrylamide gels or HPLC) before use. Protocols are well described elsewhere (*13*).

3. Tips, tubes, and all the material used in this manipulation must be very clean and at least sterilized. Wear gloves during all the manipulation of RNAs to avoid contamination of the preparation by RNAses. The same precautions should be followed in the synthesis of cDNAs. These precautions represent the lower level of protection against RNAse and it is advisable to read ref. *14* carefully. All the aqueous solutions must be filtrated through a 0.22-μm filter prior to storage.

4. We have successfully extracted total RNAs (34 μg) from punches of rat raphe dorsalis by multiplying by 5 all the volumes indicated earlier.

5. The synthesis of the ss-cDNA can be checked by the determination of the radioactivity incorporated. This could easily be done using the spin column described in Section 3.3. If I and T are, respectively, the incorporated (determine the radioactivity in the eluate of the spun column) and total radioactivity (count the column or an aliquot of ss-cDNA mixture prior desalting), the mass M of ss-cDNA is given by the formula:

$$M = (I/T) \times (\text{total dCTP mass during reaction}) \times 4$$

In the conditions used, $M(ng) = (I/T) \times 330$. Classically, the amount of ss-cDNA is 10–30% of the mass of starting mRNAs.

6. For the cloning of the 5' ends, the labeling of the ss-cDNAs is not necessary provided the activity of the enzyme has been checked before. A more informative experiment, in the cloning of 5' ends of mRNAs, is to carry out PCR with the ss-cDNAs as matrix and two specific primers. One of the primers has to be chosen close to the known 5' ends of the cDNA clones. After PCR, verify the specificity of the product by Southern blotting (Section 3.10.). This allows an accurate verification of the synthesis of a ss-cDNA long enough to pursue cloning experiments.

7. In the conditions described, these columns retain salts and small oligonucleotides (approximate length of 10 nucleotides).

8. When the pattern of the radioactivity contained in the fractions is examined, two peaks can easily be isolated. The first, which is also the less intense, corresponding approximately to fractions 7–11, contain the cDNAs. The second peak, which is more intense, starting approximately from the 15th fraction corresponds to the kinased primers and the radiolabeled nucleotides if they have not been removed before chromatography.

9. During the preparation of this chapter we have successfully tested a new method to block the 3-OH of BM 3'. It consists in synthesizing this oligonucleotide with a 3'-NH_2 end.

10. During this manipulation, prepare controls that will be used in the PCR experiments following SLIC.

   a. Prepare a nonligated sample composed of the same mixture as described in Section 3.6.2. step 1 minus the enzyme. This nonligated control will be useful to determine if unspecific amplification using the ss-cDNA as matrix has occurred.

b. Prepare another control containing the mixture described in Section 3.6.2, step 1 minus ss-cDNA and the enzyme. This control will reveal the presence of contaminant in the SLIC mix when compared to the PCR control.

11. A very small amount of BM 3' can escape the AcA 34 purification. In such case, after the first round of PCR amplification we have used a Sephacryl S-400 chromatography to purify the amplified DNA.

12. In all the PCR experiments, it is useful to prepare a control sample with the PCR mixture minus DNA. This allows the detection of contaminant in the PCR mix.

13. Be sure that the PCR product is kinased. PCR primers are classically synthesized without phosphate group at their 5' end. It is recommended to perform a limited PCR (10 cycles) with kinased primer before blunt-ending.

Acknowledgments

We are grateful to our colleagues. We thank A. Hicks for critical reading of the manuscript. This work was supported by fellowships from the Societe de Secours des Amis des Sciences followed by the Foundation Claude Bernard to J.B.D.M.E. and from the Ecole Polytechnique and the Association pour la Recherche contre le Cancer and the Programme Lavoisier to J.D. and by grants from the Centre National de la Recherche Scientifique, the Institut National pour la Recherche Médicale, the Association pour la Recherche contre le Cancer, and Rhone-Poulenc Rorer.

References

1. Mullis, K. B. and Faloona, F. (1987) Specific synthesis of DNA in vitro via a polymerase-catalyzed chain reaction. *Methods Enzymol.* **155,** 335–350.
2. Saiki, R. K., Gelfand, D. H., Stoffel, S., Scharf, S. J., Higuchi, R., Horn, G. J., Mullis, K. B., and Erlich, H. A. (1988) Primer-directed enzymatic amplification of DNA with a thermostable DNA polymerase. *Science* **239,** 487–491.
3. Delort, J., Dumas, J. B., Darmon, M. C., and Mallet, J. (1989) An efficient strategy for cloning 5' extremities of rare transcripts permits isolation of multiple 5'-untranslated regions of rat tryptophan hydroxylase mRNA. *Nucleic Acids Res.* **17,** 6439–6448.
4. Loh, E. Y., Elliot, J. F., Cwisla, S., Lanier, L. L., and Davn, M. M, (1989) Polymerase chain reaction with single-sided specificity: analysis of T cell receptor δ chain. *Science* **243,** 217–220.

5. Belyavsky, A., Vinogradova, T., and Rajewsky, K. (1989) PCR-based cDNA library construction: general cDNA libraries at the level of a few cells. *Nucleic Acids Res.* **17,** 2919–2932.

6. Frohman, M. A., Dush, M. K., and Martin, G. R. (1988) Rapid production of full-length cDNAs from rare transcripts: amplification using a single gene-specific oligonucleotide primer. *Proc. Natl. Acad. Sci. USA* **85,** 8998–9002.

7. Ohara, O., Dorit, R. L., and Gilbert, W. (1989) One-sided polymerase chain reaction: the amplification of cDNA. *Proc. Natl. Acad. Sci. USA* **86,** 5673–5677.

8. Dumas Milne Edwards, J. B., Delort, J., and Mallet, J. (1991) Oligodeoxyribonucleotide ligation to single-stranded cDNAs: a new tool for cloning 5' ends of mRNAs and for constructing cDNA libraries by in vitro amplification. *Nucleic Acids Res.* **19,** 5227–5232.

9. Dumas, S., Darmon, M. C., Delort, J., and Mallet, J. (1989) Differential control of tryptophan hydroxylase expression in raphe and in pineal gland: evidence for a role of translation efficiency. *J. Neurosci. Res.* **24,** 537–547.

10. Moseman Mc Coy, M. I., and Gumport, R. I. (1980) T4 ribonucleic acid ligase joins single-strand oligo(deoxyribonucleotides). *Biochemistry* **19,** 635–642.

11. Tessier, D. C., Brousseau, R., and Vernet, T. (1986) Ligation of single-stranded oligodeoxyribonucleotide by T4 RNA ligase. *Anal. Biochem.* **158,** 171–178.

12. Berger, S. L. and Kimmel, A. R. (1987) Guide to molecular cloning techniques. *Methods Enzymol.* **152,**

13. Sambrook, J., Fritsch, E. F., and Maniatis, T. (1989) *Molecular Cloning. A Laboratory Manual* (2nd ed.), Cold Spring Harbor Laboratory, Cold Spring Harbor, NY.

14. Blumberg, D. D. (1987) Creating a ribonuclease free environment. *Methods Enzymol.* **152,** 20–24.

15. Wilkins, R. J. and Snell, R. G. (1987) Centrifugal transfer and sandwich hybridization permit 12-hour Southern blot analysis. *Nucleic Acids Res.* **15,** 7200.

Index